C0-AWW-139
Rehabilitation of the
3 3073 00261461 6

REHABILITATION OF THE BRAIN-DAMAGED ADULT

APPLIED CLINICAL PSYCHOLOGY

Series Editors: Alan S. Bellack, *Medical College of Pennsylvania at EPPI, Philadelphia, Pennsylvania,* and Michel Hersen, *University of Pittsburgh, Pittsburgh, Pennsylvania*

HANDBOOK OF BEHAVIOR MODIFICATION WITH THE MENTALLY RETARDED
Edited by Johnny L. Matson and John R. McCartney

THE UTILIZATION OF CLASSROOM PEERS AS BEHAVIOR CHANGE AGENTS
Edited by Phillip S. Strain

FUTURE PERSPECTIVES IN BEHAVIOR THERAPY
Edited by Larry Michelson, Michel Hersen, and Samuel M. Turner

CLINICAL BEHAVIOR THERAPY WITH CHILDREN
Thomas Ollendick and Jerome A. Cerny

OVERCOMING DEFICITS OF AGING: A Behavioral Approach
Roger L. Patterson

TREATMENT ISSUES AND INNOVATIONS IN MENTAL RETARDATION
Edited by Johnny L. Matson and Frank Andrasik

REHABILITATION OF THE BRAIN-DAMAGED ADULT
Gerald Goldstein and Leslie Ruthven

In preparation

SOCIAL SKILLS ASSESSMENT AND TRAINING WITH CHILDREN
An Empirically Based Handbook
Larry Michelson, Don P. Sugai, Randy P. Wood, and Alan E. Kazdin

BEHAVIORAL ASSESSMENT AND REHABILITATION OF THE TRAUMATICALLY BRAIN DAMAGED
Edited by Barry A. Edelstein and Eugene T. Couture

COGNITIVE BEHAVIOR THERAPY WITH CHILDREN
Edited by Andrew W. Meyers and W. Edward Craighead

A Continuation Order Plan is available for this series. A continuation order will bring delivery of each new volume immediately upon publication. Volumes are billed only upon actual shipment. For further information please contact the publisher.

REHABILITATION OF THE BRAIN-DAMAGED ADULT

GERALD GOLDSTEIN

Veterans Administration Medical Center, Highland Drive
and University of Pittsburgh School of Medicine
Pittsburgh, Pennsylvania

and

LESLIE RUTHVEN

Private practice
Wichita, Kansas

PLENUM PRESS • NEW YORK AND LONDON

Library of Congress Cataloging in Publication Data

Goldstein, Gerald, 1931–
Rehabilitation of the brain-damaged adult.

(Applied clinical psychology)
Includes index.
1. Brain damage—Patients—Rehabilitation. I. Ruthven, Leslie, joint author. II. Title.
III. Series. [DNLM: 1. Brain damage, Chronic—Rehabilitation. WL354 R3453]
RC386.2.G66 616.85'8 80-15126
ISBN 0-306-40498-2

A Division of Plenum Publishing Corporation
233 Spring Street, New York, N.Y. 10013

Printed in the United States of America

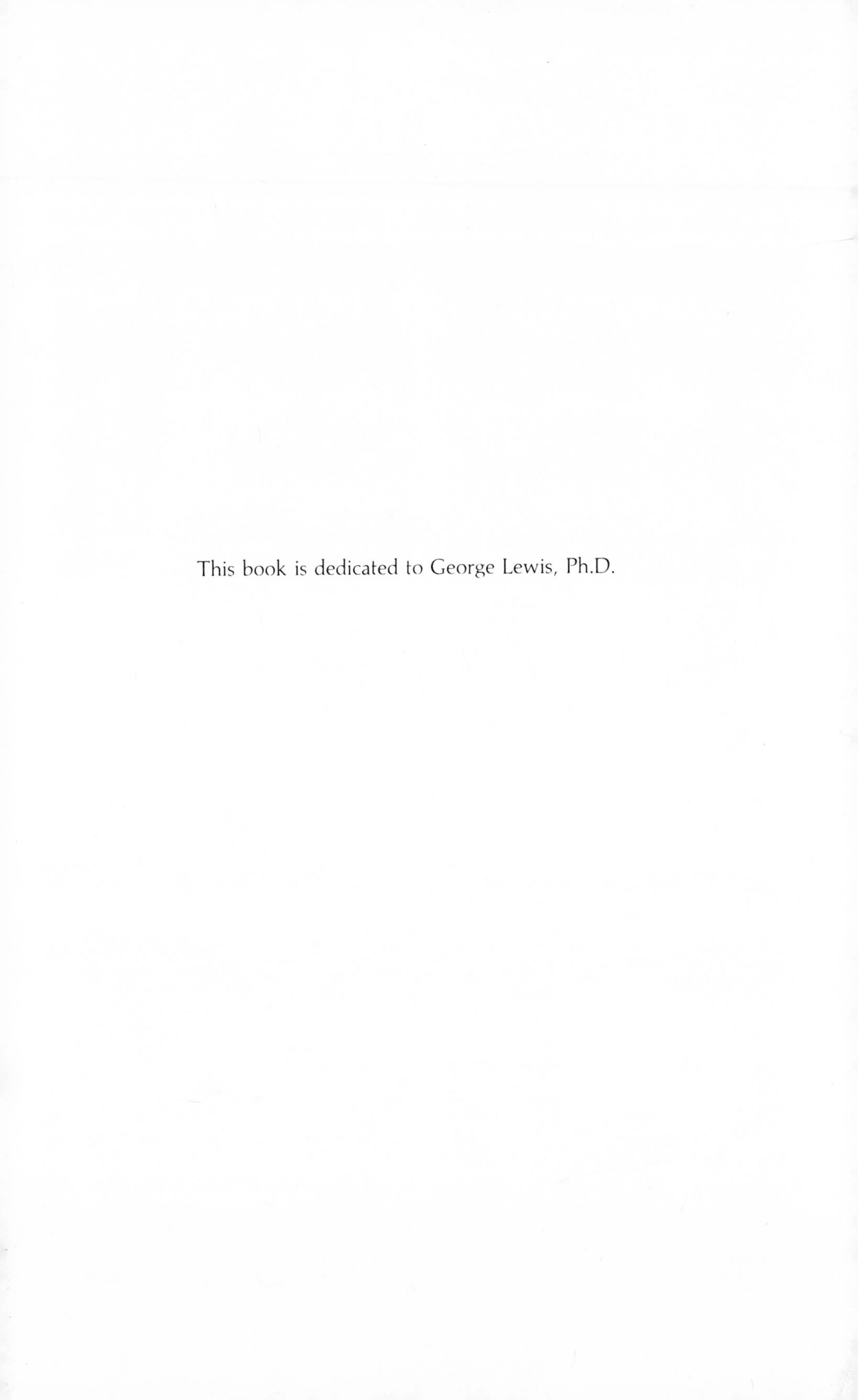

This book is dedicated to George Lewis, Ph.D.

ACKNOWLEDGMENTS

Acknowledgment is made to the National Institute of Mental Health (Grant No. 1R01 MH 18353) and to the Veterans Administration for support of research described in this book.

CONTENTS

1

INTRODUCTION

Basic Issues in Rehabilitation of the Brain Damaged

Definitions

Because of the vagueness surrounding the term *brain damage,* it is necessary at the outset to define the population to which this book may have some application. Although it is usual to speak of the brain-damaged patient in a general way, the conditions referred to cover a variety of specific disorders. In this book we will be discussing only individuals who become brain-damaged as adults. We will be addressing ourselves specifically to adults who have sustained demonstrable, structural brain damage. Those conditions in which brain dysfunction is a possible etiological agent, such as a number of functional psychiatric disorders, will not be considered. Thus the entire topic of mental retardation and early life brain damage will not be treated here, nor the many problems associated with minimal brain damage syndromes in school age children. Modern psychiatric thinking has tended to blur the distinction between the so-called functional and organic disorders (cf. Shagass, Gershon, & Friedhoff, 1977), but we would adhere to the view that the patient with structural brain damage continues to present relatively unique assessment and treatment problems.

Furthermore, the emphasis of this book will be placed on individuals with nonprogressive, chronic brain damage. Certain brain disorders are slowly or rapidly progressive diseases, and although rehabilitative attempts should certainly not be ruled out in such cases, the nature of those disorders has strong implications for rehabilitation planning. We will not be exploring those implications in a specific way. Within the general scope of nonprogressive brain damage, we will be dealing mainly with two populations: head-injured young adults and chronically institutionalized older brain-damaged patients.

With regard to brain damage in young adults, head trauma is the most common etiological agent. Although it is true that young adults may acquire brain tumors, cerebral infections, and certain forms of cerebral vascular disease, their occurrence is infrequent relative to head injury. Multiple sclerosis and Huntington's disease occur in young adults, but their progressive natures seriously affect rehabilitation. In the case of chronically institutionalized older patients, a variety of brain disorders is represented. Generally, these disorders reflect the long-term effects of some slowly progressive process, such as cerebral vascular disease, chronic alcoholism, and a group of diseases known collectively as senile or presenile dementia. These processes are not generally rapidly progressive but produce behavioral changes that often significantly impair the individual's potential for independent living in a noninstitutional setting.

The term *rehabilitation* also has some vagueness surrounding it and requires some definition. Stedman's Medical Dictionary (1972) defines rehabilitation as "Restoration, following disease, illness or injury, of ability to function in a normal or near normal manner." This definition is rather strict with reference to what is generally meant when one speaks of rehabilitation of brain-damaged patients. More frequently, reference is made to restoration of function in regard to certain reasonable goals. Such goals might be ability to function outside of an institutional setting or employment at a relatively nondemanding job. The ability to communicate simple needs may be a reasonable goal of rehabilitation in the case of some brain-damaged patients. One could think in terms of a range of goals extending from simple activity of daily living considerations to educational and vocational retraining oriented toward restoration of full participation in community life. Obviously, young adults and more elderly chronic patients lie at different points on this continuum. In the case of the young brain-damaged adult, it is often reasonable to plan in terms of restoration to some level of independent functioning in the community through educational and vocational retraining. In the case of the older chronic patient, the goals more typically place emphasis on restoration of skill in the more basic activities of daily living. Particularly in the case of the patient who has been institutionalized for a long period of time, some sort of resocialization program is also often a necessity.

In general, although that part of the definition of rehabilitation dealing with restoration of function is acceptable, the part about returning to normal or near normal function as a criterion of success is unrealistic for most brain-damaged patients. From the outset, the reader of this book should be aware that we will be offering no new treatments or devices that in any way assure complete restoration of function in patients with nonprogressive brain damage, or arrest or reversal of the pathological process operating in progressive brain damage.

The Intended Readership

The definitions offered above suggest the nature of the audience that would find this book of interest. Basically, it is written for individuals who have long-term clinical contact with brain-damaged patients. Within the medical profession, it may be of interest to psychiatrists and neurologists who provide continued care to chronic patients. In the behavioral sciences, psychologists whose practice is primarily in the rehabilitation area, or clinical psychologists who treat brain-damaged patients may find it useful. Even though the authors are both psychologists, the book was written in a manner that minimizes the need for technical knowledge of psychology, so that it may find acceptance by a multidisciplinary audience. We hope it will be useful to individuals in the various rehabilitation specialties such as speech and occupational therapy. We view this work as a clinically oriented book and not as a scholarly research reference. As we will demonstrate, scholarship in the area of rehabilitation of brain-damaged adults is highly limited, and it would be premature to produce a definitive textbook at this point. However, there is by now a reasonably extensive literature, some data, our own clinical experience and that of others, any combination of which may serve to make a preliminary contribution to this very difficult area.

Recovery of Brain Function

The literature concerning recovery of brain function has been reviewed by Stein, Rosen, and Butters (1974) and more recently by Finger (1978). Luria (1948/1963) has written specifically about recovery from head trauma, but the book was actually written immediately after World War II and so his research is somewhat out of date. In any event, the reader is referred to these three volumes for a reasonably comprehensive review of the theoretical and scientific work in this area. Also the reader should be mindful that with the exception of Luria's book most of the work done has been accomplished through the use of animal models, and that recovery from brain damage in humans has not received extensive study. A clinical folklore has developed in this area, which we will comment on later, but its scientific documentation has not been extensive.

The problem of recovery of function following brain damage is a complex and intricate matter. Numerous factors are at work, including interspecies differences, sex differences, time of acquisition of the brain lesion, location of the lesion, amount of tissue destroyed, manner in which the tissue is destroyed, and the particular form of neuropathology involved. Finger (1978) found that contemporary neuroscientists no longer accept the naive notion that since central nervous system tissue does not regenerate, the functions originally mediated by that tissue cannot return. There is now evidence (Stricker, 1979) that brain damage does not necessarily always totally destroy neurons. It can sometimes impair only the function of neurons, and that impairment of function can sometimes be at least partially compensated for by certain biochemical mechanisms. Since the early studies of Monakow (1914), it has been known that there is gradual spontaneous recovery of certain functions that become impaired immediately after brain injury. Reduction in pressure produced by temporary edema is often associated with restoration of function. In general, recovery of function is apparent on the basis of both animal and human research, as well as on the basis of clinical observation. It is now a matter of gaining fuller understanding of its nature, mechanisms, and limitations.

Important longitudinal studies of recovery of function are being conducted at the animal level by Goldman (1971, 1972, 1974) and at the human level by Lezak (1976) and by Ben-Yishay and various collaborators (Ben-Yishay, Gerstman, Diller, & Haas, 1970). These studies are aimed at discovery of the relevant parameters of recovery. Which functions recover and which do not? What are the influences of such factors as age, sex, extent of brain damage, and location of brain damage on outcome? Other research is looking at both the anatomical and neurochemical aspects of recovery. For example, anatomic studies by Butters, Rosen, Soeldner, and Stein (1975) have indicated that recovery of frontal lobe function may be associated with very specific destruction or lack of destruction of particular regions of the brain. In the case of primates, the destruction of a particular portion of the principal sulcus of the frontal lobe may rule out recovery. It may be that the basis for the neuroanatomic specificity is really neurochemical in nature. Wolfe, Stricker, and Zigmond (1978) have pointed out how recovery can take place on a neurochemical basis following catecholamine-depleting lesions. It is possible that particular neuroanatomic structures are important because they contain receptor sites for various neurotransmitters (Lipton, DiMascio, & Killam, 1978).

Recent research has emphasized the plasticity of brain function, particularly in young organisms, and has shown that many behaviors impaired by brain damage can be restored, at least to some extent. In the modern era, as early as 1951, a book was written about recovery from aphasia (Wepman, 1951). However, recovery from aphasia had been noted for some time. It is probably worthwhile to repeat a remark made by Gowers in the year 1878 and cited in Searleman (1977):

> Loss of speech due to permanent destruction of the speech region in the left hemisphere has been recovered from, and that this recovery was due to supplemental action of the corresponding right hemisphere is proved by the fact that in some cases, speech has been again lost when a fresh lesion occurred in this part of the right hemisphere. (p. 508)

Numerous explanations of the recovery process have been offered, such as the taking over of originally impaired function by intact tissue, but none of them has been equivocally proven. However, as we have indicated, there is clinical folklore about the nature of recovery and some of it is reasonably well established. Recovery is more likely in a relatively young organism, and the likelihood of recovery is greater if the area of damage is relatively small. Furthermore, the loss of certain structures is more important than that of other structures, as we have just illustrated in the case of the principal sulcus in regard to recovery of frontal lobe function in primates. There appear to be a number of compensatory factors working at neurochemical and neuroanatomic levels that tend to promote recovery. In the case of cerebral vascular disease, the development of collateral cerebral circulation sometimes aids in recovery of function. In general, brain damage produces a dynamic situation, and even though we speak of static or nonprogressive brain damage, in principle there is really no such thing. The characteristics of lesions and of preserved brain tissue continually change. Although these changes are sometimes for the worse, they are sometimes for the better. In some cases, symptoms that have been present for some time may diminish in intensity for unknown reasons. Slight changes in such variables as blood pressure or degree of edema can be associated with often remarkable changes in symptomatology. The point of these remarks is simply to illustrate that essentially all types of brain damage create a more or less fluctuating situation that may change slowly or rapidly over very variable time courses. (For the reader who wishes a more detailed and technical review of these matters, the references cited above will be useful.)

The matter of spontaneous recovery of function has very important implications for rehabilitation efforts. The crucial question has to do with effectiveness of treatment. There are those who would argue that rehabilitation is something you do while the patient is getting better by himself through spontaneous recovery mechanisms. There are others who believe that rehabilitation is primarily useful in regard to promoting spontaneous recovery. These individuals like to work with patients during the acute phase of their illnesses, and are less sanguine about working with patients whose disorders have become chronic. Still others would argue that rehabilitation efforts are worthwhile in and of themselves, and should be attempted regardless of the stage of the illness or of its progressive or nonprogressive nature. There would appear to be a strong necessity in any case to demonstrate that rehabilitation efforts have a positive effect independent of spontaneous recovery of function. The literature dealing with recovery of function tends to stress its spontaneous nature and does not generally deal specifically with the issue of treatment. There is a strong interest in the adaptive mechanisms the nervous system has developed to protect itself. Furthermore, there is at least clinical evidence to suggest that once the period of spontaneous recovery has ended, no further changes are to be expected, even when heroic treatment efforts are made. In most clinical circles, brain damage that has stabilized is often viewed as chronic, incurable disease. Indeed, the psychiatric diagnosis "chronic brain syndrome" is frequently applied to these individuals, with the connotation of irreversible brain damage. A commonly shared attitude is that nature will take its course with these patients, and that only treatment of a nursing or supportive type is possible. In psychiatric settings, so-called organic patients are often viewed as having poorer prognoses than patients with the so-called functional psychiatric disorders. The apparent absence of structural pathology seems to offer greater hope than can be mustered for individuals with structural impairment. However, we know of no data that would objectively support this view.

The Natural History of Recovery

In order to evaluate the effectiveness of rehabilitation, it is necessary to have a baseline against which the value of the treatments offered can be assessed. As indicated above, a reasonable baseline would be the extent to which the patient recovers without the intervention of any direct treatment methods aside from routine medical and nursing care. Within the realm of structural brain damage, many scenarios may be described, depending on the nature and severity of the pathological process. In the case of the degenerative diseases, there is often a slow but inexorable course of deterioration in which rehabilitative efforts can only temporarily stave off the progression of the illness, while perhaps making life more pleasant during the patient's remaining years. Such disorders as Huntington's disease and multiple sclerosis have this characteristic. There is little point in being sanguine about these illnesses because only major scientific discoveries involving direct methods of prevention and treatment will have any impact on their ultimate outcome. In the case of Huntington's disease, genetic counseling that provides the information that there is a 50% risk of offspring acquiring the disease may aid in its gradual elimination. Although we will not be emphasizing the treatment of progressive illnesses in this book, the clinician working with brain-damaged patients is highly likely to encounter such cases. Perhaps the most important thing to realize about these conditions is that acquisition of any of them is a tragic, catastrophic life event and that the crucial issue is often a matter of formulating reasonable life plans for the patient and his or her family without offering false hope. In the case of many of these disorders the patient may be expected to live a close to normal life span, and so the course of the illness from the time at which symptoms first appear may be quite lengthy.

Traumatic brain damage is a different matter. Patients who sustain head injuries with resultant brain damage are typically most severely ill immediately following the injury (or in some cases within several days thereafter), and if the critical period is survived, there is usually a greater or lesser degree of recovery thereafter. The natural history of the posttrauma course is well known, and is described in detail in many neurological textbooks (e.g., R. D. Adams & Victor, 1977). The victim frequently loses consciousness and may remain unconscious for a substantial length of time. When consciousness is regained, the patient is usually confused and disoriented. There may be memory difficulties, irritability, headache, and easy fatigability. Sometimes these patients have seizures. The memory disorder often takes the form of an anterograde amnesia in which the patient cannot recall, or has difficulty in recalling, anything from shortly before the acquisition of the trauma to the present time. This acute episode, or period of confusion and disorientation, is often seen in the stroke patient as well as in the head-injured patient.

During the acute phase of these disorders, mortality and morbidity are the major treatment issues. The goal is to keep the patient alive and prevent as much residual neurological impairment as possible. Indeed, in some cases such radical procedures as anesthetizing or reducing the temperature of the brain (Safar, Bleyaert, Nemoto, Moossy,

& Snyder, 1978) are accomplished in order to minimize permanent damage. In the case of head trauma, the first person to treat the patient is frequently a neurosurgeon who may perform a procedure known as a *debridement.* This procedure is performed when the vault of the skull has been penetrated and it is necessary to clear the area around the wound of bone fragments and other debris. If the patient is unconscious, it is also necessary to prevent infection, maintain nutrition, and monitor and maintain vital signs. During this phase of the recovery process, there is generally concern about seizures, and the patient may be placed on anticonvulsant drugs as a preventive measure. When the patient regains consciousness, we can get a little clearer picture of what the outcome will be. More or less specific symptoms may emerge, such as partial paralysis or loss of the ability to communicate (aphasia), some of which may ultimately disappear while others may remain. From the standpoint of rehabilitation, any assessment of the patient's functioning during the acute phase will probably give a distorted notion of what the outcome will be, since there is typically a phase following the acute episode in which recovery is rapid. In young patients particularly, the recovery may be very rapid.

The length of the recovery phase is somewhat variable and dependent on a number of factors. There are a number of clinical rules of thumb and some research regarding this matter. In the case of head trauma, an early study by Russell (1932) demonstrated a strong relationship between speed of recovery and duration of traumatic amnesia. Russell and Smith (1961) later confirmed this finding with a larger sample. According to R.D. Adams and Victor (1977), age is an important factor, with elderly patients remaining disabled for lengthier periods than younger patients. Russell and Smith (1961) demonstrated that older patients had lengthier posttraumatic amnesias and more signs and symptoms than younger patients. R. D. Adams and Victor suggested that improvement can continue over a period of five or more years, but recovery is more complete in children than in adults. Sands, Sarno, and Shankweiler (1969) and Smith (1981) also found an age relationship in stroke patients. Stroke patients who were younger maintained their gains and recovered over a lengthier time period than did older patients. Thus several factors, such as age, duration of posttraumatic amnesia, and length of time unconscious (Smith, 1961), affect the recovery course. According to Lezak (1976), the best rule of thumb for predicting improvement during the early stages of brain damage involves the initial rate of improvement; the greater the initial rate, the greater the further improvement.

In summary, when an individual has a head injury or stroke, there is a fairly typical course that may vary somewhat not only on the basis of the extensiveness of the brain damage but also because of a number of other factors. The major factors appear to be the age of the individual, the length of time that the individual is amnesic, and the length of time unconscious. In any event, the typical picture often involves a period of unconsciousness, a period of confusion (usually with amnesia), and a recovery period that may vary greatly in amount of time. Several questions arise with regard to the matter of rehabilitation, including the matter of when to do a rehabilitation-oriented assessment, when to initiate treatment, and how to determine prognosis.

Natural Recovery and Rehabilitation

Lezak (1976) suggested that in order to obtain a reasonable prognosis it is best to assess the patient from three to six months following the acute episode. Any attempt to make an assessment prior to that period may provide a distorted notion of what the outcome will be. Most clinicians (e.g., Porch, 1964) believe that treatment should be initiated as soon as possible, particularly before the patient loses motivation or develops bad habits. These suggestions are in the nature of "common sense" clinical opinions. However, the matter of prognosis has been studied more systematically, and there are some objective findings. Meier (1970, 1974) and Meier and Resch (1967) have done some very significant work regarding outcome from stroke. The general strategy required the correlating of a predictor neuropsychological test battery with an outcome criteria battery as well as with outcome measures derived from the neurological examination. In one study (Meier, 1970), the prediction of neurological outcome one year poststroke based on neuropsychological test indicators yielded 75% accuracy. Later, Meier (1974) presented cases of several stroke patients tested longitudinally that clearly illustrated the variety of recovery patterns possible. However, certain neuropsychological tests, notably Trail Making, the Halstead Tactual Performance Test, and the Porteus Mazes, were found to be generally good predictors. Ben-Yishay, Gerstman, Diller, and Haas (1970), utilizing an extensive battery of demographic and test variables with left hemiplegics, were able to successfully predict outcome in terms of length of stay, improvement in ambulation, and improvement in self-care. Utilizing a multiple-correlation analysis, a multiple R of .89 was obtained for all three criteria. The contribution of the neuropsychological tests used was found to be particularly noteworthy. In general, although we are only in the preliminary stages of exploration, neuropsychological tests do seem to be showing promise as predictors of rehabilitation outcome. However, other predictors, including time of initiation of treatment (G. F. Adams & McComb, 1953; Mahoney & Bartel, 1954), measures of physical function such as blood pressure (Bruell & Simon, 1960), and bowel continence (Peszczynski, 1961), should probably not be neglected either. As we have already seen, age may be an important factor, and was found to be a reasonably good nonpsychometric predictor in the Ben-Yishay, Gerstman, Diller, and Haas (1970) study.

Perhaps the major factor lacking in studies of this type is that they do not separate the differential effectiveness of natural recovery *alone* as opposed to natural recovery combined with rehabilitation. Largely for ethical reasons, it is difficult if not impossible to find a study involving an untreated control group to contrast with a group in active treatment. For example, all subjects of Ben-Yishay, Gerstman, Diller, and Haas (1970) were engaged in rehabilitation. Several solutions to this problem have been proposed. One of them is the "waiting list" control group made popular in psychotherapy research (Garfield, 1980). Patients on a waiting list are serially tested over a time span equal to before and after testing of a group in treatment. A second alternative involves the use of single-case designs (Hersen & Barlow, 1976) in which it is not necessary to have an untreated control group. The absence of attempts to do controlled studies of rehabilitation efficacy is well illustrated in a review chapter by Kertesz (1979). After reviewing the literature, he found only four studies of language therapy in which untreated controls were used. Of these four, only two (Basso, Faglioni, & Vignolo, 1975; Hagen, 1973) demonstrated the effectiveness of therapy. The study of Basso *et al.*, according to Kertesz, had numerous methodological flaws, and so the Hagen study seems to be the only one reviewed that demonstrated the effectiveness of language therapy. However, it was based on only 10 treated patients and 10 controls. More recently, a controlled study done by Basso, Capitani, and Vignolo (1979) was able to demonstrate that rehabilitation did have a positive influence on the language skills of aphasic patients over and above what could be attributed to natural recovery processes.

In the absence of substantial objective evidence that rehabilitative efforts are effective over and above natural recovery, the clinician is posed with something of a dilemma. Although it is important not to attribute to treatment what was in fact the product of natural recovery processes, the absence of objective data does not appear to deter clinicians from engaging in treatment. However, if one is going to do treatment, the decision to attempt it or not attempt it should not be directly associated with one's estimate of whether or not the natural recovery process has ended. A distinction should be made between the natural history of the illness and the decision to attempt or not attempt treatment. We would justify this view by the fact that since the unique contribution of treatment has not yet been objectively demonstrated, it is not really possible to determine the point at which it is optimal to initiate treatment or the point after which it is hopeless to attempt such initiation. The rule of thumb that speech therapy, for example, should be initiated as soon after the acute episode as possible should not be interpreted to mean that therapy should not be initiated if it only becomes available long after the onset of the patient's illness. Sufficient data are not yet available for that determination.

Although the necessary studies are again lacking, an additional justification for initiating and maintaining treatment is that even if it does not make the patient better, it may aid in preventing the patient from getting worse. To the extent that lack of treatment is associated with boredom and sensory or motor deprivation, perhaps unnecessarily accelerated deterioration may be expected in the absence of active treatment. The so-called pseudodementia of the elderly (Wells, 1979), according to Folstein, Folstein, and McHugh (1975), may reflect an interaction between deteriorating brain function and the mood changes that may be induced by isolation and withdrawal from community life. Even in the case of the so-called nonprogressive brain damage of the type associated with head injury, Walker, Caveness, and Critchley (1969) believe that there is evidence for a slow course of deterioration that may begin at some time after the head wound is sustained. Clinically, it is not unusual to see head-injured patients who have been stable for many years start having seizures or showing new signs of cognitive deterioration. Possibly the same phenomenon has been reported by Smith (1958) and Smith and Kinder (1959) in regard to psychosurgery patients. They indicate that a slow course of deterioration could emerge long after the surgery has been performed. We can report one particular case in which a patient who sustained a frontal lobotomy many years earlier showed CT-scan evidence of a large amount of cortical atrophy and ventricular dilation, apparently associated with the scar tissue produced by the surgery. It would be very important to know whether rehabilitative efforts may aid in diminishing the impact of this postulated slow, delayed process of deterioration following surgery or head injury.

Follow-up Studies

The history of head-injured individuals has been studied in much detail and on numerous occasions. The usual circumstance that allows for such a study is war. Starting with World War I, every major war has occasioned at least one large-scale study of brain-wounded veterans. Following World War I, the major studies were done in Germany under the direction of Kurt Goldstein (K. Goldstein, 1939, 1942). A. Earl Walker is probably the most prominent neurosurgeon associated with follow-ups done in the United States after World War II and the Korean conflict (Walker, 1957; Walker & Erculei, 1969; Walker & Jablon, 1959). In the studies of Walker and his colleagues, great emphasis was placed on the problem of epilepsy—but such areas as aphasia and general social and vocational adjustment were also considered. These studies tended to be in the nature of statistical follow-ups rather than basic science-oriented research. Nevertheless, they reported some highly important findings from a rehabilitative standpoint. For example, in a follow-up of a large group of head-injured men, Walker and Erculei (1969) found that there was no difference between level of functioning at five years and fifteen years after wounding. In other words, there was no deterioration over a substantial number of years. Although the mean age of the group at the time of the 15-year follow-up was about 40, and changes may have taken place later on, they apparently did not take place between young adulthood and early middle age. Another interesting finding was that the most commonly occurring posttraumatic complaint was "nervousness," defined as a tense, jittery feeling with an element of anxiety. The other very frequent complaint was headache. Neurological examinations of the follow-up cases revealed a high incidence of residual neurological deficit. Of 249 men examined, only 50 exhibited no neurological abnormality. The remainder exhibited one or more of a series of defects including mental impairment, hemiparesis, aphasia, and other abnormalities of perceptual or motor function.

One may accept the work of Walker and his colleagues as straightforward statistical studies of men who sustained head injuries. However, from the standpoint of rehabilitation, it is important to look at the implications of these studies. In a sense, they carry a double message. First, on an optimistic note, many head-injured veterans become employed and go on to live productive lives. Indeed, about 60% of Walker and Erculei's sample were regularly employed fifteen years postinjury. On the other hand, 40% were unemployed. Furthermore, of those employed, 25% felt that the jobs they held were less demanding than those they had held before the injury. Second, the other outstanding problem is epilepsy. Walker and Erculei found that in a group of 114 unselected head-injured men, 27.2% had some form of seizure problem.

In the area of head injury, there are other studies involving analyses of residual effects, but they go far beyond what we would ordinarily refer to as follow-ups. Rather, they were basic science-oriented investigations that looked in detail at the relationships among various parameters of the brain injuries and behavioral outcomes. We are referring to the body of research accomplished by Hans Lukas Teuber and his group initially at New York University (Teuber, 1959) and subsequently at M.I.T., as well as the Oxford studies conducted by Freda Newcombe and her collaborators (Newcombe, 1969, 1974). We will review certain aspects of these studies when we discuss neuropsychology in greater detail, but at this time, we mention these investigations to illustrate the point that a follow-up need not only involve statistical surveys of major and obvious sequelae of head trauma. Rather, they can involve detailed investigation of rather specific aspects of those sequelae. In any event, both the broader and more detailed studies make one point that is perhaps obvious but that should be stated. Injuries to the head sustained during wartime or peacetime are likely to leave permanent residual damage that may have strong implications for the victim's adaptation for the remainder of his or her life. Although successful surgical treatment immediately following the injury may reduce the number and severity of residual defects, it rarely, if ever, leaves the patient with no residual defects. In the case of head injury, it can rarely be said that a patient is treated to the extent that he is as "good as new." Patients and relatives sometimes develop the mistaken belief that since there has been a good cosmetic outcome and since there is no gross physical handicap, then there has been a complete recovery. Unfortunately, individuals who have suffered head injuries are often left with residual defects that involve more or less subtle changes in such areas as intellectual functioning, speech, and complex perceptual and motor abilities, even though there may be no deformities of appearance or gross physical handicaps. Furthermore, various difficulties may emerge well after the injury was sustained. From a rehabilitative standpoint the nature of these changes leads to numerous problems, some of which will be outlined.

There have been very few follow-ups of patients with nontraumatic brain damage. Perhaps the classic studies in this area were done some years ago by Chapman and Wolff (1956, 1959) with brain tumor patients. A great deal of clinical information regarding the behavior of such patients in natural settings was derived from these studies. We have already mentioned the follow-up studies of stroke patients accomplished by Meier (1970, 1974) and his group, and of aphasic patients by Basso, Faglioni, and Vignolo (1975). The group at NYU continues to conduct rehabilitation outcome-oriented follow-up studies with both stroke (Ben-Yishay, Gerstman, Diller, & Haas, 1970) and head-injured (Ben-Yishay, Rattok, & Diller, 1979) patients. In regard to follow-ups of both trauma and nontrauma patients, it is difficult to formulate any comprehensive conclusions except, perhaps, to say that one might note a trend from statistical surveys through basic neuropsychologically oriented research to treatment-outcome research. Perhaps we have progressed from the study of the natural histories of these conditions to the study of the effects of intervention on these histories.

Major Clinical Problems in Rehabilitation

Current rehabilitative therapy for neurological disorders is at its best when the patient has some specific deficit that can be directly treated by some established method—the best example being epilepsy. There are now a number of medications that are highly effective for controlling the incidence of epileptic seizures. Like most powerful medications, they must be used cautiously, and there may be side effects, such as gum hyperplasia or excessive somnolence. However, the most commonly used antiepileptic drug, diphenylhydantoin (Dilantin) is relatively safe for long-term use if taken in the proper dosage. Most physicians would prefer having their epileptic patients seizure free rather than not take the small risk of the usually relatively minor side effects of intelligent use of anticonvulsants. When these medications are not effective, surgical procedures can sometimes be used (Ojemann & Fedio, 1968; Penfield & Roberts, 1959) to remove epileptogenic scar tissue. Treatment of certain physical handicaps associated with brain damage is also frequently specific and effective (Rusk, 1977). There is a whole field called *rehabilitative engineering,* the practitioners of which are engaged in the design and utilization of prosthetic devices for the physically handicapped. There is also the field of *physical therapy,* in which attempts are made to improve motor function through exercise and a variety of treatments. Specific language deficits are also often amenable to direct treatment by speech therapists. A major problem, however, is that some brain-damaged individuals, indeed probably the majority of brain-damaged individuals, do not have specific defects of types that are directly treatable by established methods. Furthermore, individuals with specific problems, such as epilepsy, often have associated problems that are not solved by the direct treatment of the epilepsy. We often speak of the patient who "falls between the cracks"; the one who is not an appropriate candidate for physical therapy, speech therapy, anticonvulsant therapy, or other established treatment.

Aside from the problem of the patient for whom there is no established treatment, there is another group of patients that presents a particular problem for rehabilitation planning. For example, in terms of practical clinical decision making, we often think of a kind of "triage" of brain-damaged patients. There are those who are so physically and/or mentally debilitated that little can be done other than to provide supportive nursing-oriented care; there are also those who recover well, and, while they may have residual defects, these do not prevent the patients' employment and their adequate adjustment to community life. The third group, the one that provides the difficulty, constitutes those patients who do not have profound residual physical and neurological defects, but who do not make a good adjustment. Their level of adjustment cannot be fully explained on the basis of their objective physical and other neurological handicaps. A good example of this condition is to be found in Walker's 10-year follow-up of patients with posttraumatic epilepsy (Walker, 1957). It was found that 42% of the patients who were described as having "functionally normal" nervous systems were unemployed.

Numerous explanations have been offered as to why this phenomenon occurs. They may be divided into practical, psychiatric or psychological, and neuropsychological explanations, although there is a good deal of overlap. Those offering practical explanations point to the issue of compensation. They rightfully suggest that financial remuneration for disability is often a crucial matter with regard to rehabilitation planning. Some clinicians even speak of a "compensation neurosis." Individual values concerning whether or not one should seek and maintain employment when one is eligible for compensation based on disability enter the picture. From a rehabilitative standpoint the issue is whether or not the availability of disability compensation provides a detriment to the patient's ability to achieve what is optimal. When the patient is placed in a situation in which illness must be maintained in order to preserve his or her sole source of support, treatment efforts may be compromised. It is important for advocates of this explanation to separate its clinical implications from issues related to ethical considerations. Good treatment can also be compromised on the basis of the perhaps mistaken view that some particular patient is feigning maintenance of his or her illness to preserve compensation. This level of reasoning is often simplistic in that it does justice neither to the variety of motivations that may be guiding the patient's behavior nor to the impact of the social system that sets the rules for how disabled people should be compensated. Nevertheless, it is often of clinical value to know whether or not a patient is feigning illness. Lezak (1976) presents several procedures that may be used in determining whether or not a patient is feigning neuropsychological deficit. These procedures are based on the evaluation of inconsistencies in performance of a type that is incompatible with an explanation in terms of central nervous system dysfunction. For example, on a dot–enumeration task, the malingering patient may not show the expected gradual increase in response time as a function of number of dots in the array.

An explanation that overlaps with the practical one offered above is that there is a discrepancy between the patient's *actual* and *potential* level of functioning because a number of nonneurological symptoms have been acquired. Some would say that such patients have functional psychiatric disorders, while others would think in terms of psychological considerations involving emotional and motivational difficulties interfering with performance. Thus, brain-damaged patients may not be functioning up to potential not directly because of the brain damage but because of processes like disabling anxiety, depression, feelings of anxiety, and the like. Some students of brain disorders, particularly Kurt Goldstein and Martin Scheerer (K. Goldstein, 1939; K. Goldstein & Scheerer, 1941) have placed great emphasis on the manner in which the brain-damaged patient adjusts to his or her deficits. As in the case of other disabilities, there may be a whole continuum of adjustments ranging from appropriate and effective compensation to nonproductive and maladaptive reactions. K. Goldstein (1942, 1959b) has written most extensively and most wisely about the adjustment of brain-damaged individuals to their defects. More recently, Lezak (1979) and Kaplan (1979) have taken a renewed interest in this area. A major problem in conceptualization in this area is the difficulty in determining whether an emotional disorder is a reaction to a brain disorder or a product of it. As E. Valenstein and Heilman (1979) have pointed out, many central nervous system lesions can in themselves generate emotional disorders. We will return to this matter in a later chapter.

The Problem of Adjustment

Kurt Goldstein (1942) made an interesting and important distinction between two kinds of adjustment—*yielding* and *compensating*. In *yielding*, the person gives in to the defect in a way that demands little voluntary activity. *Compensation* involves building a counteracting mechanism on a volitional basis. Yielding is viewed as a more natural process and will persist if it assures adequate adjustment. The failure of either type of adjustment may lead to a catastrophic reaction, or to an inability to cope with the environment, associated with marked anxiety and distress. Thus, within K. Goldstein's framework, the brain-damaged patient seeks ordered behavior and avoidance of catastrophic reactions through a process of yielding to the defect or building some compensatory mechanism around it. More recent developments in neuropsychology have called our attention to the idea that there may be a strong interaction between the mode of adjustment and the site of the brain lesion. In particular, it has been suggested that left-hemisphere patients become depressed as a normal response to a catastrophic situation, while right-hemisphere patients develop a denial or indifference reaction reflecting an abnormal mood state. This suggestion was strongly supported by a study of Gainotti (1972) utilizing subjects with left- and right-hemisphere brain disease, but Hécaen (1969), Olsen and Ruby (1941), and Luria (1973) have all noted denial syndromes in patients with right-hemisphere disease.

It would appear that in the process of coming to terms with the environment after sustaining brain damage, the forces at work often involve interactions among various parameters of the lesion and the premorbid personality of the individual. Although brain damage does not generally change the personality radically, it tends to amplify characteristics that existed to a lesser degree premorbidly. Thus, for example, people who were somewhat compulsive premorbidly may become highly rigid and ritualistic. Individuals harboring mild underlying feelings of hostility may become excessively hostile, irritable, and belligerent. The basis for this commonly observed clinical phenomenon is not clear but is often put in terms of disinhibition. The cognitive processes that modulate and regulate behavior somehow become impaired with many kinds of brain damage.

In regard to the matter of adjustment, it often seems that characteristics generally viewed as desirable and positive in the normal individual are not necessarily the same characteristics that augur a satisfactory adjustment to brain damage. For example, individuals who are premorbidly aggressive, achievement oriented, productive, highly motivated, and active sometimes adjust poorly to their neurological defects. The aggressiveness turns into uncontrolled anger and the high energy level turns into agitation. In K. Goldstein's sense, these individuals somehow do not yield sufficiently to the defect, but, rather, attempt to compensate in ways that are not always entirely successful.

Since the time of Goldstein and Scheerer, little has been written about the problem of adjustment in brain-damaged patients. Small (1973) considered some of the relevant issues from a clinical standpoint, but his book was written for psychotherapists and is about neuropsychological assessment and various neurological syndromes. The problem of how brain-damaged patients adapt to and cope with their deficits is nevertheless, in clinical terms, a highly relevant one. Physical disability literature (e.g., Wright, 1960) may provide some helpful concepts, but brain damage has the unique characteristic of not only producing deficits but also of generating impairment of the adaptive characteristics required for coping with those deficits. There is an extensive literature in this area, some of which is reviewed in the chapter cited above by E. Valenstein and Heilman (1979). Luria (1948/1963) has also written about the problem of motivation in brain-damaged patients. The basic point is that brain damage not only affects cognitive, perceptual, and motor abilities but may also greatly affect the individual's emotional life and motivation. We will review this material in greater detail later. There are parts of the brain, notably a group of structures called the limbic system, that have a great deal to do with the mediation of affect and emotional states. The frontal lobes are also thought to have a great deal to do with the development of intentions and of goal-directed behavior. Damasio (1979) describes changes in goal-oriented behavior as the most characteristic changes resulting from frontal lobe damage. It would therefore appear that the way in which the individual adjusts to brain damage reflects some complex interaction between the premorbid personality and the nature of the damage itself.

Another issue that enters the adjustment picture is severity of damage. Severe brain damage, particularly if the onset is sudden, can abruptly change a functioning, productive individual into someone who is severely demented and incapable of working or engaging in other forms of productive, independent living. In such cases, the premorbid personality is, in a manner of speaking, lost in the severity of the insult. People who sustain very severe head injuries or cerebral anoxia often present this picture. Working with these patients, one may experience great difficulty in accepting the preillness history when it is contrasted to the patient's present level of functioning. This phenomenon can become especially poignant if the patient has achieved some high level of success. To learn, for example, that some severely demented patient on a hospital ward was a physician, engineer, or accountant some months ago can be particularly disconcerting.

Adjustment to brain damage is a highly complex matter, and the issue with which we began—the discrepancy between objectively determined disability and actual performance level—is not subject to simple explanation. Even though remunerative considerations and psychiatric disorder accompanying brain damage may play important roles, factors associated with both the actual brain damage and the mode of adjustment or coping style selected are also highly significant elements.

The individual responsible for rehabilitation of the patient has the problem of determining the source of the discrepancy between what seems to be the patient's potential level of functioning and the actual level. The point we would like to make here is that the issue should not be prejudged. It is too simple to say that it is because the patient is depressed or because he does not want to lose his pension. Certainly, brain-damaged patients become depressed and concerned about maintaining their means of livelihood. However, invoking automatic explanations of this type may lead to significant misunderstandings of the actual condition of the patient. Such misunderstandings may have significant implications for rehabilitation planning.

The remarks of Critchley in regard to the postconcussion syndrome (Walker, Caveness, & Critchley, 1969) put the matter remarkably well:

> Current thought as to the nature of the postconcussional syndrome would appear to be ambivalent, if not ambiguous. It is difficult to deny the clinical evidence as to the comparative—or apparent—rarity of severe enduring disability unless there is a factor of motivation. Likewise it does not seem possible to write off the widespread histological changes, however subtle they be. Can a compromise be achieved? Can it be that the postconcussional syndrome is the expression of minimal brain damage, which is potentially reversible, and that the picture of gross and protracted disablement is due to a neurotic overlay mediated at a level of awareness which varies from deep to shallow? (p. 6)

Neuropsychological Explanations

There is still another way of looking at the potential versus performance discrepancy. When we say that an individual with a disability is not doing as well as objective signs suggest he or she should, we generally mean a certain limited number of things. If a physical handicap is present, there may be a discrepancy between the degree of disability it imposes and what the patient is actually doing. In the case of brain damage, we generally mean that at the time of examination, the patient is found to have a functionally normal nervous system, but is not behaving as though that were the case. A functionally normal nervous system is generally interpreted to mean that

1. The mental status is grossly normal.
2. There is no seizure disorder and the electroencephalogram is normal.
3. There is no significant impairment of motility or sensation; that is, the patient has normal movement of limbs, eyes, and other moving body parts and does not have significant impairment of vision, hearing, taste, smell, or sense of touch.
4. There is no aphasia or other significant impairment of expressive or receptive aspects of language.
5. There is no disabling physical condition of other types, such as severe persistent headache, intractable pain, or severe dizziness.

It is possible, and indeed not uncommon, to see patients with histories of brain damage who meet all these criteria. In particular, in the case of head trauma, when a neurological examination is administered some years after the injury, the diagnosis may turn out to be normal and the injury can only be documented by the preillness history, or perhaps by a scar. Despite this picture of apparent normalcy as determined by the above criteria, these patients often do not do well. They may become long-term chronic patients in psychiatric institutions, or may remain at home and become unemployed, or may develop other behavioral or psychiatric difficulties. An analogous situation is seen in children who do not reveal abnormalties on the neurological examination but who show marked learning and behavioral difficulties that appear as though they may be related to some kind of brain dysfunction. Such children have been widely studied (cf. Denckla & Heilman, 1979), and there is an extensive nosology of the various disorders that may be acquired (e.g., dyslexia, hyperactivity syndrome, etc.). The current Diagnostic Manual for Psychiatric Disorders (DSM-III) uses the term "attention deficit disorder" to describe many of these conditions and explicitly points out that negative findings with regard to central nervous system dysfunction should not rule out making that diagnosis.

An explanation often proposed for these phenomena is that both children and adults who have sustained brain damage do not always exhibit the gross kinds of symptoms seen during the ordinary neurological examination, but do have neurological dysfunction of a more or less subtle nature. This kind of dysfunction, although not apparent in the physical examination of the nervous system, may have a highly significant influence on the individual's adaptation, particularly to a complex environment. In general, the dysfunction we make reference to is in the area of cognitive, perceptual, and motor skills. It exists, in other words, at the level of complex behavior but may not be detectable at the level of reflexes and sensory thresholds. It is often useful to remind ourselves that the brain is the organ of behavior and is responsible for mediating a wide range of activity, from simple reflex mechanisms to the highest of cognitive functions. A professional or scientific interest in the nature of higher brain functions is the essential feature of the field of neuropsychology, and, in recent years, a neuropsychological approach to the rehabilitation of brain-damaged patients has developed. One tenet of this approach appears to be that the apparent potential versus performance discrepancy under discussion may be a product of undetected neuropsychological deficits. In order to examine this proposition, it is necessary to provide some brief introduction to the field of human neuropsychology.

What Neuropsychology Is

In Chapter 5, we will provide a background of how neuropsychological testing may be used in the planning and assessment of rehabilitation programs. Here we will only discuss some general considerations concerned with providing some connecting links between the fields of neuropsychology and rehabilitation. There are several texts that offer introductions to various aspects of human or clinical neuropsychology. From the point of view of neuropsychological assessment in particular, Lezak's (1976) text is helpful, whereas a broad, general introduction to the entire field is provided by Walsh (1978). These two books can supply a more detailed introduction to the field than can be provided here. As a broad definition, neuropsychology is the scientific study of the relationships between brain function and behavior. Human or clinical neuropsychology traditionally is concerned with the study of brain-damaged patients, but the field has recently been expanded to include studies of normal individuals (e.g., Kimura & Durnford, 1974) and psychiatric patients (G. Goldstein, 1978). Although human neuropsychology is sometimes seen as being synonymous with neuropsychological testing of patients, it also has a basic research component that does not necessarily involve clinically oriented psychometric testing.

The rehabilitation of brain-damaged patients in most clinical settings has not been heavily influenced by neuropsychology. Typically, the fields of physical and rehabilitation medicine, and to some extent, clinical psychiatry, have played the major roles in the rehabilitative aspect of treatment of brain-damaged patients. Patients with major physical difficulties associated with their brain damage tend to be treated by physiatrists, whereas patients with chronic impairment of intellectual function tend to be treated by psychiatrists. In many hospitals, the department of rehabilitation medicine is most concerned with the physical aspects of rehabilitation and may also generally administer speech therapy. Such departments are typically headed by physiatrists and staffed by a number of specialty therapists, such as physical therapists, occupational therapists, speech therapists, manual arts therapists, and music therapists. It would be exceptional to find a neuropsychologist in a department of rehabilitation medicine, but there are some.

In the area of pediatric neurology, educational therapists, teachers, and school psychologists seem to fulfill roles equivalent to the specialty therapists in adult settings. It is only recently that some interaction has developed between neuropsychologists and rehabilitation specialists. Perhaps the clearest examples of how such interaction works are the programs at the New York University Institute of Rehabilitation Medicine and the Wichita and VA programs, which will be discussed in this book. The two fields are beginning to discover each other, and there appears to be some increase in the number of neuropsychologists found in rehabilitation settings. Rehabilitation specialists are becoming increasingly aware that a more detailed understanding of cerebral function might aid in the treatment of their brain-damaged patients. Neuropsychologists, too, are discovering that it is possible to go beyond diagnostic evaluations in the direction of recommending and monitoring treatment programs.

Because of the recency with which human neuropsychology has become identified as a specialty area, the professional qualifications and educational background of individuals in this field may not be widely known (Meier, 1981). Historically, neuropsychology grew out of the neurology and sensory physiology that developed in central Europe and England during the nineteenth and early twentieth centuries. Boring (1950) provides an excellent overview of this era. The specialty of neuropsychology developed among individuals trained in medicine or psychology who focused their interests on the relation between "brain" and "mind," or, in more contemporary terms, between brain function and behavior. The field has a branch involving animal research—with which we will not be concerned. Those involved with humans are generally described as human or clinical neuropsychologists. Individuals with medical backgrounds may describe themselves either as neuropsychologists or as behavioral neurologists (Pincas & Tucker, 1978). At this point, the professional distinctions are not clear-cut. Physicians trained in neurology who are more interested in higher cognitive functions and correspondingly less interested in reflexes, spinal cord injury, and elementary sensory processes, sometimes function, in essence, as neuropsychologists. Individuals trained in clinical psychology who specialize in assessment and treatment of brain-damaged patients are also now described as neuropsychologists. It is generally expected that those trained in psychology have a basic understanding of the nervous system, whereas medically trained neuropsychologists have some degree of training in psychology.

The Neurological Contribution

The contribution that neuropsychology can potentially make to the rehabilitation of brain–damaged patients is based on two considerations: (1) A sophisticated, scientifically based theory of brain function and of the consequences of brain lesions would seem to be essential for rational treatment planning. (2) Neuropsychology has developed, as a result of many years of research and application, assessment tools that have the capability of providing detailed analyses of brain function in individual patients. In essence, neuropsychology may be able to provide a rational approach to rehabilitation planning based on contemporary knowledge of brain function and on detailed evaluations of the behavior of individual brain-damaged patients. One criticism frequently made of clinical neuropsychology is that it tends to stop short at diagnosis and does not consider the implications of diagnosis for treatment. In the area of brain disorders, the usual tight association between diagnosis and treatment does not exist. In general medicine, the establishment of the diagnosis often implies the treatment. If diabetes mellitus is diagnosed, the patient is placed on the proper medication to restore normal glucose metabolism or is put on a low-glucose diet. If an infection is discovered, the patient may be given antibiotics. If a brain lesion is discovered and localized, the treatment is not as clear-cut, with the exception of rarely occurring cases in which neurosurgery or some specific medication shows promise. Even in the case of neurosurgery, however, there is often a major rehabilitation problem even if the surgery is successful.

Even though clinical neuropsychology has traditionally been diagnosis oriented, there is now some movement toward its becoming more involved in treatment. How this may be done brings us back to the matter of the discrepancy between objective deficits and actual level of performance in some brain-damaged patients. The issue here concerns the nature of the so-called objective deficits. If an individual who has sustained a brain lesion comes out of it partially paralyzed, blind, or grossly demented, it is apparent that such an individual would have substantial limitations of functioning capabilities in a normal environment. If epilepsy emerges as a significant problem following brain damage, seizures may be directly observed. The functioning changes related to this disorder are generally apparent. Perhaps the patient should not drive, swim, or work around dangerous equipment if the seizures are not being controlled by medication or other treatment procedures. As has been pointed out by neuropsychologists and behaviorally oriented neurologists, many brain-damaged patients do not exhibit gross, apparent defects (Reitan, 1966b; Teuber, 1959). Such patients may not be blind, paralyzed, grossly demented, or epileptic. Nevertheless, they are brain damaged and may be having significant problems with adaptive functioning. Neuropsychologists believe that, in many instances, the reason for the poor functioning is not entirely motivational or emotional in nature but rather largely the result of cognitive, perceptual, and motor deficits. Such deficits are often associated with various forms of brain damage. While they may not be apparent on the routine medically-oriented neurological examination, they may become apparent on a neuropsychological examination.

If we return to Walker and Erculei's (1969) follow-up study, their major categories of neurological examinational findings are: (1) hemiplegia, hemianesthesia, and/or hemianopsia, (2) aphasia, (3) isolated cranial nerve findings, and (4) epilepsy. We hasten to add that these investigators also used neuropsychological tests as part of the follow-up. The point is that in instances in which normal function is defined as absence of blindness, sensory loss, aphasia, cranial nerve dysfunction, and epilepsy, a great deal of significant impairment of brain functions may go undetected. It is not uncommon for an individual with a history of documented brain damage to have a normal physical neurological examination and electroencephalogram. However, such an individual may exhibit one or more cognitive, perceptual, and motor deficits on neuropsychological tests. In essence, the neuropsychological point of view may be summarized as follows: Brain damage is commonly associated with psychological as well as with physical deficits. The absence of physical deficits, such as those elicited by the routine neurological examination, does not imply normal brain function, nor does a normal electroencephalogram or related laboratory procedure.

Perhaps the best example of the consequences of the above considerations may be seen in the stroke, or "cerebral vascular accident,"[1] patient. Initially, stroke was thought of in terms of paralysis of a side of the body. Under the term *apoplexy,* an archaic term for stroke or cerebral vascular accident, the medical dictionary (Stedman, 1972) says, "A classical term for cerebral hemmorhage, thrombosis, embolism, or vasospasm usually characterized by *some degree of paralysis";* [author's italics] (p. 91). It later became apparent that patients with strokes involving the left- or dominant-cerebral hemisphere also tended to develop aphasia, with other mental changes also noted. Thus we have DeJong's (1967) definition of vascular lesions involving the anterior cerebral artery:

> a severe contralateral hemiplegia of the spastic type, the leg being affected more than the arm or face. There may be mental symptoms in the nature of memory loss, sluggishness, emotional lability, confusion and disorientation, and at times actual dementia. If the lesion is left sided, there is some expressive aphasia which may be temporary, and there is apraxia on the right side which may be masked by the paresis. If the lesion is right-sided, there is left-sided apraxia which, again, may be masked. It is stated that there may be a left-sided apraxia with a right-sided hemiplegia in lesions of the left anterior cerebral artery. (p. 882)

This partially quoted definition is quite sophisticated and, in a way, specific. The neuropsychologist, however, would criticize it and supplement it in a number of ways. First, terms like "mental symptoms" and "dementia" may not be sufficiently specific. Second, while aphasia is mentioned, the corresponding symptoms associated with right-cerebral hemisphere occlusion or rupture are not mentioned. Although the possible presence of memory deficit was mentioned, there are numerous kinds of memory disorder. For example, individuals who suffer a left-hemisphere stroke may have difficulty with verbal memory, while right-hemisphere involved patients may have their particular difficulty with nonverbal memory (Butters, 1979). Neuropsychological research has also shown that some patients have memory disorders associated with storing or recording new information, whereas others

[1]The term *cerebral vascular accident* is currently not widely used. Either the condition is specified when known (e.g., cerebral thrombosis) or the term *stroke* is used.

have particular problems with encoding or organization of incoming information (Butters, 1979; Butters & Cermak, 1976, 1980). To the neuropsychologist, the changes associated with stroke cannot be completely limited to hemiplegia, aphasia, and dementia. Even though neurologists such as DeJong (1967) are quite helpful in regard to pointing out differential consequences associated with damage to specific arteries, this precision is not maintained in the behavior area, where the tendency is to use such global terms as *dementia* and *memory impairment* in describing classes of behavior that are much more complex and differentiated than is implied by these terms.

The significance of refined assessment as a precursor to rehabilitation planning can perhaps be best illustrated with a number of examples. The first of them has to do with a problem called *inattention,* or imperception, extinction, or neglect. It was reported in detail by the neurologist Morris Bender (Bender, 1952) and has continued to be a matter of much research and clinical interest. A review of the recent literature is contained in a chapter by Heilman (1979). His general definition of the phenomenon is: "Under a variety of stimulus and performance conditions, patients who do not have elemental sensory or motor defects fail to report, respond or orient to stimuli presented to the side contralateral to a cerebral lesion" (p. 268). The phenomenon may occur in the tactual, auditory, or visual modality. Essentially it involves a failure to attend to one half of space either on the surface of the body or in the external environment. For example, patients with visual neglect might not attend to what is occurring in their left or right visual fields. The interesting point here is that these patients do not have the half-field type of blindness that is sometimes associated with stroke or other lateralized neurological disorders. Their visual fields, as evaluated by standard opthalmological tests, are normal. Thus, it would appear that the phenomenon is attentional in nature rather than a primary sensory defect. It thereby becomes a good exemplar of a neuropsychological defect that is clearly associated with brain damage but that exists in the absence of readily detectable neurological dysfunction. It may be pointed out in this regard that the visual neglect of the brain-damaged patient does not have the visual field pattern or other characteristics of the "tunnel vision" seen in some psychiatric patients.

It is apparent that the individual who does not attend to events occurring in half of the visual world can have major adaptive difficulties. There is an apocryphal story of an individual with left-field inattention who was killed by an automobile that approached him from his left side. There has also been a systematic study of accident proneness in patients with inattention (Diller & Weinberg, 1970). A second example involves a symptom known as prosopagnosia (A. Benton, 1979; Hécaen & Angelergues, 1962). It is the inability to recognize familiar faces. Individuals with this symptom generally have right-hemisphere or bilateral posterior brain damage, but often do not have detectable loss of vision or constriction of the visual fields. Again, such a symptom could produce major adaptive difficulties but may not be detected on the electroencephalogram and the neurological examination of sensory and motor function. A final example involves a mechanism that we rarely think of. How do we tell the difference between the movement of an object across our eyes and the movement of our eyes across a still object? If we think about it, it is clear that the stimulation striking the retina may be the same in both instances. However, we certainly know the difference between when the world is moving and we are still and the world is still and we are moving. Teuber (1964) postulated that what he called a corollary-discharge mechanism mediates this distinction, and he found that individuals with frontal-lobe damage demonstrated defects in this mechanism. Adaptive ability must be impaired in the individual with this subtle kind of neuropsychological defect. In order to navigate in the everyday world it would seem to be crucial to distinguish between one's own movements and those taking place in the external environment. Individuals with corollary discharge defects do not ordinarily have any gross impairment of vision or of primary motor functions, and therefore their adaptive difficulties cannot be attributed to impairment of these more basic functions.

We could go on at great length with examples of subtle, behavioral defects definitely associated with brain damage but not detectable by the routine neurological examinational methods. Such examples may come from the areas of perception, motor function, memory, cognition, language, and language-related abilities, such as reading, calculation, and numerous other areas. The essence of human neuropsychology is the study of the relationships among these complex behaviors and brain function. Elaborate and sophisticated theories concerning the nature of these relationships have been developed by such leaders in the field as K. Goldstein (1951), Halstead (1947) and Luria (1973). Rather than pursue the matter here, we will refer the reader to a number of textbooks including Walsh's *Neuropsychology* (1978), Heilman and Valenstein's *Clinical Neuropsychology* (1979), and Hécaen and Albert's *Human Neuropsychology* (1978).

To summarize the main points of the neuropsychological viewpoint, the following considerations may be offered. It is often observed that brain-damaged individuals do not function as well as might be expected on the basis of their objective, physical, neurological status. Some propose that this phenomenon concerns monetary compensation and other practical considerations, whereas others suggest that these individuals have emotional-psychiatric problems that limit their capacity to express the capabilities they possess. Neuropsychologists tend to question the idea that these individuals have functionally normal nervous systems, despite the fact that on the physical neurological examination, the electroencephalogram, and related procedures there are no positive findings. Sometimes these individuals have more or less subtle behavioral defects that may be elicited by neuropsychological examination, and that may be associated in a cause and effect way with the locus and nature of the brain lesion. The term *neurobehavioral* has recently been invented to describe these phenomena and relationships. Often individuals showing such neurobehavioral deficits without any of the more traditional neurological deficits have very well-documented brain lesions. For example, Teuber's cases with "corollary discharge" deficits had very well-defined penetrating head wounds of the frontal lobes, but had, for the most part, no gross physical neurological defects that could be called on to explain their problem in discriminating between self-induced and externally produced movement. While the corollary discharge theory remains controversial from the point of view of replication of findings concerning the frontal lobe localization, it nevertheless is a plausible theory of the process taking place in the brain for mediating the function involved.

Why has this voluminous store of neuropsychological information not found its way into rehabilitation? Clinical neuropsychology as an independent discipline is relatively new and is only beginning to become well known and accepted. Clinical psychologists rarely find their way into rehabilitation medicine departments, and when they do, they frequently perform the more traditional duties of clinical psychologists such as personality assessment and psychotherapy. There is a relatively small group of psychologists specifically interested in rehabilitation (Wright, 1960), but their interests are spread over the wide range of general physical disability such as blindness or paralysis. Thus, one part of the answer may relate to matters of a professional nature.

Beyond professional considerations, there seems to be a more powerful reason for the minimal work done in behavioral approaches to the brain-damaged patient. Concisely, it can be described as the aura of pessimism—the belief that little or nothing can be done for the patient who has sustained structural brain damage. The damage is permanent and irreversible, and so are its consequences. Earlier in this chapter we presented arguments for the view that there may be recovery of function following brain damage, as long as the pathology is not progressive in nature. However, even those that accept this view may believe that whatever recovery occurs will happen naturally, and any treatment done is simply an accompaniment to the patient's recuperation. Nevertheless there are so many people with behavioral difficulties who have nothing wrong with their nervous systems and are therefore more likely prospects for significant improvement or recovery.

This point of view becomes particularly puzzling when one considers the relationship between outcome in brain injury and in various functional psychiatric disorders. As indicated above, most head-injured men of Walker and Erculei (1969) were working fifteen years after injury. Meier (1974) has been successful not only in documenting recovery from stroke but in predicting pattern of recovery on the basis of neuropsychological test performance. Contrary to these somewhat optimistic findings, the potential for recovery from such functional disorders as alcoholism or schizophrenia is not particularly favorable. Thus, the prognosis for recovery from certain commonly occurring kinds of brain damage is better than it is for some of the major functional psychiatric disorders. It is not our purpose here to convince clinicians that they really should be working with brain-damaged patients but rather to comment on the view that the absence of identifiable organic pathology in a psychiatric patient is necessarily associated with a better prognosis than is the case for a patient with organic pathology. We know of no evidence for such a belief and would raise some question concerning its corollary; that therapeutic effort is better spent on patients with "functional" disorders than on patients with "organic" disorders.

An Attempted Integration of Approaches to the "Potential versus Performance" Problem

As indicated, there are divergent opinions as to why some brain-damaged patients fall short of realizing what is regarded as their potential. In practical terms, we are talking about (1) failure to pursue educational goals, (2) failure to work, and (3) an unduly lengthy stay in various kinds of institutional settings. Our attempt to integrate these views will not be elaborate, but is based on the assumption that practical, psychiatric, and neuropsychological factors may be at work in the same patient. The point can probably be best made in the form of a number of examples.

With regard to chronically institutionalized patients, one is often hard put to answer questions regarding their persistent institutionalization following resolution of the acute illness that originally brought them to the hospital. This problem area, as it relates to psychiatric patients, has been considered extensively and will not be reviewed here. We will concern ourselves only with how it relates to the brain-damaged patient. Such patients are often diagnosed as having "organic brain syndromes," and they may remain on wards of mental hospitals for extended periods of time. When inquiry is made as to the reasons for their length of stay, several findings often emerge. Frequently, the patient and/or his family are economically dependent on continued hospitalization. The patient feels he is unable to work, or even if he can work, the money he can earn may not exceed his continued hospitalization-contingent pension. In addition to these considerations, the patient may have acquired a functional psychiatric illness. He may have become chronically depressed or schizophrenic or may have developed habit disorders, such as alcoholism or addiction. He may have become severely phobic about going outside the institution. Finally, examination of these patients may often reveal significant mental impairment associated with their brain damage. The interplay among these factors may result in a diminished capacity or willingness to cope with life's vicissitudes. As Zubin (1977) has pointed out in the case of schizophrenia, coping failures may often promote the development of psychopathology. He makes the distinction between *coping* and *competence*, the latter referring to abilities and skills, and the former referring to the attitude and motivation of the person who is faced with a task or some life stressor. In the case of brain-damaged patients, the brain damage may seriously alter competence but it may require a combination of environmental, motivational, and emotional factors to affect coping effort. Thus, the patient with the same brain lesion may cope with it effectively or may experience a coping failure. A long history of coping failures may engender significant secondary psychiatric problems. Thus, the long-term institutionalized patient may suffer from a history of failure in coping with the vicissitudes of life. Since there are individuals living in the community who have extensive brain damage equal to what is seen in many institutionalized patients, the brain damage itself is probably not the only reason for the continued institutionalization. We can place the additional factors under the heading of *coping effort*. The brain-damaged patient living in the community may cope better than the individual with similar brain damage who continues to live in an institution, or alternatively, the environment may have been kinder to the community dweller than to the institution resident.

A second example is the young brain-wounded war veteran who recovers well from the medical consequences of his brain damage but who does not make a satisfactory adjustment to community living. Carey, Young, Rish, and Mathis (1974) followed a group of subjects who sustained brain wounds during the Vietnam conflict and found that 40% of these subjects were not working or going to school 2–4 years after wounding. Perhaps more interestingly, 70% of the subjects who rated themselves as "moderately impaired" were not working or in school. It is understandable that severely disabled individuals would not be in school or at work, but the low level of productivity of these types in individuals with only moderate disability is puzzling. We may remind ourselves that these subjects are young, healthy men who have very little in the way of medical problems other than those associated with their brain wounds. If we may accentuate the positive, 30% of the moderately disabled subjects of Carey *et al.* were working or in school—although 70% were not. What makes the difference?

The necessary research has not yet been done in order to arrive at a reasonable answer to the above question. To do so, we would have to objectify what is meant by *mildly*, *moderately*, and *severely disabled.* Such objectification could be based on the results of a number of diagnostic and examinational procedures including the neurological examination, neuropsychological tests, a speech and language evaluation, the electroencephalogram, computerized tomography, and so forth. These tests could form an objective index of disability. We might add a psychiatric and psychosocial evaluation to determine whether or not the person is suffering from a functional mental disorder or living in a highly stressful life situation. We could then go on to assess the individual's current situation in terms of level of productivity. It may then be possible to formulate an index of the discrepancy between the examination results and the individual's level of productivity. No doubt, it would be found that some individuals show little discrepancy while others show a great deal. After ruling out those individuals who have severely disabling physical impairments, such as paralysis or blindness, there will no doubt still be many individuals left who are not functioning according to their potential. Some may have functional psychiatric problems; some may have subtle neuropsychological deficits; some may be living in untenable situations; some may be content with, and protective of, their disability compensation. In other words, it may be demonstrated that the young brain-wounded veteran may not be functioning according to potential because of numerous considerations that may vary from case to case. More research is needed to make a more precise determination of what actually goes on, but the likelihood is strong that the neurological and neuropsychological deficits will not tell the whole story. Walker (1957) sums up this situation nicely in his statement, "The total picture must be examined, for the disturbance is more than the sum of the hemiplegia and the epilepsy" (p. 1636).

In summary, rehabilitation outcome following brain damage is a complex matter. It involves not only the brain damage itself but also the individual's capacity to cope with the deficits acquired and the environment's ability to adjust to the individual's altered state. Examination of the patient from neurological, neuropsychological, psychiatric, and sociological standpoints is important, but any of these examinations taken individually may not be sufficient to accurately predict outcome. It seems essential to know the specific combination of neurologically induced deficit, in addition to available coping skills and social setting.

Issues Related to Treatment and Rehabilitation

If one accepts the reasoning presented above it would follow that treatment or rehabilitation may consist of resolving the neurological and neuropsychological deficits, increasing the individual's coping ability, or modifying the environment. In fact, all three alternatives are often used jointly in the treatment of brain-damaged patients. We will briefly examine each of them separately.

Resolving the Deficit

To what extent have modern medicine and behavioral science devised methods for the treatment of neurological conditions involving brain function? In the case of epilepsy, great strides have been made. It is the exceptional case that does not respond to one or some combination of the various anticonvulsant drugs. On the other hand, such behavioral deficits as impaired conceptual abilities, amnesic disorders, or visual-spatial deficits remain recalcitrant to treatment. Various somatic and pharmacological therapies have been attempted, and we will examine some of them in the next chapter. Generally, behavioral methods constitute the most widely used form of treatment for cognitive and perceptual disorders. The science of speech pathology and the clinical practice of speech therapy represent the most distinct treatment-oriented disciplines in the area of brain disorders, but there is a growing interest in the application of memory training and in cognitive and perceptual retraining in general.

Those involved in this approach to treatment maintain the belief that some behavioral consequences of brain damage are recoverable to a certain extent. Great emphasis is placed on diagnosis and evaluation, since it is necessary to identify the deficit before it can be treated. Thus, there is a great interest in extended aphasia examinations and in neuropsychological testing. Ideally, the identification of the deficit is the key to treatment. If the patient has a memory deficit, give him memory training; if he has aphasia, offer him speech therapy; if he has a combination of deficits, tailor-make a program that deals with the deficit pattern. This form of treatment tends to be "school-like" in nature and involves specific instruction, exercises, repeated practice, and so forth. Its main difficulty, in and of itself, is that it assumes normal motivation; that is, it is dependent on the incentive of the patient to recover from his disorder. This consideration raises the complex issue of motivation in disorders of the central nervous system. Luria (1948/1963) has martialed much evidence supporting the view that patients with frontal-lobe lesions acquire substantial disintegration of motivation, while patients with lesions elsewhere in the brain do not do so to nearly the same extent. The clinical picture of the apathetic frontal-lobe patient who essentially cannot maintain sustained, goal-directed behavior is not at all uncommon. This lack of capacity to sustain purposive behavior has major rehabilitative implications for the frontal-lobe patient, since many rehabilitation programs require the expenditure of long hours of arduous practice. Since Luria's consideration of the problem of motivation was written shortly after World War II, major discoveries have been made concerning the relationships between certain structures in the central nervous system and various emotional states. E. Valenstein and Heilman (1979) reviewed much of this material, pointing out that lesions in various portions of the limbic system have major implications for emotional and motivational states. Interestingly, certain lesions can increase some drives (such as sexuality), while other lesions may decrease other drives. In any event, particularly in the cases of frontal-lobe and limbic-system lesions, the brain damage itself appears to produce deviations from normal motivation. States of apathy, inability to sustain goal-directed behavior, and rapid fatigability are commonly observed among many kinds of brain-damaged patients, and so it is often necessary to work with coping effort as well as with specific deficits.

Improving Coping Ability

Zubin (1977) makes an analogy between coping ability and the electrical voltage of a well-functioning motor. Although the motor may be intact, it will not do its job unless the proper amount of voltage is applied. It is interesting to note that too much or too little voltage may be disruptive. This principle, when applied to learning situations, is closely akin to the concept of motivation. The relationship between motivation and learning has been scientifically documented certainly since the time of early behaviorism and has been elaborated on in numerous ways within the context of the development of learning theory. The relationship between learning and reinforcement is clearly crucial here. It is unfortunate that not until recently has some association been made between reinforcement theory and the problem of brain damage as it appears clinically. In this regard, the heavy emphasis on the medical model in combination with the great emphasis on the impairment of cognitive, perceptual, and motor skills associated with brain damage has led to the relative neglect of the role that motivational factors may play in influencing brain function in general and recovery in particular. We mean something more specific here than the somewhat vaguely defined relationship between the desire to get better and the course of recovery. Few would deny that it is helpful, if not crucial, for the patient to want to get better if improvement is to occur. Rather, the point we wish to support here is that systematic attempts to manipulate motivational states and environmental contingencies in a manner that promotes learning may be beneficial with patients whose inner strivings to recover for one reason or another do not provide sufficient incentive for rehabilitation. It seems apparent that the form of treatment commonly known as behavior modification or behavior therapy may show promise in this regard.

A Rationale for the Use of Behavior Therapy in Rehabilitation of the Brain Damaged

Modifying the Environment

It is not our intent to provide a lengthy description of behavior therapy. The reader is referred to a text by Bellack and Hersen (1977) for an introduction to this form of treatment. Even though it has numerous historical antecedents (cf. Yates, 1970), since the 1960s behavior therapy has been rapidly emerging as a significant branch of clinical, rehabilitation, and educational psychology, having its own literature, journals, and centers. Its various methods have been applied to a wide variety of clinical and educational problems, and it has many advocates who use it as the major tool in their work with students or patients. It has its own technologies, instruments, and theories of behavior. Bellack and Hersen (1977) point out that there have been several attempts at defining behavior therapy, but they all make reference to the point that it is an empirically based form of treatment derived mainly from laboratory-type research in experimental and social psychology. In our view, most would admit that the strongest influence remains learning theory, although the earlier strong commitment to "behaviorist" or "S-R" learning theory has abated among some behavior therapists, and alternative conceptualizations such as Bandura's (1969) social learning theory have emerged. Even though popular views of behavior therapy may continue to regard it in terms of the use of reinforcement in the modification of behavior, such a conceptualization would now be viewed as an oversimplification.

While behavior therapeutic techniques have been applied to numerous populations over the course of many years, they have not been applied to any great extent to brain-damaged adults. Many of us may be familiar with the early token-economy-oriented work of Ayllon and Azrin (1968) with chronic psychiatric patients; the work of Lovaas and Newsom (1976) with autistic children; and the work of Baer and Wolf (1970) with the mentally retarded. More recently, behavior therapy techniques have been applied to alcoholics (Nathan, 1977), and a new field called "behavioral medicine" (Collins, 1981) has emerged. The question to be considered here is whether or not the successes gained in the above areas can also be achieved with brain-damaged patients. Later in this book, we will be presenting preliminary data that may answer this question.

There is a traditional belief that brain-damaged individuals have difficulty with new learning. There is much research evidence suggesting that many brain-damaged patients do indeed have difficulties in incorporating new information because of various forms of memory disturbance (Butters, 1979; Hécaen & Albert, 1978). This matter raises an important issue regarding the application of behavior modification techniques at least to some brain-damaged patients. The difficulty is an obvious one in the case of frankly amnesic patients. They may not be able to recall their reinforcement histories—praise or other reward for some desirable behavior given on one day may not be recalled at all on the following day. Indeed, the entire training session may be forgotten. Although such dramatic amnesic disorders are relatively rare, the extent to which symptoms of memory impairment are found in many, if not most, brain-damaged patients remains an open question. Thus, one difficulty is provided by a possible defect in the learning mechanism itself. There may be a comparable defect in the motivation mechanisms. We have pointed out that lesions in the frontal lobes or the limbic system may have important consequences for motivation. Our own experience with frontal-lobe patients has been similar to that of Luria (1973) in that these individuals may be extremely apathetic, and it becomes exceedingly difficult to find effective reinforcers.

On the other hand, the laws of learning are felt by many behavioral scientists to have some degree of universality, and so the applicability of those laws to brain-damaged patients becomes an empirical question. In regard to achieving the empirical demonstrations needed, we will be presenting the viewpoint that they can be accomplished most productively through a liaison between the fields of clinical neuropsychology and behavior therapy. The expertise of the neuropsychologist can be used to assess the cognitive, perceptual, and motivational deficits of the patient while the behavior therapy specialist has available a repertoire of potentially effective techniques, methods of assessment that differ from neuropsychological assessment, and research design skills that particularly suit behavior therapy studies (Bellack & Hersen, 1978; Hersen & Barlow, 1976). In any event, it is possible that the successes of behavior therapy in many educational and clinical applications may be repeatable in the case of brain-damaged patients, and it is felt that the prospects of success may be enhanced through input from neuropsychology.

Numerous examples of how this liaison may work in practical clinical situations will be given throughout this book. It will become clear that neuropsychological assessment can provide a quick and thorough description of assets and deficits to the behavior therapy specialist. In this manner, these procedures can be used not only in their traditional diagnostic role, but in rehabilitation planning as well. Typical questions that may be answered by neuropsychological assessment are whether the client should be approached with verbal or nonverbal stimulation, whether he has a memory or learning problem that necessitates frequent repetition of the same information, or whether his level of intellectual functioning is so impaired that instruction has to be carried on at a very basic, elementary level. Issues of this type can readily be assessed by the neuropsychologist who can then apply the obtained test findings to rehabilitation planning. Traditional clinical neuropsychology has provided careful and detailed analyses of behavioral deficits as they are found in patients with brain lesions of various types and in different regions of the brain. It has also dealt to some extent with the natural history of brain lesions from the point of view of recovery of function. However, it has not generally looked at deficits as potentially remediable behavior. It has not typically proceeded from observed deficits to programs for remediation for such deficits. With some important exceptions to be reviewed in the next chapter, neuropsychology has not been greatly involved in the development of training techniques and materials for treatment of brain-damaged patients. As we have already indicated, a major exception is the group of language-related disturbances known as *aphasia,* but this exception is largely owing to the fact that for many years there has been an established speech therapy profession.

Perhaps it would be useful to provide an example of our point. Let us take the case of a head–injury patient who has lost some of the functions of his right arm. The neurological examination and the neuropsychological tests suggest that the arm is normal with regard to sensation, muscle tone, and reflexes, but that the patient is disabled with regard to producing normally rapid, skilled movements of his right hand. This condition is known as *apraxia* (Heilman, 1979), which is usually defined as a disorder of learned skilled movement that is not produced by weakness, tremor, poor comprehension, or causes other than a brain lesion that specifically produces a skilled-movement defect. In other words, an individual can be apraxic because a limb is so unsteady or weak that it cannot be used to execute movements with normal skill level. Such a condition is not apraxia. With apraxia the involved limb may be quite strong and steady, but the individual still cannot perform skilled movements. We need not go into the neurological details of this condition, except to say that the problem is thought by some neuropsychologists to involve a disconnection of the motor areas of the brain from the areas in which so-called *engrams* for purposeful, skilled movements are contained. An engram can roughly be defined as a hypothetical representation in the brain of some unit of behavior. The formation of engrams is a way in which it is thought that the brain becomes modified by experience. In any event, apraxia is sometimes seen in patients who have had focal vascular lesions or head injuries. From the point of view of treatment, the more traditional physical therapy approaches to this kind of condition would probably not be useful, since the problem is not one of increasing strength or tone but of restoring skilled, dextrous movement. Thus, the recommended treatment might involve utilization of a pegboard or pursuit rotor or similar device with which the patient can practice, in the hope of improving capacity to perform skilled movements. Heilman (1979) recommends the teaching of alternative strategies. He gives the example of a patient with agraphic apraxia who could not write with a pen, but who could type.

Once the determination of the neuropsychological deficit is accomplished, behavior therapy can enter the picture with the aim of establishing a program that can optimize the patient's capacity to achieve maximal restoration of function in the briefest time. The role of the behavior therapist would be threefold: to conduct a behavioral assessment (Hersen & Bellack, 1981), to design a treatment program, and to implement and monitor that program as it progresses. The behavioral assessment might focus on those factors in the patient and his environment that could be related to treatment. One consideration might involve looking at environmental stimuli that may be associated with increases or decreases in the severity of symptoms: Does the patient become more apraxic when he is being watched by the doctor? Another aspect of the assessment might involve a search for appropriate reinforcers. In some cases it might be determined that simple instruction is sufficient, because the patient is aware of his disability and is highly motivated to achieve maximum restoration of function. In such cases, providing the patient with a training device, giving appropriate instructions, and requesting him to practice regularly may be all that is needed. In other cases, some external source of motivation may be required. Sometimes simple knowledge of results is sufficient reinforcement, and the patient's efforts will be sustained if he can see improvement over time. Knowledge of results may be combined with some form of goal setting. In the case of our apraxic patient, the following arrangement may be suggested. Let us say that the patient is right-handed, and that right-handed people do 20% better on the task being practiced with their right hand than they do with their left hand. Thus, a reasonable goal can be formulated in terms of normal performance or the performance of the patient's own unimpaired left hand. For example, the goal may be to make the impaired right hand as good as the unimpaired left hand.

The matter of whether the type of program described above will be successful is unresolved at present. Although there have been a number of isolated research studies, there has not been sufficient application of behavior therapeutic techniques to brain-damaged patients to allow for any conclusions. Retraining attempts have been reported in specific areas such as memory (Lewinsohn, Danaher, & Kikel, 1977), but a comprehensive, systematic behavior-therapy program for patients with a variety of deficits has not been accomplished. The results of the programs we will be reporting on later in this book represent preliminary attempts at implementing such research.

Psychopharmacological Approaches as an Adjunct to Behavior Therapy

It is not necessary to view behavior therapy as an alternative treatment to the by now more traditional pharmacological therapies. Stern (1978) has shown how various psychotropic medications can be used in productive combinations with behavior therapy in the treatment of psychiatric disorders. The effectiveness of many psychotropic agents, even when used independently of behavior therapeutic efforts, has been demonstrated in a voluminous psychopharmacological literature (cf. Lipton, DiMascio, & Killam, 1978). Although these drugs have their dangers, and some of them have adverse side effects, it would appear that their clinical usefulness in regard to treatment of anxiety, agitation, and depression has been well demonstrated. In our experience, most behavior therapists who work with psychiatric patients do not find drug withdrawal desirable, but only want to stabilize the drug situation such that it does not interfere with the behavioral treatment or its evaluation.

The problem with administering these medications to brain-damaged patients is that they do not seem to produce the same therapeutic benefits in such patients as they do in nonbrain-damaged depressed, schizophrenic, and anxiety-disorder patients. Indeed, Himmelhoch, Neil, May, Fuchs, and Licata (1980) reported that lithium, a highly effective medication for manic-depressive disorder, made mildly demented patients worse. Nevertheless, brain-damaged patients frequently become depressed, anxious, and disorganized, and it would be helpful if various psychopharmacological agents could be used as a means of increasing the coping ability lost because of those conditions. All that can be said at this point is that more research is required to determine what medications, in what dose levels, can be productively and safely used with brain-damaged patients. There is a clinical folklore that brain-damaged and elderly patients develop adverse side effects more readily than others, and that it is therefore necessary to prescribe psychotropic agents in very small dosages for them. We will be reviewing the matter of psychopharmacological approaches in greater detail in the next chapter. The only point to be made here is that behavior therapy and psychopharmacological approaches need not be mutually inconsistent. The major problem in using these agents with brain-damaged patients is that of hypersensitivity to many of the commonly used medications such that they become overly sedating, or contraindicated because of particularly adverse side effects or other medical considerations. The matter of drugs specifically developed to improve cognitive function will be discussed in the following chapter.

Rational Rehabilitation

The major theme of this book is that rehabilitation of brain-damaged patients can be more effective than it has been in the past if it is planned on what we would term a rational basis. The choice of this term does not imply that current rehabilitation efforts are irrational. The point is, rather, that they may be based on assumptions concerning brain function that are not completely accurate or sufficiently sophisticated, and that they are organized more around the availability of particular facilities and personnel than around the specific treatment requirements of the patient. By rational rehabilitation we mean rehabilitation that is planned on the basis of accurate, scientific knowledge of brain–behavior relationships, and that is designed to achieve some reasonably specifically articulated treatment goal. It begins with detailed assessment, because without assessment it is not possible to understand the nature of the patient's deficits. It is our view that neuropsychological assessment appears to be the best way of obtaining these kinds of data. However, other forms of behavioral assessment are needed to determine such matters as the patient's level of motivation, complicating emotional difficulties such as depression, and the psychosocial situation in which the patient exists. In addition to assessment, it is also necessary to develop an educational technology designed to remediate the variety of deficits that may be associated with brain damage. Training devices, instructional methods, and evaluation instruments must be invented or borrowed from other areas. Rehabilitation planning then becomes a matter of integrating the assessment material so that the technology available can be utilized to implement a treatment program addressed to the specific pattern of the patient's assets and deficits.

The above considerations clearly imply that there can be no such thing as a single method of rehabilitating individuals with brain damage. One of the unsophisticated assumptions alluded to above is that brain damage or "organic brain syndrome" is a single entity. Indeed, in glancing through a relatively recent text on rehabilitation (Hirschberg, Lewis, & Vaughan, 1976) the only passage we could find having to do with management of the patient with organic mental impairment contained these sentences: "(1) . . . the patient must be given an adequate amount of stimulation and (2) he must not be subject to excessive demands" (pp. 265–266). It seems clear that the term *organic mental impairment* implies a single entity, and the advice given implies a single treatment method. We are taking essentially the opposite viewpoint; the term organic mental impairment encompasses an enormous variety of entities, and a large number of treatment programs can be developed.

Numerous conceptual models developed within the framework of neuropsychology have implications for treatment. In one way or another, all of these models have been derived from investigations of patients with various brain disorders. One major area of such investigations has pointed out that there is a great discrepancy in behavioral outcome depending on whether the major brain damage is in the right- or the left-cerebral hemisphere (Dimond & Beaumont, 1974; Kinsbourne & Smith, 1974). The clear implication of these studies is that rehabilitation of patients with left-hemisphere brain damage is an entirely different matter from rehabilitation of patients with right-hemisphere damage. Basically, left-hemisphere patients tend to have language related problems whereas right-hemisphere patients tend to have problems with spatial orientation and constructional abilities. We will elaborate on this matter in detail later on. Luria's (1962/1966, 1973) theory is that brain damage results in the destruction of one or more functional systems. A functional system is a multicomponent organization that underlies some complex behavior or mental activity. Rehabilitation planning involves assessment of the functional systems that are preserved and impaired, and devising a method of restoring those behaviors that were affected by destruction of the relevant functional systems. As Golden (1978) has pointed out, this restoration can be accomplished by several means:

1. Getting another part of the brain to take over mediation of the impaired behavior.
2. Substituting a set of simple operations for a complex skill.
3. Substituting a complex skill for a basic skill.
4. Finding a new way to perform affected tasks that does not depend upon the damaged area of the brain.

We will not treat these points here but do want to show how a particular model of brain function (in this case Luria's) can provide a conceptual basis for rehabilitation planning.

It is our view that neuropsychological approaches to rehabilitation (such as Luria's) often do not provide a sufficient conceptual model. They tend to pay insufficient attention to the matter of motivation, and do not do justice to the patient's psychosocial status. When motivation is considered, it tends to be treated in neurophysiological terms, such as in the cases of limbic system and frontal lesions discussed previously. Motivation is rarely considered in purely behavioral terms, utilizing reinforcement-related concepts. For example, motivational constructs derived from the "learned helplessness" theory of depression (Seligman, 1975) may be quite applicable to brain-damaged patients. One often hears clinicians suggesting that brain-damaged patients do not remember because they do not want to remember. They do not conceptualize because they do not perceive the need to do so. Although neuropsychologists may scoff at this level of explanation, the problem of motivating brain-damaged patients to persist in often arduous rehabilitation programs remains a clinically significant one. Our attempts to unite behavior therapeutic with neuropsychological approaches to rehabilitation were largely instigated by our observation that brain-damaged patients frequently did not have the same level of desire for restoration of function that is commonly seen in patients with physical disabilities. The sources of motivational lag are generally unclear. They may be secondary to the patient's intellectual deficits, they may be directly related to structural or physiological alterations of the brain mechanisms underlying motivation, or they may be related to the same phenomena that reduce motivation in non-brain-damaged individuals, such as "learned helplessness" related matters. Nevertheless, a motivational problem is often present, and it often imposes severe limitations on what can be achieved through a purely didactic approach to rehabilitation.

A related issue involves the psychosocial situation. The patient's emotional status and the quality of the environment in which he finds himself are often crucial considerations. As we will see in Chapter 8, many brain-damaged patients have severe emotional difficulties, if not frank functional psychiatric disorders. Depression is often a major complication of structural nervous system disease. R. D. Adams and Victor (1977) pointed to the increased incidence of suicide in patients with progressive neurological conditions. Whittier (1977) traced relationships between Huntington's disease and psychopathology. He cited the following sentence by Huntington himself, referring to the disease bearing his name. "The tendency to insanity, and sometimes that form of insanity which leads to suicide, is marked" (Huntington, 1872) (p. 320). Thus, the relationship between certain neurological disorders and what we would now refer to as "functional psychiatric disorders" or "psychogenic conditions" was surmised at least in the nineteenth century. In addition to the patient's emotional or psychiatric status, the milieu is often a crucial consideration. Living in a situation of intellectual deprivation can apparently make matters worse than they have to be on the basis of the neurological disorder itself. Long-term institutionalization or living in a noninstitutional environment in such a dependent manner that all needs are satisfied without problem solving or other kind of effort can be expected to aggravate the deficits produced by brain damage. As in other areas of rehabilitation, a nonconstructive milieu can provide a serious handicap, and returning the patient to such a milieu can reverse some of the successes achieved in the rehabilitation program. Thus, the role of the family or the patient's caretakers in the absence of a family is often critical in regard to treatment outcome.

We have attempted to indicate that a rational rehabilitation program has to be above all based on a detailed assessment of the patient's deficits and preserved abilities. Neuropsychological testing is an ideal tool for providing such an assessment. However, issues related to educational methods and motivational considerations are also important. Ingenuity must be exercised to plan a program of treatment to remediate the observed deficits, and often it is necessary to plan some means of motivating the patient to persist in trying to meet the rehabilitation goals. Often, such efforts may be compromised by the presence of a serious emotional disorder and/or a psychosocial situation that may be working at odds with the treatment effort. In the remainder of this book, we will provide discussions of all these areas as they impinge on the rehabilitation of structurally brain-damaged adults.

2

APPROACHES TO THE REHABILITATION AND TREATMENT OF THE BRAIN-DAMAGED PATIENT

Introduction

As in most health-related fields, there is a great deal of variation in the ways in which brain-damaged patients are treated and rehabilitated. These variations are associated with a number of factors, including differences in treatment setting, professional training, and theoretical orientation. In this chapter, we will first discuss these variations under the general headings of treatment settings, the rehabilitation disciplines, and philosophical and theoretical orientations. This discussion will be followed by our attempt to develop a multidimensional approach to rehabilitation. We will include reviews of several components of the research and clinical literature regarding rehabilitation of brain-damaged adults. As pointed out in Chapter 1, the term *brain damage* covers a wide variety of disorders, so one can expect various interactions among patient variables and treatment settings, relevant professional disciplines, and approaches taken to treatment. These interactions will be discussed throughout the chapter.

The Treatment Setting

Brain-damaged patients have varied fates depending on a number of factors not the least of which may be chance. Chance is often a consideration in regard to whether or not an individual becomes institutionalized; for example, just happening to be observed in engaging in deviant or peculiar behavior in the community may lead to institutionalization in a psychiatric facility. Certain communities may be more or less tolerant of particular behaviors, and so whether or not one becomes institutionalized may depend more on community standards than on the condition of the individual, or the need for institutional treatment. The presence of a supportive, caring family or social network (Hammer, Makiesky-Barrow, & Gutwirth, 1978) may prevent institutionalization, while the opposite may promote it. Thus, the decision to treat a patient in an institutional or noninstitutional setting does not rest solely on clinical considerations. A number of extraneous factors enter the picture—many of which are fortuitous.

Thus, institutional treatment of brain-damaged patients need not be necessitated by the nature of their conditions but may be based on a number of socioeconomic and other environmental considerations. From a rehabilitative standpoint, the question should be that of determining where the best place to treat and rehabilitate the individual may be. Should it be at home, where he has the support of family and friends, or in a hospital where expertise of a large professional staff and an armamentarium of equipment and other treatment facilities are available? Perhaps a day hospital or sheltered workshop, in which the patient can live at home and go to the institution only for therapy, is the best solution. We are not raising this question so that we can say that any one of the alternatives is superior. Rather, we are initiating a discussion of those conditions under which each of the alternatives may be the most appropriate treatment setting. As many of us know, trends in the mental health movement have gone from the accentuation of hospital treatment to a preference for treatment in the community. There has also been a growing interest in continuity of care, such that the patient is followed through the course of acute treatment, rehabilitation, and follow-up by the same treatment team, or at least by the same agency or institution.

When brain-damaged patients become institutionalized, they generally go either to psychiatric hospitals or nursing homes. If the patients' symptomatology involves behavior that is bizarre, deviant, or disruptive, they are generally hospitalized in a psychiatric facility. If the difficulties are more physical than behavioral, placement is generally made in a nursing home or a chronic neurology ward in a large general medical hospital. Although such institutional facilities often give excellent, humane care, they are, unfortunately, often viewed as terminal facilities, with the explicit or implicit assumption that the patient will remain there for the rest of his or her days. We generally think of these institutions as facilities for the aged, though it is not uncommon to find relatively young brain-damaged patients on a ward consisting of mainly elderly patients. Social consequences of this situation can be awkward, but there is little alternative, since institutional facilities for young, adult, brain-damaged patients are essentially nonexistent. Generally, the only viable alternative is that of placing such patients on general psychiatric wards. This obviously has its drawbacks.

Whereas there is really no research literature that specifically addresses itself to the problem of institutionalization of brain-damaged patients, there is an enormous literature on institutionalization of both the elderly and the psychiatric patient. Since the early work of Stanton and Schwartz (1954) and Goffman (1961) it has been recognized that institutions, particularly what Goffman first described as total institutions, have significant consequences for the people living in them. In general, and particularly in the case of psychiatric patients, these consequences tended to be described as negative in nature. It now seems clear that the social structure of institutions appears to have significant effects on the symptomatology of the illness for which the patient is institutionalized. For example, Wing and Brown (1970) demonstrated that improvement in the ward environment was associated with clinical improvement in some schizophrenic patients. It was made clear in their study that it was not the patients' symptomatology that produced an impoverished ward atmosphere, but rather that an unstimulating ward atmosphere led to social withdrawal by the patients. It has also been shown that brief hospitalization is no less effective than long-term hospitalization in regard to successful treatment outcome (Caffey, Galbrecht, & Klett, 1971; Herz, Endicott, & Spitzer, 1975, 1977). Thus, it would appear that institutional life in unstimulating, monotonous environments can have detrimental effects on individuals, and that psychiatric patients, at least, can be treated as effectively with brief periods of hospitalization as they can with continued long-term hospitalization. These considerations have encouraged clinicians to think in terms of whether the advantage of sending a patient to an institution outweigh the negative influences institutionalization may have. The community mental health movement has come about partly because of the iatrogenic consequences of institutional life. As most of us know, in recent years there has been a large reduction in bed occupancy at state and federal neuropsychiatric hospitals, and a corresponding increase in the number of community mental health centers. Where possible, community treatment is generally viewed as the treatment of choice. When hospitalization is required, efforts are made to make it as brief as possible. Some authorities consider the problem of institutionalization as being so potent that institutionalization itself is described as a disease (Eisdorfer, 1980).

The consequences of the de-institutionalization movement are becoming known as we are able to study patients who have recently been discharged from institutions. Some years ago, for example, Michaux *et al.* (1969) did a longitudinal study of a group of psychiatric patients during their first year out of the hospital. Although these kinds of studies have provided reason for optimism (Bleuler, 1972), it would be fair to say that the major problem of rehospitalization has emerged. There have been several studies of the problem of relapse, particularly in schizophrenia (Hogarty, Ulrich, Goldberg & Schooler, 1976; Vaughan & Leff, 1976), which focus on the various conditions that promote and discourage relapse. Many psychiatric patients return to the hospital, and it is important to evaluate the consequences of such returns. Perhaps the patient does better by being hospitalized only intermittently, but it is also possible that the cumulative effect of having to return to the hospital periodically may ultimately be more harmful and disruptive than continued hospitalization.

Another problem that would appear to be particularly relevant for brain-damaged patients is that of relocation. Although long-term institutionalization can promote excessive dependency and loss of social skills, de-institutionalization can produce a great deal of stress. Since the classical studies of Holmes and Rahe (1967) we have become very aware of the influence life change may have on morbidity and mortality. In the case of elderly people in particular, life change may have dire consequences. The morbidity and mortality figures for the first year in a nursing home are shocking. Apparently, the move from community life to life in an institution may have very unfortunate consequences (Lieberman, 1969; Lieberman & Lakin, 1963; Lieberman, Prock, & Tobin, 1968). Thus, the attempt to move the patient "for his benefit" may not turn out to be at all beneficial. Even though more research is needed to determine whether relocation *per se* or relocation involving going from a better to a worse life situation is the culprit, it seems that life change in elderly and impaired people may (in and of itself) produce stresses that may not be properly dealt with. We are aware that the early "stressful life events" literature has been criticized recently on methodological grounds (Dohrenwend & Dohrenwend, 1978). However, the criticisms offered would not appear to alter the conclusions one would draw from the studies of relocating elderly people to nursing homes. We would only add that relocation of the long-term institutionalized patient to the community may also be quite stressful. It is important to consider the *potential* impact of discharge on a long-term patient.

Therefore, in determining the optimal treatment setting for a patient, one should weigh the disadvantages that may be engendered by institutionalization and relocation against the advantages the setting in question has to offer. On the one hand, if placement at an institution is being considered, questions should be raised concerning such matters as estimated length of stay and how well the patient will be able to tolerate the stresses induced by the contemplated life change. On the other hand, the advantages offered by institutions should not be neglected. If the patient is living at home but is not doing anything in the way of work, education, or treatment, and he can go to an institution in which the potential is present for his becoming a healthier, more productive individual, then perhaps he should go to the institution. We are not trying to convey the impression that community life is all good whereas institutional life is all bad.

Many of these considerations apply to the brain-damaged patient as well as to the psychiatric, general medical, or elderly patient. However, there are some special problems to be considered. Of course, the patient with acute brain damage should of necessity be treated in a hospital. A real choice emerges only when the *acute* condition has subsided, and the patient is in need of some form of continued treatment. Should he be sent home with arrangements made for appropriate outpatient follow-up, or should he be sent to a convalescent hospital, to a nursing home, or to a rehabilitation center? In the case of the brain-damaged patient, there may be a continued need for general medical treatment and for special modalities, such as speech or physical therapy, all of which may not be available in the patient's community. In such cases, the institutional setting has some appeal. As indicated, expertise in the area of rehabilitation of brain-damaged patients is not widespread. We often find such patients and their relatives or friends looking nationwide, if not internationally, for a promising treatment facility.

Sometimes the treatment setting is not a matter of choice but is determined by where we find the patient. In the case of the chronic patient who has been institutionalized for a long period of time, although it may be desirable to discharge him from the institution and rehabilitate him in the community, doing so immediately is unrealistic. The rehabilitation process must usually be initiated within the institution. In such cases, rehabilitation generally has two goals: to restore functioning and to return the patient to the community. This situation varies greatly from the one in which the patient is a community resident, develops brain damage suddenly, perhaps as the result of a stroke or head injury, and begins a rehabilitation program in the hospital immediately following the acute phase of the illness.

These considerations may help in deciding on the appropriate treatment setting (when there is a choice) and in outlining a program of appropriate treatment within a given setting when there is no choice. Some of the following ideas may be useful. If the patient is at home and treatment in an institutional setting is being contemplated, it is important to weigh the advantages of being in an institution against the possibly harmful consequences of institutionalization and relocation. Treatment programs that might be appropriate for noninstitutionalized patients may not work well with long-term institutionalized patients. For example, programs that deal almost exclusively with cognitive and perceptual retraining may work well with postacute patients who have been hospitalized only briefly, but may have little effect on long-term patients, without psychosocial intervention involving a more general orientation to community living. It is hoped that we have learned our lesson and that we no longer keep people in hospitals for lengthy periods of time without some specified treatment goals and plans. However, this regrettable situation continues to exist in many institutions that no longer maintain a long-term treatment policy in general, and in a few institutions that for some reason or other have not changed their philosophy. In many cases, large numbers of the patients in this remaining group are brain-damaged patients living in psychiatric hospitals. Such patients are often nonresponders to the usual forms of psychiatric treatment, and because they do not improve, they remain hospitalized.

Various efforts have been made to provide viable alternatives to institutional care. One important development has been the hospital-based home-care system. In this arrangement, the patient is treated at home by a visiting professional staff, which may include nurses, rehabilitation specialists, physicians, psychologists, and so forth. Various prosthetic devices are often installed in the home to assist with any physical and/or cognitive handicaps the patient may have. These programs also frequently involve training of family members in the treatment and care of the patient. For example, the NYU group (Diller, Ben-Yishay, Gerstman, Goodkin, Gordon, & Weinberg, 1974) has a program for training spouses to improve the speech of aphasic patients. Hospital-based home care has the advantage of combining the technical expertise needed for much rehabilitation work with all the advantages of a home environment. Other solutions have been the halfway house, the sheltered workshop, and the day hospital. Halfway house programs are particularly good for patients who have been institutionalized for long periods of time and need a gradual transition to community living. The day hospital is often a good setting for the patient who is currently unemployable but does not require round-the-clock nursing care. Sheltered workshops, as the name implies, can provide a good rehabilitation environment for the individual who can work but who, on account of some mental or physical disability, cannot compete adequately in an ordinary work situation.

The Rehabilitation Disciplines

Numerous health-related disciplines become involved with treatment and rehabilitation of brain-damaged patients. Recent emphasis on interdisciplinary approaches has encouraged joint effort by members of various professional groups, but this goal is not always optimally achieved. In this section, we will briefly review what we see as the major contributions of the various disciplines to the care of the brain-damaged patient, and proceed to make some remarks concerning interdisciplinary approaches.

Neurology and Neurosurgery

What is a neurologist and what does a neurologist do? It is not easy to answer these questions without some degree of stereotyping, but we will make every effort to do so. A neurologist is a physician who has received a minimum of three years of additional training in diagnosis and treatment of diseases of the nervous system. In private practice, the neurologist acts as a consultant to physicians in general practice. In a hospital setting, the neurologist is generally associated with a neurological service that may have both inpatient and outpatient facilities. In both cases, a large amount of the neurologist's time is spent on evaluation, diagnosis, and follow-up. A neurological evaluation can be a very elaborate procedure, including the taking of a history, the administering of a specialized physical examination of the nervous system, and the utilizing of results of a wide variety of laboratory procedures, notably the electroencephalogram and various radiological techniques. The neurologist's role in direct treatment varies with the illness under consideration. There is no know cure for a large number of neurological illnesses, and treatment recommendations in these cases largely involve supportive measures that may prolong life or reduce discomfort. In cases in which cure or arrest is possible, the neurologist may refer the patient to a neurosurgeon or prescribe any of a large number of medications or other therapies. The use of medication for neurological disorders is most advanced in the area of epilepsy, but progress is being made in the treatment of cerebral vascular disease, Parkinson's disease, and other disorders.

The neurologist has several roles in regard to rehabilitation. He is generally the person responsible for making the medical diagnosis and for placing the patient on appropriate medication. The neurological examination often generates data that may aid in the patient's treatment and rehabilitation. The neurologist may discover that the patient needs new glasses or a hearing aid or may find that the patient has some more general medical disorder that is producing the neurological symptoms. By examination of the patient's motor system, he may be able to make specific recommendations to the physical therapist. The neurologist may be the first person to identify a speech disorder and recommend that the patient be seen by a speech pathologist. He may find, as he often does in psychiatric settings, that symptoms that appear to be neurological in nature do not have an organic basis. In such cases, treatment would be different from what would be the case for the neurologically impaired patient. In general, the neurologist coordinates the general treatment effort based on his evaluation of the case and his diagnosis.

The neurosurgeon has training in neurology with additional training and expertise in surgery. Neurosurgeons may operate on any of the major components of the nervous system, including the brain, the spinal cord, and the peripheral nerves. With regard to brain-damaged patients, the neurosurgeon may be the first physician to treat them. For the most part, this occurs in the case of penetrating head wounds where *debridement* is often employed to clear the wound of bone and other tissue fragments to prevent infection. This procedure is usually followed by treatment with antibiotics and anticonvulsants. Occasionally, the neurosurgeon may be the last physician to treat the patient. In these cases all more conservative treatment efforts have failed, and surgical intervention provides the last hope. In addition to performing operations, neurosurgeons spend a great deal of time in consultation, with the aim of determining whether or not the patient is an appropriate candidate for surgical treatment.

In general, the services of a neurosurgeon are most useful in the treatment of head trauma, brain tumors, and certain cerebral vascular disorders. Sometimes surgery is effective in the treatment of hydrocephalus, through a shunting procedure; sometimes surgical procedures are also used to treat psychiatric illness. This latter procedure, known as *psychosurgery*, is rarely used in current practice but has gained a great deal of notoriety because of its relatively common use in the past. Psychosurgery has been amply discussed elsewhere (E.S. Valenstein, 1973), and the topic will not be pursued further here. Neurosurgical procedures are also sometimes used to treat epilepsy (Bogen & Vogel, 1962; Ojemann & Mateer, 1979).

It should be pointed out that while the neurosurgeon may play a crucial role in treatment, rehabilitation—as distinct from treatment in general—often begins only after the surgeon has finished attending to the patient. With few exceptions, as in surgical treatment of Parkinson's disease, a surgical patient does not come off the operating table with complete restoration of normal function. A patient (and even persons close to him) may develop a magical belief about surgery in which the doctor is able to "fix my head" so that everything will be as good as new. For example, a patient who is about to undergo *cranioplasty*, a procedure that involves repair of the skull defect only, may develop the idea that this procedure will remove functional defects. One aphasic patient we worked with had the idea that a prospective cranioplasty would make him speak correctly again. It is often very important for the rehabilitation staff, including the neurosurgeon, to work with the postsurgical patient to aid him in understanding that, although the pathological procedure causing his difficulties may have been substantially slowed down or stopped, *restoration of function* is a different matter which may take substantial time and hard work to achieve. Often, the goal of neurosurgery is survival, and if the patient does live, then it is necessary to work with whatever capabilities he has remaining.

Neuropsychology

Neuropsychology is a broad field covering the entire spectrum of brain–behavior relationships. At this time we can only concern ourselves with *clinical neuropsychology*, which is the branch that is concerned with working with patients in regard to their diagnosis and treatment. Our question, then, is what is a clinical neuropsychologist? By training, a clinical neuropsychologist is usually a person who has received a Ph.D. degree in psychology, generally with an area of specialization in clinical or physiological psychology. There are a few programs in which it is possible to obtain a Ph.D. degree in neuropsychology. By virtue of their training, clinical neuropsychologists may have expertise in the areas of neuroanatomy, neurophysiology, neuropathology, and behavioral techniques for the assessment of patients for the sequelae of brain lesions. Some neuropsychologists are stronger in assessment, others are stronger in neuroscience. By and large, clinical neuropsychologists spend most of their time doing some form of evaluation of patients with documented or suspected brain lesions. When a lesion has been documented on the basis of structural evidence, the interest is primarily in the extent and pattern of functional disability. In the suspected cases, the neuropsychologist serves as a member of an interdisciplinary group charged with the making of a proper diagnosis. The techniques used vary greatly in content and theoretical orientation, ranging from the individualized and highly flexible neuropsychological investigation of Luria (1973) to the objective and standardized examination used by Reitan and his associates (Luria & Majovski, 1977; Reitan & Davison, 1974). The general philosophical differences go back a long way in psychology, perhaps to the days of Allport's distinction between nomothetic and ideographic methods (Allport, 1937) and to Meehl's discussion of clinical versus statistical prediction (Meehl, 1954). We will not pursue this matter here, except to point out to the nonpsychologist that what is termed a *neuropsychological assessment* in one setting may look nothing like a neuropsychological assessment elsewhere.

In addition to the contributions that they have already made to our current understanding of psychological deficit and relationships between brain and behavior, neuropsychologists can make a highly meaningful contribution to the rehabilitation arena. Chapter 5 contains information regarding how this task may be accomplished. In general, the neuropsychologist, working at a behavioral level, can identify the patient's deficits and areas of intact ability. On the basis of the analysis of this pattern, it is possible to sketch a program of rehabilitation that is rational, or appropriate for the status of the individual patient (Golden, 1978).

The neuropsychologist may enter the picture at various stages of neurological disorder. He may work with a neurosurgeon in regard to establishing a diagnosis on an acutely ill patient. On the other hand, he may work in a clinical setting in which the bulk of the patients have chronic, long-standing disorders. If the neuropsychologist does rehabilitation work, the ideal time for him to be involved is as the acute phase of the illness is being resolved. For all intents and purposes, this stage occurs after the patient has regained consciousness (if he had been unconscious) and is no longer completely bedfast. Activities such as speech therapy can even be begun while the patient is still confined to bed, as long as he is sufficiently alert.

Psychiatry

A psychiatrist is a physician with (usually) three additional years of training in mental disorders. At one time there was a strong relationship between neurology and psychiatry, and the term *neuropsychiatry* is still used in some settings. In recent years, however, most psychiatrists have focused their efforts on the so-called functional disorders such as schizophrenia and depression. Nevertheless, many psychiatrists work with brain-damaged patients, particularly in institutional settings. We will limit our discussion here to the relationship between psychiatry and such patients. In many psychiatric hospitals, a psychiatrist may be the administrator of a ward that houses brain-damaged patients. Most patients are generally placed in programs with other kinds of psychiatric patients. Thus, of necessity, much of their treatment resembles the kind of treatment given to individuals with functional psychiatric disorders. Individualized aspects of treatment such as medication and assignment to some specific type of therapy may vary, but the treatments used are generally of the same type given to the various kinds of psychiatric patients.

Like the neurologist, the psychiatrist may contribute to the treatment and rehabilitation of the brain-damaged patient in a number of ways. First, in his traditional role as a psychiatrist, therapy may be offered to help solve the patient's emotional and interpersonal problems. The psychiatrist may offer psychotherapy or various types of medication with the aims of reducing anxiety, alleviating depression, or improving thinking. Brain-damaged individuals frequently have emotional and interpersonal difficulties that are sufficiently pronounced to require psychiatric treatment. The psychiatrist also frequently has the job of administering a ward of patients, some of whom may be brain-damaged. Typically, such patients do not have acute, recently acquired brain lesions but do have chronic brain disorders of various types. In these cases, the job is primarily that of establishing a program for such patients, including getting them on the proper medications, arranging an activity program, and setting the tone for the ward milieu.

Psychiatrists and other physicians have been looking for some kind of medication to improve the cognitive and perceptual functioning of brain-damaged patients. Results with the standard psychotropic drugs such as phenothiazines, antianxiety agents, and antidepression agents, while successfully used with many kinds of psychiatric patients, have been disappointing with the brain-damaged patient. Psychopharmacological approaches to rehabilitation will be discussed in the next section and elsewhere in this book.

Psychopharmacology

Pharmacologic treatment of brain-damaged patients is based largely on research in two areas: the pharmacology of aging and of memory and learning. Both of these areas have been reviewed in a volume by Lipton, DiMascio, and Killam (1978), to which the reader is referred for technical information. We will lean heavily on the reviews presented in that volume for our discussion here.

Several approaches to discover the neurochemical mechanisms of cognitive processes have been developing recently. There has been, for example, a great deal of interest in the *cholinergic system* in relation to memory. Essentially, the cholinergic system is a set of neurons that interact chemically by means of choline and its various derivatives. Drachman (1978) reports that scopolamine, an anticholinergic agent, could impair memory and that dementia could be associated with some disorder involving cholinergic neurons. This information, however, cannot be applied in an uncomplicated manner (e.g., simply giving cholinergic drugs to brain-damaged patients). Investigators are also interested in the role of stimulants. The most well-known research in this area concerns the use of Ritalin (methylphenidate) with minimally brain-damaged children (Conners, 1974; Satterfield, Atoian, Brashears, Burleigh, & Dawson, 1974). Stimulants such as Cylert (magnesium pemoline) and Dexadrine have also been used (Plotnikoff, 1968, 1971) to enhance memory and learning abilities. According to Ban (1978), Ritalin and other CNS stimulants have been used to treat elderly psychiatric patients, with mixed results. A stimulant drug called pentylenetetrazol was found to enhance discrimination learning in mice, but according to the results of several studies, Ban suggests that this compound does not appear to improve mental function in the elderly patient. According to Lehmann and Ban (1975) a stimulant drug called Pipradrol seems promising. There have not been enough controlled studies to draw conclusions.

Another approach to the problem involves attempts to dilate the cerebral blood vessels of elderly atherosclerotic brain-damaged patients, thereby increasing blood flow. The most commonly used drugs are papaverine (Pavabid), cyclandelate (Cyclospasmol), and dihydrogenated ergot alkaloids (Hydergine). Studies have shown that these medications do appear to improve mental functioning, but it is not clear whether they do so by increasing cerebral blood flow (Ball & Taylor, 1967; A. R. Taylor, 1971). The findings with Hydergine have been particularly favorable. Ban's review (1978) indicates that there were 12 double-blind studies in which Hydergine had significantly greater therapeutic effects than placebo. In two studies, the effects of

Hydergine were found to be superior to those obtained with papaverine. Nicotinic acid has also been used to increase blood flow, and in one study (Lehmann & Ban, 1975) it was found that favorable response to nicotinic acid was associated with improvements in perceptual, cognitive, and psychomotor function.

Another seldom-used alternative concerns altering cerebral metabolism. Various hormone preparations are used for this purpose. One such medication called Pyritinol was found to improve short-term attention in some patients, according to Ban's review. This compound affects glucose consumption and could be effective because of the role carbohydrate metabolism may play in senile dementia. There is also some evidence that fluoxymesterone and corticosteroids (two hormone preparations) are effective to some extent in the treatment of patients with memory disorders or dementia.

The aim of psychopharmacological treatment of brain-damaged patients is not only the improvement of mental functioning but is also often directed toward relief from various psychiatric symptoms, such as depression, anxiety, and agitation. Thus, tranquilizers, antianxiety agents, and antidepressants are sometimes given to brain-damaged patients. Numerous problems are thereby introduced. With regard to tranquilizers, or antipsychotic agents, Epstein (1978) shows that this type of medication tends to be overused in the case of elderly patients with brain syndromes. He cites a study by R. Barton and Hurst (1966) in which nursing personnel could not distinguish between elderly patients taking chlorpromazine (Thorazine) and those on placebo. In general, the effectiveness of antipsychotic drugs with elderly demented patients is questionable. Furthermore, many of these drugs have side effects that are particularly undesirable for the elderly and brain-damaged. If these medications are used, it is generally felt that smaller doses than what is indicated for younger, nonbrain-damaged patients should be used.

The antidepressants tend to have significant cardiovascular side effects. Use of these drugs may be associated with cardiac abnormalities and with undesirable alterations in blood pressure. For example, the tricyclic antidepressants (e.g., Elavil) can cause postural hypotension, a sudden lowering of blood pressure on standing up (Bigger, Kantor, Glassman, & Perel, 1978). Sovner and DiMascio (1978) report that tricyclic antidepressants can sometimes increase confusion. There is, however, one drug for affective disorders that shows hope for the brain-damaged patient—lithium. Lithium is a potentially dangerous drug that may have some seriously adverse effects, but there has been one clinical study by Williams and Goldstein (1979) which shows that it could have a dramatic effect on mental function in severely demented patients. Ten cases were studied representing such conditions as Korsakoff's syndrome, stroke, and cardiac arrest. Of the ten, eight were rated as having improved with lithium. The study was not controlled, but the initial results seemed promising. L. D. Young, Taylor, and Holmstrom (1977) reported that lithium treatment of manic patients with organic brain syndromes controlled the mania without making the organicity worse. On the other hand, lithium has been reported to cause confusional episodes in some patients (Agulnik, DiMascio, & Moore, 1972). Himmelhoch, Neil, May, Fuchs, and Licata (1980) reported that lithium was found to be contraindicated for manic-depressive patients with dementia. It is clear that the literature now available gives mixed reviews with regard to use of lithium for organic brain syndrome. However, the Williams and Goldstein (1979) study calls for further investigation.

With regard to the antianxiety or anxiolytic drugs, perhaps the first thing to be said is that the barbiturates should be avoided because they may have a paradoxical effect and produce excitement instead of sedation. Such drugs as diazepam (Valium) are better, and seem to reduce anxiety and hyperactivity in elderly patients with organic brain syndromes (DeLemos, Clement, & Nickels, 1965). However, it should be pointed out that long-term use of drugs of the diazepam type may have negative consequences for all people, particularly when the medication is withdrawn. The withdrawal symptoms from diazepam and related compounds are similar to those related to barbiturates or alcohol, and include convulsions, tremors, abdominal and muscle cramps, vomiting, and sweating.

We might add a brief comment about the antiseizure drugs. Obviously, treatment of epilepsy is a major aspect of the rehabilitation of many brain-damaged patients. The proprietary names of the most commonly used compounds are Dilantin, Mysoline, and Tegretol. A new antiseizure drug, sodium valproate, has recently become available in the United States. Sometimes antiseizure drugs do more than control seizures. Dilantin is thought to bear some relation to memory in that it blocks the impairment of memory produced by ouabain, a digitalis-like drug that is used clinically to improve heart action. Tegretol is said to have antidepressant properties. Despite these characteristics, however, the primary indication for these drugs is epilepsy and it is usually not beneficial or wise to administer them to patients who do not have epilepsy.

At this point, some general comments concerning the psychopharmacological treatment of brain-damaged patients are in order. First, most of the human research has been done with elderly organic brain syndrome patients or with hyperkinetic children. Thus, the applicability of the findings to younger adult individuals is questionable. Perhaps the vasodilators may not be of particular value to young patients with nonvascular brain lesions (e.g., head trauma). Second, in evaluating the research done with these medications, it is crucially important to look at the populations studied. As indicated above, the research with the vasodilators has been positive. On the other hand, many clinicians believe that vasodilators are not of much help to their patients; it may be that they are effective for early vascular disease, but become less helpful as the degree of atherosclerosis advances. Third, many of the drugs discussed are powerful compounds, and may have side effects that are more detrimental than their direct effects are therapeutic. Some of these drugs are downright dangerous when used with elderly and brain-damaged patients. In general, caution should be the watchword, and these drugs should not be used without careful medical supervision. Finally, it is clear that we do not yet have the pharmacological "miracle cure" and many of the beneficial effects produced by these drugs may be quite modest. One must therefore always weigh the benefits against the possible dangers when using these medications.

Despite what appear to be discouraging findings thus far, the search for effective and safe memory and learning enhancers does go on. The compounds currently undergoing intensive study are the peptides, especially a substance known as ACTH 4,10; a new drug not yet available in the United States called Piracetam or Nootropil; an experimental compound called vasopressin; and lecithin, a choline agonist. Various investigations are underway at this time to evaluate the effectiveness of these agents with demented patients. It is far too early to offer an opinion about any of them, but it may be mentioned that the most recent findings with lecithin have been rather discouraging (Corkin, Growdon, Sullivan, & Shedlack, 1982).

Nursing

Nursing care of the brain-damaged patient often involves a combination of psychiatric nursing skills with a high level of competence in regard to general medical nursing skills. Obviously, the brain-damaged patient has mental symptoms requiring understanding from a psychiatric or behavioral standpoint, as well as the physical illness that produced the brain damage. The problem of epilepsy comes immediately to mind. The nurse must know what to do when patients have seizures and what *not* to do. Similarly, the patient with cerebral vascular disease is likely also to have generalized vascular disease, and so there must be careful monitoring of cardiovascular status. Because of the combination of mental and organic symptoms, the nurse must be particularly skilled in regard to understanding the often unclearly stated complaints of the patient.

As in most kinds of illness, it is the nursing staff that has the most contact with the patient. Nursing personnel are often responsible for implementing various treatment programs and for administering certain therapies. For example, they may be involved in behavior therapy programs or in conducting reality–orientation therapy. The stereotype of the nurse dispensing medication, recording vital signs, and looking in on the patient does not really apply in psychiatric settings. Even though nursing personnel do these things, their major efforts are generally directed toward various treatment programs. Thus, they may lead groups, counsel individual patients, carry out rehabilitation training, or perform any of a variety of treatment-oriented activities.

With regard to care of the brain-damaged patient, nursing staff members are involved at all stages, from assisting in surgery to following up the patient in the community. Rusk (1977) uses the term *rehabilitation nursing* and outlines the role of the nurse in regard to coordinating treatment and the more technical aspects of rehabilitative care, such as instructing in activities of daily living and working with prostheses and other treatment-related mechanical devices. Frequently it is the nurse who works with the family regarding care of the patient at home.

Physical Medicine

The terms *physical medicine, physiatry,* or *rehabilitation medicine* refer to a relatively recently established medical specialty that focuses on problems related to rehabilitation. This specialty was essentially founded by Howard A. Rusk, who early in his career was impressed by what he viewed as a neglect of the rehabilitative aspects of treatment by the medical profession, that tended to look on rehabilitation as an extracurricular, adjunct activity of medicine; something dealing with social work and vocational training, but of little concern and having few implications for medicine (Rusk, 1977). Now, many years later, there is a growing specialty of physiatry, comprising physicians with advanced training in rehabilitation methods. There is also a growing field of rehabilitation engineering that has to do with the development of protheses and related devices for the physically handicapped. It would probably be fair to say that physiatry concentrates primarily on physical disabilities, but there is a strong interest in the brain-damaged patient and his behavioral difficulties. Physiatry has also had a strong impact on the psychiatric hospital. Many such hospitals now have physiatrists on their staffs, and, frequently, these individuals coordinate the work of a number of different kinds of rehabilitation specialists, such as occupational therapists, speech therapists, and physical therapists.

Rusk (1977) sees physiatry as the profession responsible for planning and coordinating the multidisciplinary rehabilitation effort, as represented in Figure 1. In their everyday work, physiatrists examine and consult with patients to determine the nature of their disabilities and to recommend one or more of a large variety of treatment modalities. For example, if a patient is found to have a gait disturbance, he would be sent to a physiatrist who would examine him and recom-

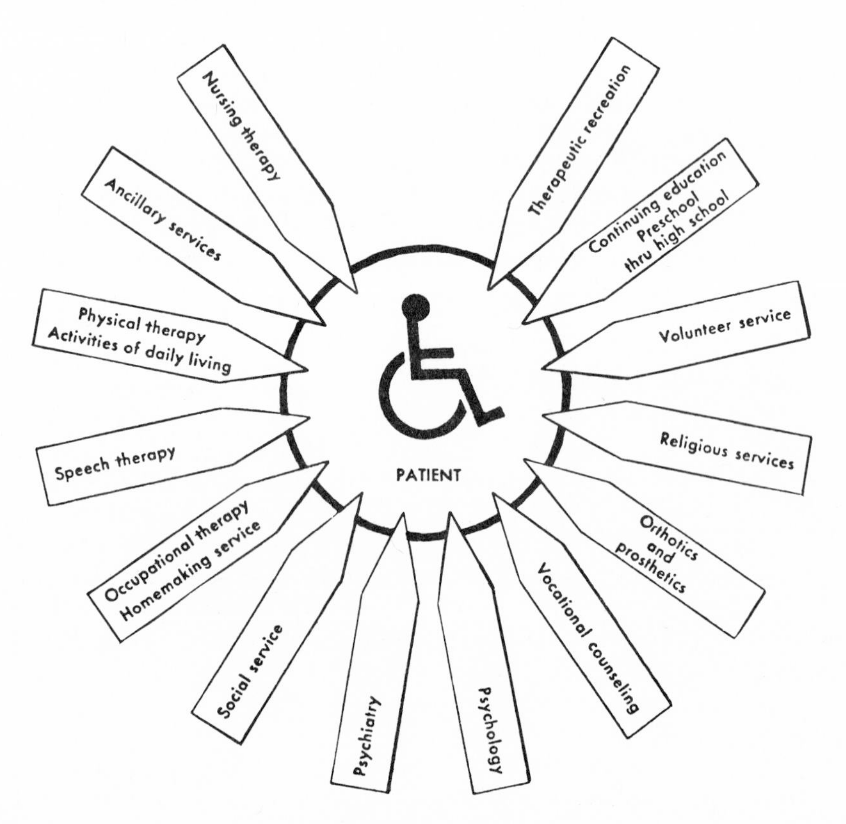

Figure 1. The rehabilitation program.

mend some treatment modality, such as physical therapy. The physiatrist also works with physical therapists, occupational therapists, and other rehabilitation specialists to describe the nature of the patient's condition and to ascertain the form of treatment needed. Physiatrists are particularly interested in muscles and the motor system. They are well trained in *electromyography*, a technique used for evaluating muscle status and the generation of muscle charts, a mapping out through physical testing of the strength of individual muscles.

Recourse to physiatry generally follows the acute phase of the illness, because its role is primarily that of restoring normal function following trauma, surgery, stroke, or similar events. It also plays a role with chronically ill patients; primarily that of retaining and prolonging function in so far as is possible. Thus, the physiatrist and the team of rehabilitation specialists may work with patients with progressive illnesses, such as multiple sclerosis or Huntington's chorea, with the aim of retaining functioning for as long as possible. The rationale for the application of rehabilitation medicine in psychiatric settings is of a somewhat different nature. Of course, psychiatric patients can develop physical disabilities requiring rehabilitation of the traditional type. However, these disabilities account for only a small portion of the rehabilitation-medicine effort in many psychiatric settings. The various rehabilitation specialists have developed rationales and specific treatment techniques for various psychiatric syndromes. Such modalities as physical therapy or music therapy are now directly applied in the treatment of schizophrenia, depression, and organic brain syndrome. The therapeutic professions will be described in the section on rehabilitation specialties. It is not unusual to see extensive rehabilitational facilities, including gymnasiums, swimming pools, manual arts therapy shops, and occupational therapy shops, in psychiatric hospitals. Admittedly, this approach to psychiatric treatment is controversial and not universally accepted, because some feel that these facilities really serve only recreational and time-occupying purposes, and do not contribute substantially to treatment or rehabilitation. As previously pointed out, the problem appears to be primarily one of utilizing the therapies and facilities available in a rational way. They can be merely entertainment or beneficial therapy, depending on how they are used.

Speech Pathology

There are two levels of professional practice in the field of speech pathology. Individuals with master's level training, who are generally known as *speech therapists,* usually work directly with patients to assess and treat speech problems. Those with Ph.D. degrees are usually known as *speech pathologists,* and assess and treat patients, but often do research and supervision as well. Speech therapists and speech pathologists are generally competent in evaluating hearing disorders, and may perform and interpret audiological examinations. In the area of brain disorders, speech therapists and pathologists have particular expertise in the assessment and treatment of aphasia. Those who specialize in this area are now characterized as *aphasiologists.* There are very important links between speech pathology and the basic sciences of audiology and linguistics. Those branches of linguistics called *psycholinguistics* and *neurolinguistics* are of particular significance for speech pathology.

It would be both demeaning and inaccurate to describe the profession of speech pathology only in terms of administering hearing tests and evaluating and treating patients for speech difficulties. Speech pathologists have made major contributions to the scientific study of both normal and pathological language, of communication processes, and of relationships between language and the brain. Although the profession has applications well beyond the area of aphasia, neurologists, neuropsychologists, and speech pathologists share a common interest in it. Extensive tests for aphasia have been constructed by several speech pathologists (e.g., Eisenson, 1954; Holland, 1980).

It is generally desirable to initiate speech therapy as soon after the acquisition of the brain lesion as possible. Therefore, the speech therapist prefers to enter the picture at the very beginning of the recovery process. Therapy with a patient who has been aphasic for some time is generally problematic. Natural recovery processes are no longer at work and so cannot be enhanced by therapy, and the patient may have developed poor speech habits. For the reader who may be unfamiliar with the area of aphasia, the textbook by Goodglass and Kaplan (1972) may be helpful.

The Rehabilitation Specialties

In addition to the retraining of language abilities, efforts are often made to restore a number of other functions. These efforts are typically conducted by a professional group known as *rehabilitation specialists.* Some rehabilitation specialists, such as *physical therapists,* deal largely with restoration of physical function, while others, such as *occupational therapists,* concentrate more on purposive behavior. Sometimes the distinction is vague, but it would probably be fair to say that physical therapists stress exercises whereas occupational therapists work with various media, such as crafts of various sorts and the kinds of materials and objects the patient may encounter in everyday life. Activities of Daily Living (ADL) is a core concept, with the aim of both physical and occupational therapists being that of retraining the patient to engage in such activities to the greatest extent possible. These therapists also train patients in the use of such prosthetic devices as artificial limbs and braces. A field that is closely related to occupational therapy is *manual arts therapy*. Manual arts therapists are typically skilled in some vocational specialty as well as in clinical management of patients. Thus, they may be accomplished carpenters, machinists, photographers or similar vocational specialists, in addition to being therapists. There are also therapists who use the performing arts as media, including music or dance. Such therapists are generally accomplished artists who use singing, dancing, or playing of a musical instrument as therapeutic devices in various applications.

Like the speech therapist, the rehabilitation specialist would prefer to have a patient who is coming out of the acute phase of his illness. In the case of prosthetic training, it is often important to work with the patient even before he receives the prosthesis, in preparation for it. The occupational therapist has a major role with the brain-damaged patient, even if the patient does not have a gross physical disability. Occupational therapists can deal with problems of perceptual-motor coordination and other perceptual and cognitive difficulties commonly seen in brain-damaged patients without marked physical disabilities.

Ophthalmology and Audiology

An *ophthalmologist* is a physician with specialized training in diseases of the eye and visual system. Examination of the eye is of particular interest in the case of brain-damaged patients for several reasons. First, certain types of blindness are caused by damage or disease involving the brain rather than the eye itself. The visual field examination can help in localizing this brain damage. Second, direct inspection of the eye with a slit lamp or ophthalmoscope may aid significantly in diagnosing several kinds of brain disorders, including brain tumors, cerebral vascular disease, and multiple sclerosis. Third, impairments seen particularly in children, such as a reading disability, may be found by the ophthalmologist not to be cognitive impairments but rather visual defects. Prescription of the proper eyeglasses may sometimes solve the problem.

Audiologists may either be physicians or sometimes Ph.D.s. As mentioned previously, speech pathologists are generally trained in audiology and audiometry. The audiologist contributes to rehabilitation planning in much the same manner as the ophthalmologist. Some hearing disorders relate to brain damage rather than specific damage to or disease of the ear or auditory nerve. Sometimes difficulties that may appear as psychiatric or intellectual in nature may in fact be the result of loss of hearing. What is sometimes understood as a failure to understand may be a failure to hear or to hear well enough to understand.

Interdisciplinary Approaches

Treatment by an interdisciplinary team is clearly becoming common practice in institutional health settings. There are essentially two models for how these teams function. In one, a group of professional specialists brings to bear various areas of expertise on the problems of the patient. Thus, for example, the psychiatrist makes his contribution, the neuropsychologist his, the ophthalmologist his, and so forth. Perhaps these specialists meet at a case conference to see the patient, review the record, integrate the findings, and make plans. In the other model, the traditional professional roles are not as clear-cut, and individuals from a number of disciplines may perform common tasks. This approach, a more controversial one than the first, is sometimes known as *blending*. The focus is on the problem of the patient rather than on the contributions of the individual specialities and professions. This approach is often reflected in the change from so-called source-oriented to problem-oriented medical records. The controversial aspect of blending-type programs is that staff members may perform activities that are not within the traditional purview of their professional disciplines. A nurse may administer psychological tests, or a psychologist may do the sorts of things usually done by occupational or physical therapists.

The blending approach would appear to have increasing difficulty as licensing, certification, and health insurance considerations continue to define and limit various professional practices. Nevertheless, in areas like treatment of brain-damaged patients, where no one group can be said to have extensive expertise, there would seem to be some advantage in a diffusion of roles, activities, and responsibilities. There are obvious limitations to blending; speech therapists do not perform neurosurgery and neurosurgeons do not give aphasia examinations. The matter of blending as opposed to maintenance of traditional roles within the context of a multidisciplinary team has not been evaluated, and so we really do not know which system is more beneficial. It would be premature to condone or condemn either one. Whereas the blending system could conceivably encourage individuals to go beyond their professional limitations, the more traditional system could produce rigidities that may have negative consequences for the patient. In any event, an interdisciplinary approach to assessment and treatment of brain-damaged patients requires extensive staff development, primarily because of the newness of the area. We will deal with this matter in detail in Chapter 3.

Philosophies of Rehabilitation

The conduct of rehabilitation programs reflects the beliefs and values of the people who administer those programs. In the case of rehabilitation of the brain-damaged adult, we would say that systematic treatment philosophies have not yet emerged and that it is necessary to borrow from other fields. The extent to which any of these other fields constitutes an appropriate model for treatment of the brain-damaged patient is open to question and will be discussed further. For our present purposes, a distinction is being made among what will be called the physical disability, psychiatric, behavioral, and neuropsychological models. This distinction, although not really theoretical in nature, is an attempt to reflect styles of current treatment of brain-damaged individuals.

The Physical Disability Model

In a literal sense, brain damage is a form of physical disability in that it involves dysfunction of an organ of the body. Furthermore, many types of brain damage are associated with what we more normally think of as physical disabilities such as paralysis and blindness. In real life, the brain-damaged patient is often treated in a setting that contains large numbers of physically disabled patients. For these reasons, rehabilitation efforts for brain-damaged patients may be modeled after what is done for physically disabled individuals. When one reviews the rehabilitation literature, it soon becomes apparent that the stress is on treatment of physical disability. Even in the case of neurological disorders, the emphasis is on such things as gait, spasticity, prevention of decubitus ulcers, and the like.

The physical disability model can be viewed in a more general way by asking whether or not brain-damaged patients are like other people with disabilities. As has become clear in recent years, disability is multifaceted. Naturally, there is the matter of the specific defect itself, but there are a host of issues surrounding disability that go far beyond this defect. These areas are thoroughly explored in a book by Wright (1960) in which there is a great deal of material concerning such matters as the ways in which people experience their disabilities and the ways in which disabled people are viewed by others. Wright's book also deals with the various ways in which individuals adapt to disability, and how interactions between disabled people and people close to them affect development and formation of the self-concept. Wright's work was greatly influenced by the somatopsychology of Barker (1948) with whom she collaborated for many years (Barker & Wright, 1952, 1954; Barker, Wright, & Gonick, 1946). *Somatopsychology* is concerned generally with the ways in which the physique affects a person's psychological situation, and specifically, with the ways in which disabled people cope with their life situations.

Regarding the brain-damaged patient, the social psychological situation is clearly a pertinent factor. Nondisabled individuals tend to have characteristic attitudes of both positive and negative natures toward the disabled. For example, the disabled in some respects are treated like minority group members and tend to be afforded an inferior status position. On the other hand, people with disabilities are often viewed as more conscientious than nondisabled people, and those who have coped successfully with major disability are often greatly admired. Feelings of admiration, pity, fear, aversion, and attraction may all be engendered by the disabled person. The brain-damaged patient, particularly if he has a visible disability, is not unlike other disabled individuals in this social psychological respect. The pertinence of these factors to rehabilitative effort is readily apparent.

Even though brain-damaged people are disabled and share the life situation of all disabled individuals, there is an important difference between the two groups. The difference is based on the fact that the brain is not like any other organ of the body. Damage to the brain characteristically alters the nature of experience, and this alteration has numerous implications. The brain-damaged patient often has a diminished capacity to intellectually understand and emotionally experience the nature and extent of his disability. This one distinction should drastically alter what one does in rehabilitation and how one does it. Even in the case of physical rehabilitation, the effective treatment of individuals who are physically disabled because of brain damage (e.g., the stroke patient) might be quite different from how patients with physical injuries not produced by brain damage are effectively treated. In dealing with the brain-damaged patient, it is important to give up the assumption that the patient's feelings about his illness, motivation, and capacity to integrate experience are the same as one's own.

In general, the physical disability model is pertinent but incomplete in the area of brain damage. It is particularly relevant from the point of view of the impact of the disability on others, and on how the attitudes, beliefs, and behavior of others can influence rehabilitation. However, this model must be supplemented by a psychology of the brain-damaged individual, provided for by the psychiatric and neuropsychological models.

The Psychiatric Model

Psychiatry, from its earliest beginnings, has had a long-standing interest in brain damage. The role of the psychiatrist in the treatment of the brain-damaged patient has already been discussed in this chapter. Here, we will be more concerned with the theoretical stance of what we are describing as a psychiatric rehabilitation model. In psychiatry, a distinction is made between disorders attributable to some specific organic factor involving the brain, and disorders that cannot be attributed to any such factor. The latter conditions are known as *functional disorders,* some of which may not have an organic etiology and some of which may have an organic etiology that has not yet been definitively established (e.g., schizophrenia). The key concepts in the psychiatric model revolve around the terms *delerium, dementia,* and *organic brain syndrome.* Since delerium is usually a self-limiting condition that does not require rehabilitative efforts, our discussion will focus on dementia and organic brain syndromes.

What is the nature of the psychopathology of the brain-damaged patient? How can we describe his illness and his mental condition? Within the framework of modern psychiatry, a kind of "breakthrough" in regard to this kind of inquiry was achieved by Kurt Goldstein (1939) during World War I. He believed that the various deficits seen in brain-damaged patients could be described largely in terms of a single underlying concept: loss of the abstract attitude. The underlying neurology for K. Goldstein's view was holistic in nature and corresponded with his view that loss of the abstract attitude could occur as a result of damage to any area of the cerebral hemispheres. Moreover, it could manifest itself in a number of modalities including the language area (K. Goldstein, 1948) and the area of problem-solving ability (K. Goldstein & Scheerer, 1941). Since the appearance of Goldstein's writings, brain damage in patients has very frequently been described in terms of the loss of abstraction ability or its consequent characteristics, concreteness and rigidity.

Modern psychiatry characterizes the brain-damaged patient by his impairment in the area of abstract reasoning, but also by defects commonly seen in memory, judgment, orientation, and learning ability. There are also frequently observed secondary symptoms of irritability, depression, and socially maladaptive behavior. The course and treatment of the condition are dependent on the nature of the organic disorder that produced the brain damage. However, the clinical phenomenology of dementia is quite similar across a number of different etiological conditions. As in the case of the physical disability model, there is a great deal of interest in adjustment. Patients with the same kind of brain lesion may respond in widely varying ways depending on motivation, premorbid personality, premorbid intellectual and educational level, and related considerations. Thus, within a psychiatric framework the brain-damaged patient may be described along three parameters: the nature and extent of the mental deficit that can be associated with a structural lesion, the secondary symptoms associated with this defect, and the interaction between the patient's mode of adjustment and his premorbid personality. How does this framework relate to rehabilitation? As discussed previously, the psychiatrist is often the professional with major responsibility for treatment of the chronically brain-damaged patient. How is such treatment conceptualized? The first approach is often an interdisciplinary one that entails evaluating what can be done medically. Can the pathological process be reversed or at least arrested by medication, surgery, or other form of somatic therapy? Following or accompanying this attempt, there may be recourse to what can be found in the armamentarium of the psychiatrist, psychologist, or other mental health specialist. Kurt Goldstein was a strong believer in individual psychotherapy as treatment for brain-damaged patients. The point, of course, was not to reverse or remove the brain damage, but to support or assist the patient in coping with life situations. In the case of the brain-damaged patient, the philosophy of many psychiatrists is that while the damage may not be reversible, the patient may be relieved of various types of discomfort. In order to do so, a variety of treatment modalities may be used, including various types of medication, psychotherapy, and milieu therapy. For some psychiatrists and clinical psychologists the philosophy of treatment can be captured in the phrase "brain-damaged people have emotional problems too."

It would not be fair to say that the psychiatric model offers only palliative or custodial treatment. Within the limits imposed by the structural brain damage, functioning may improve when the patient is relieved of depression and anxiety. Elsewhere in this book the problem of emotional disorder in brain-damaged patients is discussed in detail, but we can mention here that a large proportion of these patients have, in addition to their brain syndromes, a diagnosable psychiatric disorder. The use of psychotherapy and medication for treatment of these disorders is often effective. The only cautionary note is that the psychiatric drugs often do not work in the same way with brain-damaged patients as they do with psychiatric patients in general. Therefore, it is often necessary to deviate from the usual indicated medication for the condition being treated, or from the usual dosage level, when medicating brain-damaged patients.

The psychiatric model may be said to have the attractive feature of treating the patient as a whole human being. It deals with the patient's feelings, motivations, interpersonal relationships, social behavior, and related areas, as well as with the cognitive and perceptual problems. However, like the physical disability model, it is incomplete. Although there is an awareness that brain damage alters normal experience, there is no theory or technology for thoroughly assessing and remediating the specific alterations that occur. With regard to assessment, the psychiatrist's major instrument is the interview in combination with a review of the history. Recent developments in psychiatric interviewing (e.g., Spitzer & Endicott, 1973) have provided the field with greater structure and objectivity, but the major contributions have been in the area of the functional psychiatric disorders. Furthermore, psychiatry has not developed a technology for remediation of specific alterations of behavior that occur as a consequence of brain damage. The discussions of behavioral and neuropsychological models are better equipped to deal with this problem.

The Behavioral Model

Since this model will be discussed in great detail in a subsequent chapter, we will be brief here. G. Goldstein (1979) has outlined a proposed interface between behavior therapy and human neuropsychology, the point in common being the rehabilitation of the brain-damaged patient. Golden (1978) has also suggested that behavior therapy-oriented approaches may be useful in rehabilitating brain-damaged individuals. However, to the best of our knowledge, there has been no detailed consideration of the specific application of behavior therapy to brain-damaged patients comparable to what has been done for such groups as alcoholics (P. M. Miller, 1978; Nathan, 1977), phobics (Leitenberg & Callahan, 1973), patients with anorexia (Agras, Barlow, Chapin, Abel, & Leitenberg, 1974), and the mentally retarded (e.g., E. S. Barton, Guess, Garcia, & Baer, 1970). There is very little data indicating whether behavior therapy in any of its many forms is effective with brain-damaged adults. Nevertheless, adoption of a behavioral model may provide a new way of looking at brain-damaged patients that may have many positive implications for assessment and treatment.

The history of behavior therapy and presentations of its basic principles may be found in many sources (cf. Bellack & Hersen, 1977; Yates, 1970, 1975). The only question to be raised at this point concerns whether or not a treatment approach based largely on principles of learning can be effective in a diagnostic group whose members characteristically have severe impairments of the ability to learn. It has been well documented that severely cognitively impaired individuals *can* learn—particularly since the appearance of the behavioral literature in the mental retardation area (Corte, Wolf,& Locke, 1971; Giles & Wolf, 1966; Groves & Carroccio, 1971). Furthermore, Bellack and Hersen point out that even when a behavioral disturbance is found in an individual with an organic illness, the disturbance may not be caused by the illness. Rather, it may be under environmental control and if so, then it is possible that it can be placed under systematically achieved stimulus control. Not all of the maladaptive behaviors of brain-damaged patients are necessarily attributable to the brain damage. Some of them may be more strongly associated with reinforcers coming from the environment. Thus, while it may not be possible to alter the pathological condition of the patient, it may be possible to alter environmental contingencies.

The problem of the brain-damaged patient presents a significant challenge to behavior therapy. There are patients who are so densely amnesic that they may have no recollection of their reinforcement history. Other patients have such extensive intellectual impairment that it is difficult to utilize such secondary reinforcers as tokens. As indicated in the previous chapter, the brain mechanisms for motivation may have been damaged or destroyed, particularly if the frontal lobes or limbic system are involved in the brain damage. We will propose that it may be possible to effectively meet these challenges through an alliance between behavior therapy and clinical neuropsychology. The neuropsychologist can efficiently assess the patient's pattern of impaired and preserved abilities, identify the problem to be treated, and suggest what is possible and not possible in view of the nature of the neurological disorder. The behavior therapist may then design an appropriate treatment program. We will also propose that behavioral assessment (Bellack & Hersen, 1978; Cone & Hawkins, 1977) provides a means of evaluating brain-damaged patients that complements what is obtained from neuropsychological tests, and that the single-case research design, that is largely associated with behavior therapy (Hersen & Barlow, 1976; Paul, 1967), may be an invaluable method of implementing and evaluating rehabilitation programs.

The Neuropsychological Model

In recent years, neuropsychologists have become increasingly involved in systematic efforts to rehabilitate brain-damaged patients (Golden, 1978). In the past, clinical neuropsychologists tended to stress the diagnostic and assessment aspects of their practices; now, they are looking more at the implications their testing procedures may have for direct intervention. The challenge inherent in this modern view is a difficult one to meet in that it calls for recommendations for treatment of many conditions that have been declared untreatable. Since there is no known cure for many of the brain disorders, what more can be done beyond making a diagnosis and providing humane nursing care? Now, many neuropsychologists (as well as other professionals) view this question as simplistic and naive. There are many reasons for this being the case. First, many brain disorders are treatable and sometimes curable. Patients recover from strokes with a great deal of restoration of function, and neurosurgeons often achieve treatment success for patients with brain tumors, Parkinson's disease, and other disorders. When one speaks of the lack of effective treatment for brain damage, the discussion should be limited largely to brain damage of the chronic or slowly progressive type. Even in the case of chronic brain damage of certain types, however, active treatment is often effective. Second, extensive efforts are now underway to find cures for certain disorders that were formerly viewed as incurable, and to improve the quality of life during the remaining years for those individuals who do, in fact, have incurable disorders. With regard to incurableness, it has been found that certain forms of dementia may be produced by slow-growing viral infections (Gibbs & Gajdusek, 1978). The discovery of the responsible organisms may ultimately lead to methods of direct treatment. Improvement of the quality of life is being approached pharmacologically and psychosocially. In the pharmacological area, some investigators are attempting to treat Alzheimer's disease patients with cholinergic agonists such as choline and lecithin

(Growdon & Corkin, 1980), with some modest degree of success. Eisdorfer (1980) and Lawton (1980) have outlined psychosocially oriented treatment programs for patients with dementias of the senile and presenile types. It is thus necessary to maintain a balance between being overly optimistic about the course of many brain disorders and not being so pessimistic that treatment efforts are either not made or only made halfheartedly. The attitude that the patient can only be kept comfortable and treated humanely while he is dying can be counterproductive. This viewpoint has been most effectively supported by the Committee to Combat Huntington's Disease, who have provided an extensive popular literature regarding the care of individuals with this illness (e.g., Guthrie, 1979). Patients with Huntington's Disease may live for as long as fifteen years after appearance of the first symptoms. The point made strongly by the Committee (and by others) is that ways should be found to make these years as creative and productive as possible, even if a cure is not available.

Several models have emerged within neuropsychology concerning recovery and restoration of function. However, they have several points in common. Most crucial, perhaps, is the heavy reliance on assessment, although there may be some disagreement as to the nature of that assessment (G. Goldstein, 1980). Neuropsychological assessment is generally a thorough survey of cognitive, perceptual, and motor skills achieved through utilization of a wide variety of tests. The rehabilitation-oriented neuropsychologist usually maintains the goals of identifying the patient's deficits, primarily through the assessment and direct treatment of those deficits. Neuropsychologists tend not to emphasize acceptance and relief from stress as treatment goals. Neuropsychologically oriented treatment can be demanding and arduous.

The first rehabilitation model discussed is the one employed by the group at the NYU Institute of Rehabilitation Medicine. Fortunately, a group member, Leonard Diller, has made an explicit presentation of his group's rehabilitation model. It is definitely a neuropsychological model, in that it attempts to place treatment into the general framework of what is known about brain–behavior relationships. However, it places a somewhat different emphasis from what is commonly found among clinical neuropsychologists on assessment. Diller makes the point that there is a particular relationship between diagnosis and treatment, and that the diagnostic frame of reference differs depending on whether or not one is going to engage in treatment efforts. The diagnostic information gained from a rehabilitation-oriented assessment has to be specific, since one must determine just what skill needs to be remediated. The remediation model proposed includes several components: identifying the to-be-remediated skill, selecting tasks appropriate for training in that skill, analysis of those tasks in terms of performance of daily and nondaily activities, and seeking neurological correlates. The skills chosen for remediation programs tend to be perceptual and cognitive skills, apparently because of their wide range of potential generalizability. The NYU group has devised a number of innovative ways of retraining brain-damaged patients, and these techniques are designed in accordance with certain principles. Because of the nature of the effects of brain damage, stimuli that normally serve as cues for certain behaviors tend to lose their saliency, and the trainer then has to call attention to certain cues as part of the remediation procedure. In effect, cues have to be made more salient for the patient. Although the patient must be made aware that he has to solve a problem, this process should be carried out in a manner that minimizes failure experiences. Offering clues to solution is one way of reducing failure.

The NYU group has reported on several projects related to their work with rehabilitation of hemiplegic and aphasic patients. One of them concerned training hemiplegics to pass the Block Design subtest of the Wechsler Adult Intelligence Scale. A second project involved studying and attempting to rehabilitate scanning behavior in hemiplegic patients. Particular emphasis was placed on visual-spatial neglect of one half of the visual field (Weinberg, Diller, Gordon, Gerstman, Lieberman, Lakin, Hodges, & Ezrachi, 1977). A third project involved training spouses to improve the functional speech of aphasic patients, using a teleprompting system in which suggestions and feedback were given to the spouse through a headphone. Although some of this training may appear academic in nature, the investigators were quite interested in the relationships between the training and adaptive functioning. They found, for example, that the Block Design training was associated with the organizational aspects of eating; namely, maneuvering of food, stacking, accidents, and so on.

In a monograph summarizing their work up to 1974 (Diller, Ben-Yishay, Gerstman, Goodkin, Gordon, & Weinberg, 1974), this prefatory remark was made:

> Changes in mental life as a result of brain damage are frightening to experience and to observe. Scientific investigations in the past century, particularly in the field of neuropsychology, have been concerned with trying to understand these changes. Very little has been done with regard to the question: Granted the fact of brain damage, what can be done to improve mental function? (p. 1)

In their research, the NYU group has demonstrated that their training techniques are of benefit to patients, and that the specific training done generalizes to other areas. For example, scanning training and Block

Design training were found to be associated with improvement in occupational therapy. In effect, their data have successfully countered the view that the functional consequences of brain damage are irreversible, and that treatment efforts can only be palliative in nature.

An alternative model was proposed some years ago by Luria (1948/1963), who had been actively involved in rehabilitation work. His general philosophy concerning rehabilitation (Luria, 1962/1966) is nicely captured in this passage:

> According to these dynamic ideas, a disturbance of a function arising as a result of a local brain lesion no longer appears so permanent and irreversible as was hitherto considered. In order to restore the disturbed function (or, more accurately, the disturbed functional system), a careful analysis must first be made of the disturbance as it stands, the primary defect which caused the functional system to become impaired must be found, and then an attempt must be made to replace the disturbed link by an intact link, *thus reconstructing the functional system as a whole and enabling it to operate with the aid of new, intact links.* (p. 66)

In the same work Luria went on to give some interesting examples of retraining such functions as handwriting, and cited the work of several Soviet neurologists and psychologists who had developed techniques for the retraining of various cognitive and perceptual abilities in brain-damaged patients. To the best of our knowledge, except for the mateial provided in an early work (Luria, 1948/1963), Luria's methods are, unfortunately, not well known in the United States, primarily because of the absence of English translations. It is clear, however, that Luria, like other neuropsychologists, placed a heavy emphasis on careful examination of the patient in order to identify and delineate the nature of the disturbance of function. Such identification and delineation may be accomplished by a variety of procedures and tests, but all neuropsychologists emphasize the crucial significance of some carefully accomplished form of assessment.

In some of our own work we also developed a neuropsychologically based model for rehabilitation. It also stressed neuropsychological evaluation and focused on perceptual and cognitive processes. However, it contained two unique components. First, it placed great emphasis on neuropsychological evidence of impaired cognitive functions of one or the other cerebral hemisphere. Thus, patients with primarily left-hemisphere brain damage were seen as spending their time most productively in language or language-related programs. Patients with right-hemisphere brain damage were assigned to visual-spatial or constructional activities. Consequently, the principle of functional asymmetry of the cerebral hemispheres was applied to rehabilitation planning and implementation. Second, it tackled the problem of motivation. As we mentioned in the first chapter, brain-damaged patients, for various reasons, are not always ideally motivated for rehabilitation. Either they do not comprehend the nature of their loss or there is some defect in the neural mechanisms that underlie motivation, as in the case of patients with frontal-lobe or limbic-system lesions. We were impressed by the promising early results stemming from the application of behavior therapy to psychiatric patients (e.g., Ayllon & Azrin, 1968) and felt that perhaps some of the same methods could be applied to brain-damaged patients. What resulted from these considerations was an attempt to combine clinical neuropsychology with behavior therapy, thereby taking advantage of both the expertise concerning brain–behavior relationships held by neuropsychology and the capability of behavior therapy in regard to the design of retraining programs. Our preliminary research (done some years ago) was a project designed to implement and evaluate a number of programs planned on the basis of neuropsychological assessments and the application of some of the behavior therapy techniques that were available at the time. As stated in an earlier work (G. Goldstein, 1979), "The point of view we would like to propose is that neuropsychological assessment can form a productive alliance with behavior therapy in regard to the planning, implementation, and evaluation of individual rehabilitation programs" (p. 35).

Specific Neuropsychological Models

Some neuropsychologists interested in rehabilitation have limited their interests to more or less specific areas of perceptual and cognitive function. Probably the two most widely studied areas are language function and memory. These neuropsychologists have proposed specific ideas about how neuropsychological information can be used in the planning and implementation of retraining programs. In the area of language retraining, neuropsychologists have formed alliances with speech pathologists. We will first discuss various models for language therapy, then move on to memory retraining.

Not all language therapists employ the same treatment philosophy. Holland (1979) has described three possible models. The first is didactic in nature, and involves exercises and instruction that are used to reteach lost language skills. The second model is based on the assumption that one should attempt to impart to healthy tissue the ability to mediate functions originally mediated by destroyed or impaired tissue. For example, the aphasic patient may be asked to do a great deal of writing with his left hand, in order to get an abundance of language-related material into the right cerebral hemisphere. Theoretically, at least, such retraining aids the right hemisphere in taking over the control of language abilities. On the other hand, some language therapists feel that in some cases it is more appropriate not to attempt to restore the lost language skills but to teach the patient to communicate through alternative channels. The so-called visual communication or VIC system developed by H. Gardner, Zurif, Berry, and Baker (1976) is an example of this method. Patients are taught to communicate through the use of pictures. The general idea is to teach the patient to get the message across, or to receive the message, regardless of how it is done.

Without going into detail about the various types of aphasia and related language disorders, we will simply point out that the treatment approach taken may be closely related to the type of aphasia the patient has. Thus, the language therapy used for a patient with so-called fluent aphasia (cf. Goodglass & Kaplan, 1972), in which speech is smooth and extensive, would be different from what would be done for the halting, impoverished speech of the nonfluent aphasic. In the case of the fluent aphasic, a major difficulty is with language comprehension which (it is thought) secondarily produces fluent but generally unintelligible speech. In the case of the nonfluent patient, comprehension may be almost normal, but the patient is impaired in the ability to produce sustained, smooth speech. Simply stated, the major rehabilitation task for the fluent patient is to learn to comprehend better, while the nonfluent patient must learn to speak more extensively and smoothly, or to develop some alternative method of communicating. Luria (1948/1963), for example, points out that in the case of fluent or sensory aphasia, the patient may lose the ability to differentiate the sounds of speech. It then becomes necessary to reteach the patient that a change in only one sound of a word may alter its meaning. The patient may be shown a series of pictures paired with words that makes the point clear.

Although classification of the aphasias has been longstanding, a classification of the memory disorders has only recently emerged. There are apparently several types of amnesic conditions, and it now seems clear that, to be successful, memory training must take cognizance of this typology. For the reader particularly interested in the matter of the different types of amnesia, a chapter by Butters (1979) and a book by Butters and Cermak (1980) should be consulted. Again, without going into detail, in certain types of amnesia the major impairment is in short-term memory, and as one goes into the increasingly distant past, memory improves in a gradientlike fashion. Other amnesic patients do not show this so-called anterograde-retrograde gradient and have uniformly poor memories across time. Certain patients with lateralized lesions may have memory problems for specifically verbal or nonverbal material, depending on the laterality of the lesion. Some patients with amnesic disorders may benefit from cues, whereas others may not. Some may benefit by providing mnemonics, perhaps through using visual imagery, whereas others may not.

On the basis of this relatively newly acquired information concerning the amnesic disorders, it may be possible to design training programs that are specifically suited to the deficit pattern of the patient. Prior to doing this, however, it is again necessary to do a careful assessment. Although the standard instruments for evaluating memory, notably the Wechsler Memory Scale (Wechsler, 1945), may not serve to identify different types of amnesias, the research that has provided the information regarding the different kinds of amnesia has also provided new tests that may ultimately prove to be invaluable to rehabilitation planning.

Integration

The Reality of the Brain-Damaged Patient

We will now elaborate on the point that while we admittedly have a neuropsychological bias, we do not treat the terms *rehabilitation* and *cognitive-perceptual training* as synonymous. Rehabilitation is a much more global term, and should ideally encompass the reality of the patient. More accurately, we should describe the matter as the reality of the patient and his situation. Thus, in an attempt to provide an integrated rationale for rehabilitation of the brain-damaged patient, we will lean heavily on clinical considerations and try (to use the vernacular) to "tell it like it is."

One aspect of this reality can be put most succinctly by saying that the patient is a whole person. Although the focus may be on cognitive and perceptual defects in the case of brain damage, such defects do not exist separately from a human personality. Particularly in the case of the brain-damaged patient, the cognitive, perceptual, and motor defects are often so striking that we tend to neglect other aspects of function. Indeed, we name these patients after their defects, calling them *aphasics* or *hemiplegics*. It is only in recent years that clinical researchers have paid much attention to affective disorders in brain-damaged patients (E. Valenstein & Heilman, 1979). Problems associated with depression, agitation, and mania are not at all uncommon and frequently require treatment. A severe depression can provide a significant obstacle to rehabilitative efforts, as can anxiety and other "nonorganic" symptoms. One thing that is clearly learned when one is involved with the total treatment of brain-damaged patients is that psychiatric difficulties are the rule rather than the exception. The image of a brain-damaged patient as a normal, healthy individual with no difficulties other than those associated with his brain lesion is largely a myth. There are both premorbid and postmorbid factors that greatly color the situation.

Let us take the case of the young, adult patient who sustained a head injury. From a rehabilitation standpoint, it is worthwhile to raise some question as to why the patient had a head injury. Did it involve a car accident while the patient was intoxicated? Was it a matter of poor judgment in a situation in which the patient became injured because of lack of knowledge of the consequences of his action? Was the patient injured in a fight or because of some impulsive action on his part? Following the head injury, the clinical phenomenology of the patient should reflect the patient's premorbid personality and capacity. Postmorbidly, one also has to consider the reaction of the patient to his deficits as he perceives them. There is (almost necessarily) a devaluation of the self-concept. Brain-damaged patients often feel that they have become worthless to themselves and to others because of their brain disorder.

Another reality is that the brain-damaged individual frequently is an unemployed person. Often, heads of families lose the breadwinner role. Homemakers often lose the capacity to manage a household and care for their families. The economic aspects of unemployment frequently cause problems, and the patient must depend on insurance, public welfare, or the support of others who might have been supported by him premorbidly. One of the psychological consequences of unemployment is development of the feeling that one is no longer needed. Typically, from the time the brain-damaged patient regains consciousness he is taken care of, and this situation sometimes continues long after acquisition of the brain lesion. He becomes a dependent for a greater or lesser period of time. In the case of brain damage, as compared with other disabilities, the patient tends to be viewed as incompetent as a caregiver to others.

Typically, the brain-damaged patient is a family member: a parent, sibling, spouse, or child. The acquisition of brain damage often significantly alters the family network of relationships, and the characteristics of these alterations may be significantly related to rehabilitation outcome. The acquisition of a lesion frequently ushers in a period of chaos for the family—out of which comes some kind of reorganization in which a number of productive and nonproductive reorganizations take place. Perhaps the most unproductive reorganization is *rejection*, whereby the family will have nothing to do with the patient and leaves his care in the hands of some institution. The opposite pole involves overprotection and overindulgence, in which the family views the patient as completely incapacitated and feels that everything must be done for him. We believe that the most positive reorganization occurs when the family becomes, in essence, a part of the rehabilitation team and participates in the program established for the patient. In any event, the way in which the family and other acquaintances of the patient respond to him is a significant determiner of rehabilitation outcome.

A Multidimensional Approach

We must, as members of various professional disciplines, learn from each other. Without attempting to be critical, we would say to the physician that a stroke is not merely paralysis of a side of the body. To the dynamically oriented psychologist and psychiatrist, we would say that the cognitive deficits of the brain-damaged patient are not necessarily resistance, denial, or the consequences of depression but are often direct manifestations of structural brain damage. To the neuropsychologist we would offer that whereas the patient may have dramatic cognitive and perceptual deficits, he may also have a rejecting family and a serious depression, both of which have to be considered in his treatment. It would seem useful to take a field theoretical approach to the rehabilitation process, in which the patient is at the focus of various forces impinging on him. Family therapy and "Network Therapy" advocates have strongly supported this position. In psychiatric and general medical settings there is currently a strong emphasis on systems theory and the Problem-Oriented Medical Record System (POMR) (Weed, 1970), both of which focus on the complex of problems of the individual patient rather than on the diagnosis or the roles of the different treatment specialties. Approaches of this type, although seemingly overly theoretical, remind us that there may be many countervailing forces going on in the patient's life space, and that all of these forces may not always be productive. Thus, retraining efforts in the clinic may be nullified or diminished in effectiveness by overprotection or rejection by the family. The POMR helps to focus on the patient's problems and through the so-called SOAP system forces us to: hear his complaints (S = subjective complaints), look for objective signs of difficulty (O = objective symptoms), make some assessment and evaluation (A = assessment), and formulate a plan (P = plan). Typically, the problem list calls for a multidisciplinary approach in that on the list "family conflict" may share equal status with "confusion," which may in turn share equal status with "hypertension." All of these matters are examples of problems, and ideally they should all be resolved. In current health practice there is an emphasis on comprehensive "nose to toes" care that follows the patient from the hospital to the outpatient clinic to the community. The rehabilitation of the brain-damaged patient clearly calls for care having this scope.

Thus far, we have discussed the contributions of various disciplines to rehabilitation, the nature of interdisciplinary treatment, and various treatment models. In actual rehabilitation, it is necessary to somehow organize the disciplines and treatment model or models into some operational framework. We would recommend a multidimensional framework, especially since brain damage commonly pervades all areas of life. It is unlike may other illnesses in that it cannot generally be treated within the context of a normal ongoing life. One can be diabetic and maintain normal function by taking insulin, or one can have a physical handicap and continue to function with the aid of some prosthesis. However, it is not usual to see patients with brain lesions whose treatment can take place in the context of an ongoing normal life. There are, of course, such individuals, but they are exceptional and are not usually presented as candidates for rehabilitation. Therefore, the pervasive effects of brain damage typically require multiple forms of treatment involving many disciplines.

Medical Care

Many brain-damaged people are sick. (Naturally they are sick from their brain damage, but that is not what is meant here.) What we mean is that the brain damage may cause or be caused by some organic difficulty that requires continued medical observation and treatment. The brain is not removed from the rest of the organism. Rather than go into detail about this matter, we will provide some examples. Patients with cerebral vascular disease typically do not only have *cerebral* vascular disease; they have *generalized* vascular disease. They may have impaired cardiac function or hypertension or diabetes or any of a number of cardiovascular difficulties. These conditions should be monitored and, if possible, treated medically. In the case of head injury, the reverse process may take place. The traumatic brain damage may produce symptoms outside of the brain in a previously healthy individual. Sometimes head trauma produces hydrocephalus or other conditions that can lead to such symptoms as incontinence. We have recently seen a case of this type in which a premorbidly healthy young man sustained a severe head injury following which bowel function became chronically abnormal. His gastrointestinal difficulties required continued management, observation, and treatment. Thus the physician should maintain continuous participation with the rehabilitation group in order to fulfill these three basic functions:

1. To implement new developments in medical research as they may pertain to treatment of the patient
2. To monitor the progress of the brain lesion and its consequences
3. To advise what type and level of activity would be prudent for the patient to engage in given his physical condition

Work Therapy

Clearly, a major goal of any rehabilitation effort is that of making the patient as productive as possible. Whereas many brain-damaged patients are so aged or so impaired that return to remunerated work is not a realistic goal, the majority of young brain-damaged individuals can probably return to work or school under the proper circumstances. This aspect of a multidimensional approach pertains to such patients. A rehabilitation team can productively aid in getting the patient back to work. From the point of view of restoration of self-concept, relearning of impaired skills, and avoidance of developing a debilitatingly sedentary life pattern, work is fine therapy.

The issue of employment of brain-damaged patients has some unique characteristics. First of all, there is the medico-legal matter of employment of epileptics. Epilepsy can limit one's ability and opportunity to drive, work around potentially dangerous equipment, and generally work at any job in which insurability of the employer is an issue. One must consider, in this complex issue, the rights of the epileptic as well as the dangers he may present to himself and to others. In the case of the nonepileptic individual with brain damage, there is often another serious problem. It revolves around the question of why the patient is not working. Brain damage is most often not a visible disability, and there are usually no physical limitations that (in and of themselves) would prevent employment. Yet there is a high rate of unemployment among young, brain-damaged individuals (Carey, Young, Rish, & Mathis, 1974). There seems to be a poorly understood discrepancy between what these individuals are seemingly capable of and their actual levels of productivity. With regard to rehabilitation, the point is that if one could create a structure and environment that make work *possible* for these individuals, they would then derive many of the ego-enhancing, educational, and physical benefits that can be acquired from work. Thus, the sheltered workshop or prevocational training center can be an ideal setting for a multidisciplinary rehabilitation approach.

Retraining

Work, in and of itself, may be beneficial in rehabilitation, but the therapeutic benefits of work can be greatly enhanced by assigning the patient to an appropriate task. The appropriateness of a task is largely a question of what the patient's deficits and remaining abilities are. Thus, a program of productive work can mean not only work but cognitive-perceptual retraining as well. Such an arrangement can be developed through the cooperation of neuropsychologists and rehabilitation specialists. In such programs, patients can develop such skills as the capacity to maintain attention, problem-solving ability, conceptual thinking, spatial relations ability, and language skills through rational selection of work tasks. It is sometimes necessary to retrain a skill to a certain level before training can take place at the level of productive work. We refer to such training as *prevocational*, in the hope that the patient can eventually graduate to some actual work task. In our terminology, the work of the NYU group is prevocational in nature, although many of their patients are too aged and infirm to return to paid employment.

The planning and implementation of retraining of brain-damaged patients is a process requiring innovation and creativity. It can benefit from input from many disciplines. The point that this planning should be rational bears repetition. Training programs should not be based on the availability of facilities or of skilled personnel in particular areas but on the needs of the individual patient. One major contribution of the NYU group was the introduction of scientific concepts, developed by neuropsychologists, into the planning of retraining programs. One should not treat left- and right-hemiplegics in like manner because the behavioral deficits suffered by members of each of these groups are very different from each other. One should not simply send a patient to OT, for example, until the occupational therapist, the neuropsychologist, and members of other disciplines have devised a plan for *what the patient should be doing in OT.*

Recreational Activities and Environmental Exposure

Most normal, healthy people make a distinction in their lives between work and play. In the case of the chronically ill individual, this distinction can become blurred, as there really is not a time when one is being task oriented and another time when one is just having fun. The inclusion of recreational activities in rehabilitation programs can aid in normalizing the structure of the patient's day, thus aiding him in readjusting to a normal environment. Additionally, recreational activities can serve at least three other functions. The first benefit is that, if the recreational task is appropriately chosen, it can become a part of the cognitive-perceptual retraining. The judicious choice of verbal or nonverbal games can make game playing a training activity. Indeed, the Block Design training of the NYU group (Ben-Yishay, Diller, Gerstman, & Gordon, 1970) has a gamelike quality and is generally enjoyed by patients. In the area of speech therapy, Wilcox and Davis (in press) have developed a program called Promoting Aphasic Communicative Effectiveness (PACE) that is designed to improve ability to communicate meaning through a "twenty-questions" type game. The second benefit is that recreation of the gross motor type, such as participation in calisthenics or sports, helps reduce the effects of sedentary life that often go along with institutional living. Many hospitals have recreational therapy programs allowing for regular participation in sports and other avocational activities. Although there is no direct research evidence that such activities are particularly beneficial for brain-damaged patients, there have been numerous studies involving the effectiveness of running and other exercise for psychiatric patients, medical patients, and normal individuals with health problems (Cantwell, 1978; Clausen, 1977; Reischel, 1977). The third benefit is that recreational activity encourages social interaction with patients and staff.

Many brain-damaged patients rarely leave their homes or the institutions in which they reside. They may therefore develop a kind of sensory deprivation that could exacerbate their symptoms. If at all possible, institutionalized and homebound individuals should be taken on field trips regularly. They may go to parks, zoos, concerts, museums, sports events, or other cultural or recreational facilities and activities. These activities need not be merely recreational; they could prevent some of the detrimental effects of perceptual and motor deprivation, as well as aid the institutionalized patient in readjustment to community life. Avedon (1974) reports that in the Netherlands holiday cruises are offered for chronically ill institutionalized individuals, and sightseeing tours are also provided on specially equipped buses. He also nicely points out that there is nothing new about the use of recreation in treatment by quoting this passage from Browne (1837), which describes treatment of the sick by the priests of ancient Egypt:

> [The patients] were required to walk in the beautiful gardens which surrounded the temples, or to row on the majestic Nile. Delightful excursions were planned for them under the plea of pilgrimages. Dances, concerts, and comic representations occupied a part of the day, as constituting the symbolic worship of some divinity. (pp. 141–142)

Mental Health

As we have indicated, the view of the brain-damaged patient as a normal individual except for the lesion is a stereotype that seldom exists in reality. Aside from the behavior changes that can be directly attributed to the lesion, there is usually some kind of emotional difficulty. Often it is a depression, but agitation, anxiety, impulse-control problems, and frank psychotic disorders are also frequently seen. It is sometimes exceedingly difficult to differentiate between that which is attributable to the lesion and that which either existed before the lesion, or is an emotional reaction to the impact of becoming brain damaged. Some very intricate interactions can take place. For example, we worked with a severely head-injured patient who was incontinent of feces. The neurologist could not determine whether or not the incontinence had a neurological basis, but there may have been some degree of loss of control of the anal sphincter. In the course of a behavior modification program, it became apparent that the patient could control his bowels—at least to a certain extent. It was therefore likely that he was using the incontinence (at some level) to manipulate his environment. Thus, the patient's impaired social judgement, his emotional reaction to his illness, and the possibly direct effect of his neurological problem may have all been interacting to produce the incontinence.

We believe that very few people who actually work with brain-damaged patients on a day-to-day basis would not vouch for the need for psychotherapeutic intervention in many (if not most) cases. Such intervention may consist of individual or group psychotherapy, medication, or other treatments in the armamentarium of the mental health professions. As we have indicated, the psychiatric model of rehabilitation may overemphasize this aspect of the complex problems associated with brain damage, and this overemphasis may lead to such things as overuse or inappropriate use of psychotropic drugs. On the other hand, one cannot afford to turn overemphasis into neglect. In some cases, the major adaptive problem is not the organic brain syndrome but the accompanying chronic depression or anxiety. We have seen institutionalized, severely brain-damaged patients who could function in the community were it not for disabling anxiety. We have also seen patients who would not cooperate willingly in rehabilitation efforts because of feelings of hopelessness and futility. Sometimes a course of treatment with antianxiety or antidepressive drugs would help substantially in getting these patients rehabilitated.

The application of psychotherapy to brain-damaged patients is controversial in that many clinicians feel that such patients, by virtue of their intellectual impairment, would not benefit from such treatment. For example, the amnesic patient may not be able to remember from one therapy session to another. The aphasic patient may not be able to articulate his thoughts and feelings by speaking, and may not comprehend the speech of the therapist. The patient with generalized intellectual impairment may not be able to cope with the symbolic nature of much that goes on in psychotherapy. In our experience, whereas formal individual psychotherapy is rarely the treatment of choice, other forms of psychological treatment are useful. One of them simply involves providing the opportunity for counseling and receiving advice on an ad hoc basis. Supportive-type group psychotherapy is also often useful in that it encourages social interaction and provides the patient with an opportunity to exercise his sometimes impaired ability to communicate. A third behavioral approach that will be elaborated on in detail later is behavior therapy.

Mainstreaming

The term *mainstreaming*, which has become very popular in education circles, has some applicability in the present context. It means that students with various educational disabilities—be they physical, cognitive, or emotional in nature—are placed in classes with normal children rather than in "special" schools or classes. In the case of the physically handicapped in particular, great efforts have been made, and are continuing to be made, to make educational and other public facilities more accessible. Much of the current educational philosophy seems to be promainstreaming. It is felt that individuals with various disabilities do better educationally and socially if they mix with normal children and children with other kinds of disabilities, rather than being segregated into special programs. The question is whether or not it is advantageous to the rehabilitation of brain-damaged patients to have them placed with other patients with other forms of physical, perceptual, and cognitive handicaps.

Unfortunately, this specific question cannot be answered scientifically at present. However, there are data on psychiatric patients in general. In 1979, Ellsworth, Collins, Casey, Schoonover, Hickey, Hyer, Twemlow, and Nesselroade reported on the results of a large VA cooperative study, in which it was demonstrated that patients hospitalized on mixed wards containing acute and chronic patients had better treatment outcomes than patients on wards with more narrowly defined populations. Using our terminology, it can be said that patients who were mainstreamed did better than patients who were not. Fairweather (1964) reported essentially the same finding; a mixture of acute and chronic as well as active and passive patients was associated with an increase in the effectiveness of treatment. One consideration here is that Ellsworth *et al.* indicated that staff working with a combination of newly admitted (less than six months of hospitalization) and long-term patients spent slightly more time with the newly admitted patients, and it was these patients who were used by Ellsworth *et al.* to evaluate program effectiveness. It is possible that the chronic patients also benefited from being on wards with acute patients, but that was not documented by the study.

A presently unanswered question concerns whether or not it is advisable to mainstream brain-damaged patients, in the sense of treating them in the same facilities that treat psychiatric and/or physically disabled people. Some clinicians are of the opinion that to do so is inadvisable, particularly in the case of mixing brain-damaged patients with patients with functional psychiatric disorders. It is sometimes felt that brain-damaged patients, by virtue of their impaired intellectual functioning, develop some of the maladaptive behaviors of the psychiatric patients, without benefiting from being with people with other kinds of disabilities. In one program that will be described we placed brain-damaged patients together with mentally retarded and learning-disabled individuals in a sheltered workshop setting. We did not study the matter systematically, but there did not appear to be any adverse effects on either group. One possibly effective strategy may be that of treating brain-damaged patients in a separate setting during the early rehabilitative stages and gradually allowing them to try out some of their newly relearned skills in the more complex environment that may be provided by a mixed-patient population.

The Need for Comprehensive Programming

Why does a rehabilitation program have to be multidimensional? Often, it does not. However, in the case of the brain-damaged patient it often does because of the pervasiveness of the condition. Thus, failure to attend to one area may greatly compromise the effectiveness of excellent treatment given in some other area. We are, therefore, strongly in favor of the development of comprehensive programs in which one individual or team is responsible for coordination and implementation of a total treatment effort. Although this recommendation may sound vague, it does have specific implications—some of which are clinical, some of which are administrative in nature. In dealing with the administrative aspects of the problem, the program coordinator must have control of all aspects of the program. He need not be a dictator, but he should also not be in a position in which his decisions are regularly vetoed by some higher authority. Functioning in this area often runs into medico-legal difficulties, particularly when the program leader is not a physician. There are several solutions to this problem. One possibility is for the leader to work outside of a medical setting, receiving medical consultation as required. Obviously, a second solution is for the leader to be a physician. However, the most reasonable solution within a medical setting, if the leader is not a physician, is to establish working relationships with physicians such that the leadership situation is clear. In any event, administration of a comprehensive, multidimensional rehabilitation program requires control in at least these areas:

1. Admissions and discharges
2. Medication
3. Access to various clinics and treatment facilities
4. Access to consultation and various diagnostic procedures
5. Authority to write and cancel orders
6. Supervisory authority over the treatment team insofar as their participation in the program is involved

These comments should not be taken to mean that we are encouraging nonphysician program directors to practice medicine. It does mean that the director is responsible for reaching some satisfactory arrangement such that the medical aspects of the total treatment are coordinated with that effort. It also implies that a program-oriented organizational framework is probably better than a discipline-oriented framework when comprehensive treatment is desired. Again, the most relevant documentation of this claim comes from work with psychiatric patients. Two examples are to be found in the areas of crisis intervention and psychosocial treatment of chronic patients. Lieb, Lipsitch, and Slaby (1973) have written a book entitled *The Crisis Team* in which it is clearly pointed out that the fundamental working unit is the treatment team, which is always multidisciplinary in composition. Much of the book describes a comprehensive program for assessment and treatment of individuals in crises, implemented in an integrated manner by the multidisciplinary team. There is a well-outlined program going from intake, through obtaining a history, to prescribed treatments and ultimate disposition. The authors of this work do not present systematic data attesting to the effectiveness of their approach, but they do present a number of illustrative cases. There are, however, much data regarding program-oriented treatment of chronic patients in mental hospitals. A major contribution to this area was made by Paul and Lentz (1977), who clearly demonstrated the advantages of an organized social learning-oriented program over traditional milieu programs. *The Token Economy* (Kazdin, 1977) is another classic example of an organized program approach, in this case based on operant-conditioning principles.

Aside from research and clinical evidence of the effectiveness of program-oriented approaches to rehabilitation and treatment, ordinary observation of discipline-oriented health service facilities often reveals a number of problems. There is frequently a matter of dual loyalties, in which, for example, the nurse may work with a ward administrator but also be under the supervision of the director of the nursing service. It should be emphasized that the administrative aspects of rehabilitation programs may have significant consequences for those programs. An equivocal leadership pattern in which delegation of responsibility is not clear or a situation in which the program coordinator's decisions are commonly vetoed or "second guessed" by others can easily lead to failure.

On the basis of the degree of success achieved with comprehensive, multidimensional programming with psychiatric patients, there is reason to believe that such an approach may be useful with brain-damaged patients (at least generally). It would appear that this approach has two aspects: the introduction of a variety of modalities and the implementation of some conceptual model. The use of multiple modalities is often necessitated by the fact that patients tend to have a variety of treatment needs. The absence of a model could produce unorganized, disjointed programming. As a model, the token economy allowed for the coherent organization of treatment programs in a number of applications. One important point is that the model should be made explicit. If it is not, it may be difficult to achieve consensus among the treatment team members regarding its potential efficacy and general acceptability. For example, some people may not be sympathetic to behavior therapy. They may object to it on clinical, ethical, or scientific grounds. Although everybody is entitled to his opinion, it is not fair to the staff member or ultimately to the patients to place someone opposed to behavior therapy in a behavior–therapy-oriented program. The philosophy of the program should be explicitly stated, and the potential staff member should be able to choose to participate or not participate.

The Goals of Rehabilitation of the Brain Damaged

In this chapter, we initially tried to present several approaches to rehabilitation along three dimensions: setting, discipline, and philosophical orientation. We then attempted to integrate this material. One final component of this integration revolves around the matter of treatment goals. Do different people mean different things when they use the term *rehabilitation?* Given the anatomical fact that central nervous system tissue does not regenerate, the pathological fact that many brain disorders are presently incurable terminal illnesses, and the clinical fact that many of the physical and behavioral symptoms associated with brain lesions are exceedingly difficult to treat, what can we hope for? In what sense can we reasonably use the term rehabilitation? If one were to hold a conference of informed individuals in this area, it is likely that a variety of views would emerge. Wright (1960), for example, provides an extended discussion of normalcy as a rehabilitation goal. She is critical of the concept that normal function should be held up as an ideal but points out that it does have some value as a potent motivator. In rehabilitation, however, she suggests that the nature of the disability in combination with the preserved abilities may be a better guide to what to do than is the standard of normal function. For example, individuals with gait disturbances may learn to locomote far more effectively through some unconventional procedure than through attempting to relearn normal walking. It is interesting to reflect on the goal of normal function in the case of rehabilitation of the brain-damaged person, particularly when one considers that goal as one being held by the patient, those treating him, and his family and acquaintances. From the point of view of the brain-damaged patient, the whole question of personal goals is greatly colored by the patient's cognitive ability to form goals at all. Unlike the rather literate people that Wright (1960) draws her examples from, most seriously brain-damaged people have a significantly diminished capacity to articulate general goals to themselves or others. The extent to which brain-damaged individuals compare themselves unfavorably to normals and develop consequent feelings of shame and inferiority is very much open to question. Indeed, the extent to which such processes as defense mechanisms, compensation for inferiority, and development of guilt operate in individuals with severe brain damage is not really well understood. Obviously, the goal of normal function, as established by individuals other than the patient, is often unrealistic and may have unfortunate consequences. In speaking with friends and relatives of patients, it often appears that this goal sometimes seems to be based on magical thinking. Somehow, through some unknown mechanism, the affected spouse, parent, or child will return to the way he used to be. Thus, it would appear that the ideal of normal function is a goal that is usually not clearly articulated by the brain-damaged patient, and that is more or less unrealistic as a goal set for the patient by others.

An alternative set of goals may be described as educational in nature. The aim is to teach the patient something, and if he learns it, then the goal has been achieved. For example, if the NYU group's Block Design training is successful, and if the success is demonstrated by improvement in the patient's ability to eat and engage more effectively in other behaviors requiring coordination in three dimensions, then a goal has been achieved. The patient may still be hemiplegic and may still have significant intellectual and perceptual deficits, but the rehabilitation effort has demonstrably improved his functioning, albeit in a limited area.

In these two examples of possible goals we can see an enormous discrepancy in what is meant by rehabilitation when different people use the term. Still another view is expressed by those who set deinstitutionalization as a goal. The goal of treatment is to get the patient out of the hospital. Such a goal may be established in a variety of ways, including use of retraining techniques, medication, and working with the family. In any event, if the patient is discharged and does not return to institutional life, then he can be viewed as rehabilitated. This goal is a particularly pertinent one for patients with chronic disorders and long histories of institutionalization. It is an objective goal and is easily measured. One can, of course, raise the question of what the patient is doing outside of the institution. If his life style is relatively unchanged or worsened, then whatever was achieved by his de-institutionalizing is open to question. We have learned, in recent years, that getting the patient out of the hospital does not solve all the problems. For example, relapse and rehospitalization are not infrequent (Goldberg, Schooler, Hogart, & Roper, 1977; Hogarty, Goldberg, & Schooler, 1974).

We could continue to dicuss alternative goals and their implications at great length. Perhaps what is most important is to have explicit goals, regardless of what they may be. The process of setting goals, evaluating their feasibility, and rejecting those that are unrealistic on the basis of experience rather than prognosis may be what really produces progress. The only proviso is that scholarship and expertise should play some role in the process, particularly in regard to defining the limits of what is possible. We are all familiar with cases in which a handicapped person walked or performed some other major function again despite the fact that a doctor told him that he would never do so. However, when a doctor tells a patient that some function will never return, he is usually right. We tend to hear about the exceptions and not the rule.

From a practical standpoint, what can be expected from individuals with nonprogressive brain damage in the way of recovery of function? This question is an empirical one and can best be answered by research. Unfortunately, however, little research has been done, and what we can report is limited. It is clear that the NYU group has been successful in improving adaptive behavior in hemiplegic patients. Such areas as functional speech, eating, dressing, accident avoidance, and grooming apparently can be improved. It is therefore not unrealistic to train the stroke patient in certain cognitive and perceptual skills and to expect demonstrable improvement in activities of daily living. Seeing aphasic patients improve with speech therapy is not at all uncommon. In our clinical experience, we have seen institutionalized patients with severe amnesic and language disorders recover sufficiently, following an intensive treatment program, to leave the hospital and function independently for extensive periods of time. It may be pointed out that, in all these instances, the patients involved had substantial, irreversible brain damage. Thus, in our view, some degree of recovery of function is a reasonable goal, even when treatment is initiated long after the acute phase of the brain disorder has subsided, if there was such a phase. It also seems clear that we are looking at genuine treatment effects and not spontaneous recovery of function. Thus, there seems to be some basis for expecting some degree of functional recovery following rehabilitative treatment efforts applied to individuals with nonprogressive brain damage. Such recovery may be associated with improved capacity to perform activities of daily living, and, in some cases, with increased potential for employability.

In view of these considerations, two basic points may be made concerning goal setting in the rehabilitation of brain-damaged patients. First, it is important to set and communicate reasonable, explicit goals so that the patient, the patient's relatives, and the treatment team are all aware of the nature of the program. These goals may be ambitious but should take cognizance of the physical limitations imposed by the patient's neurological condition. Second, while normal function is commonly viewed as an ideal treatment goal, available research findings are probably a more reasonable guide to goal setting. Our present state of knowledge is limited with regard to what the best approaches are to rehabilitating brain-damaged patients and also with regard to what is and is not possible for the brain-damaged patient. As we have seen, in the absence of direct knowledge, clinicians have borrowed heavily from models of treatment originally intended for patients with physical disabilities and psychiatric disorders. The recent application of behavior therapy and neuropsychology to rehabilitation of the brain-damaged patient has introduced some new concepts to this area, but further investigation is needed to evaluate their potential more fully. Although a call for further research has probably become the most commonly used cliché in scientific writing, we confess to not being able to find a more appropriate way to end this chapter.

3

STAFF DEVELOPMENT

Whereas staff education is important in all clinical endeavors, it is particularly important, for a number of reasons, in regard to treatment of brain-damaged patients. Perhaps most significantly, rehabilitation-oriented treatment of brain-damaged patients is a somewhat new field, and little is known about it. Otherwise well-trained members of multidisciplinary treatment teams may be relatively untutored in regard to rehabilitation of individuals with brain dysfunction. They may know a great deal about the medical, physical, and psychiatric aspects of these disorders, but they may not know a great deal about *neuropsychology;* in our view the scientific specialty area that currently has most to say about various relationships between brain dysfunction and behavior. As we have already indicated, clinical neuropsychology or behavioral neurology is a relatively new field, and approaches to brain-damaged patients from a treatment standpoint, using these fields as a knowledge base, have only been attempted in a few settings throughout the world. It is therefore usually necessary to provide the treatment staff with a great deal of technical information in an area that may be quite new to them.

A second reason for the importance of staff development has to do more with beliefs and attitudes. Working with brain-damaged patients is generally an arduous task, and the rapid, dramatic results sometimes seen in general medicine and psychiatry are not often seen in these patients. We rarely see complete cure and recovery. Progress is usually slow, and much effort must be made to achieve sometimes minimal gains. Many brain-damaged patients have terminal illnesses, and, despite our best efforts, the patient may deteriorate before our eyes. Issues of death and dying (cf. Kübler-Ross, 1969) are often prominent. On the other hand, it would seem to be important that the staff have some degree of optimism regarding treatment outcome. If the aura of pessimism regarding treatment of brain-damaged patients is overly pervasive, then one is likely to see many self-fulfilling prophecies. The belief that the patient cannot recover can turn into the fact that he does not recover, only because insufficient or inappropriate treatment was given.

In addition to the problems associated with treating brain-damaged-patients, a host of problems may arise when the program director elects to place an emphasis on behavior therapy. Hersen and Bellack (1978) have provided an extensive analysis and literature review of staff training and consultation in the behavior–therapy area. There is an extensive literature on training of behavior therapists (e.g., J. M. Gardner, 1972; Loeber, 1971; McKeown, Adams, & Forehand, 1975). The emphasis is generally on the paraprofessional staff, such as students and nursing assistants, since, as Hersen and Bellack indicate, such individuals may have "life-or-death" power over the effectiveness of behavior–therapy programs. In effect, staff training and development would appear to be crucial when entering a field in which there is relatively widespread pessimism concerning outcome, by utilizing a method that remains somewhat controversial.

Leadership

Regardless of the treatment philosophy adopted, be it behavior–therapy-oriented or otherwise, it would seem crucial that the leader of the program be well versed in the area of brain–behavior relationships. Generally, such individuals are neurologists with specialized training in behavioral neurology or psychologists with specialized training in clinical neuropsychology. For purposes of this discussion, we will refer to such individuals as *neurobehavior specialists,* since they may be neurologists, psychologists, or members of other professional disciplines, such as psychiatry, speech pathology, or physiatry. The important point is that the leader has substantial technical and theoretical knowledge of what is known concerning how the brain mediates behavior in humans, and what the behavioral consequences are of the various forms of brain damage. Aside from scientific considerations, of course, the team leader also has to function as a clinician and as an administrator, who is sensitive to, and able to deal effectively with, a patient's problems of a social, interpersonal, or emotional nature, in addition to the specific rehabilitation problem. Ultimately, one is always treating a whole human being rather than an isolated deficit. With regard to being an administrator, the success or failure of a program is often determined by the competence with which it is administered. An intolerable authoritarian style, a neglect of financial and other administrative matters, a failure to communicate with other organizations in the same institution in which the program is located, and other administrative errors may lead to the downfall of the program.

Where are such ideal leaders to be found? Unfortunately, there are no specific training programs in brain–damage rehabilitation, either in medical schools or in psychology departments. Thus, beyond the M.D. or Ph.D. degree, it is necessary to obtain additional training in the neurobehavioral sciences. There is now a small but growing group of professional individuals who identify themselves as *clinical neuropsychologists* or *behavioral neurologists.* Many of these individuals are rehabilitation oriented, but not all of them have received formal training in rehabilitation. For such individuals, we would recommend, as a start, attendance at one of the workshops in cognitive and perceptual retraining sponsored by the NYU Institute of Rehabilitation Medicine. For those interested in adopting a behavior–therapy approach, postdoctoral training in that area is readily available. One may wish to correspond with the Association for Advancement of Behavior Therapy. There are also several workshop programs available in the area of neuropsychological assessment. For those interested in predoctoral training in human neuropsychology, there are several programs in existence. The ones at the University of Oklahoma, University of Minnesota, and University of Victoria (British Columbia) are among the most prominent of these programs.

We confess to knowing no way of making a person a "good clinician" or effective administrator. With regard to the clinical skills needed, all we would offer at this point is that the team leader should have some degree of expertise in clinical psychiatry or psychology. Rehabilitation efforts are often compromised by considerations concerning poor interpersonal relationships, motivational difficulties, or emotional disorders. Thus, the team leader should be sensitive to such matters as personality conflicts among staff and patients, or the existence of a psychiatric disorder in a patient aside from the neurological problem. The recognition of depression is crucial. The relation between organic mental disorders and depression have become a matter of great interest (Wells, 1980), particularly with regard to the matter of so-called pseudodementia in the elderly—severe difficulties in cognitive function determined to be associated with an affective disorder rather than with diffuse progressive brain impairment.

The Staffing Pattern

We have already described the rehabilitation-related disciplines, and here we will briefly indicate what can be viewed as a core-staffing pattern for the team. In this discussion, the emphasis will be on inpatient treatment teams, but there are obvious applications to the outpatient situation. For many reasons, a physician should certainly be a member of the team (if not the leader). Of course, in medical settings, there is always the question of medico-legal responsibility. While it is true that rehabilitation can be viewed as an educational process as well as a medical one, a realistic appraisal would suggest that brain-damaged patients quite frequently have medical problems associated with the brain damage itself, with age, or with other considerations. Therefore, it usually does not make sense to separate the medical aspects of rehabilitation from the educational aspects. The particular training of the physician is also an important matter. Ideally, specialized training in neurology, psychiatry, or physiatry can be a definite asset. Aside from purely professional and legal matters, use of medication is often an important part of the treatment process, since without a physician, use of medication is impossible.

It is highly desirable to have rehabilitation specialists as team members rather than as consultants. The presence of an occupational therapist, a physical therapist, and a speech pathologist in the treatment area (or at least at team meetings) can be helpful, because it allows for direct communication regarding activities in speech therapy or in the OT or PT clinics to the other members of the team. As we will see later in our material on treatment of chronic patients, sustained interaction among the rehabilitation specialists, the nursing personnel, the program physician, and the neuropsychologist is really the essence of rehabilitation planning and evaluation.

While it is obvious that nursing personnel are necessary in an inpatient setting, we would point out that in behaviorally oriented rehabilitation programs, the nursing staff does far more than the traditional tasks performed in general medical settings. For example, Ayllon and Michael (1959) described the psychiatric nurse as a "behavioral engineer," and illustrated how nursing personnel could be taught various principles of operant conditioning, such as those involving extinction and selective reinforcement, in order to deal with problematic behaviors of chronic psychiatric patients. In general, the nursing staff in a medical setting or the teaching staff in an educational setting constitutes the group that has the most direct contact with patients or students. In many respects, these individuals work as extensions of other professional staff members to implement program plans. We will refer to these individuals as the *line staff.*

Numerous examples could be given of how line staff and other staff interrelate. If the patient is on some kind of exercise program, a line-staff member may receive instructions and guidance from the physiatrist or physical therapist concerning the nature of the training to be given and the limitations of the client's capacity. However, the line-staff member may do the actual training. Similarly, the occupational therapist may aid the line-staff member in regard to selection of media and level of task complexity; the speech pathologist can provide specific language-retraining procedures; the behavior therapist can provide a single-subject design program to be implemented. In all these cases, the implementation is often accomplished by the line staff. Sometimes, the entire line staff is involved. Types of programs in which the client must be consistently treated with a particular attitude, or those that operate within the framework of a particular comprehensive system, such as a token economy (Kazdin, 1977), generally require continued participation by the entire staff.

In summary, the staffing pattern of a rehabilitation program should include a neurobehavioral specialist, a physician, a group of rehabilitation specialists including a speech pathologist, occupational therapist, and speech therapist, and a line staff that is trained to perform more than the traditional nursing or educational duties. Such training may be in the area of behavior therapy, or more specifically in the application of specialized rehabilitation techniques and strategies to brain-damaged patients. Beyond the training requirements for each of the professional disciplines involved, there obviously has to be some specialized training concerning the specific nature of the rehabilitation program in which the team will be operating. It is this topic that we will address next.

Staff Training

There is unfortunately no body of research literature describing elegantly designed studies of training methods in brain–damage rehabilitation as there is in the field of behavior therapy (Hersen & Bellack, 1978). However, it would appear that Hersen and Bellack's general conclusion that a mixed approach involving some lecture material and some "hands on" supervised experience is likely to be more productive than a purely didactic educational methodology. Rather than speculate further on what may be the "best" way to train a rehabilitation staff, we will concentrate on what we feel the content of the training should be. The only additional methodological comment we will make is that we have found it useful to conduct a formal orientation for about a two-week period prior to initiation of a program. Even though such orientations tend to be largely didactic in nature, they can provide a great deal of essential information in a very efficient, concentrated manner.

The content areas for training in brain–damage rehabilitation are defined by the term itself. One has to understand brain damage and rehabilitation. A technical understanding of the consequences of brain damage requires some knowledge of the sciences of neuroanatomy, neurophysiology, and neuropathology. In other words, one must know something about normal brain structure and function, and something about brain diseases and other disorders. Furthermore, one has to know something about how the brain mediates behavior, namely, *neuropsychology*. We will proceed to mention briefly certain basic facts concerning each of these areas, not so much to provide a training syllabus as to acquaint the reader in a very general way with some of the basic concepts.

Neuroanatomy

It is important to know that the brain is divided into three major portions: the brain stem, the cerebellum, and the cerebral hemispheres. We will concern ourselves here with only the cerebral hemispheres. In general, but with some major exceptions, the right-cerebral hemisphere controls the left side of the body and contains the centers for vision in the left half of space, or the left visual field. The two hemispheres are connected by three commissures, the major one being a structure called the *corpus callosum*. Each hemisphere is divided into four major lobes: the frontal, temporal, parietal, and occipital lobes. As another generalization, the frontal portion of the brain has to do with mediation of movement, while the posterior portion mediates various modalities of perception. Thus, pathways descending from the frontal lobes control and regulate skilled movement, whereas pathways ascending to the temporal, parietal, and occipital lobes deliver messages regarding auditory, somatosensory, and visual information. This basic neuroanatomy explains why, for example, patients with damage to the left hemisphere may become paralyzed on the right side of their body, whereas patients with damage to the posterior portion of the brain may have various visual difficulties, such as the loss of ability to recognize objects or faces.

Neurophysiology

Neurophysiology is the study of how the brain functions. As we know, the brain consists of an astronomical number of individual cells or neurons, and neurophysiologists are generally interested in how these neurons function, individually or collectively. For our purposes, there are two important branches of neurophysiology, *electrophysiology* and *neurochemistry*. Electrophysiology is closely associated with the electroencephalogram (EEG) and related procedures. The brain generates a normal pattern of electrical activity that varies with differing stages of waking and sleep. Electrophysiologists are therefore particularly interested in this normal electrical activity in regard to such matters as the relationship between particular wave forms and the structures in the brain that generate them. Electroencephalographers are interested in the detection of abnormal patterns that may be associated with a variety of neurological disorders. The area of neurochemistry has proliferated in recent years primarily because of the discovery of a group of substances known as *neurotransmitters*. There is now thought to be a number of systems in the brain for the various neurotransmitters, and one may hear such terms as the *GABA* or the *DOPA* system. These substances seem to be responsible for communication and storage of information. For example, it is now thought that the poorly understood mechanism for memory may be the modification of these systems on the basis of experience. That is, the so-called memory trace may be a neurochemical event.

Neuropathology

Neuropathology is the study of disorders in neuroanatomy or neurophysiology. The neuropathologist is often viewed as the individual who does the brain cutting and examination of the autopsy. However, we are using the term in a broader sense here, and really mean the study of the wide variety of things that can go wrong with the brain. There are several general principles. As an organ of the body, the brain can acquire many of the same disorders acquired by other organs. It can acquire a disorder because of dysfunction of some other organ or organ system, just as it can produce disorders of other organs or systems because of a disorder of its own. For example, heart disease can impair the oxygen supply of the brain, whereas a defect in the hypothalamus can set off a series of neuroendocrinological events that could produce hypertension and eventual heart disease. Typically, neuropathology is classified into a number of accepted entities. Gen-

Table 1
Types of Neurological Disorders

Type	Brief description
Brain malformations	Developmental anomalies in brain structure
Trauma	Injury to the brain, generally produced by an accident or war injury
Vascular disorders	Damage to the brain produced by impairment of blood circulation
Neoplastic disorders	Generally, cancer of the brain but cysts are also included
Demyelinating diseases	Multiple sclerosis and rarer disorders in which there is progressive erosion of the coating or myelin sheath of neurons
Neuronal degenerative disease	Disorders such as Huntington's or Alzheimer's Disease in which the neurons themselves degenerate
Infectious diseases	Often called encephalitis or meningitis; in these conditions, the brain becomes infected
Toxic and metabolic disorders	Alcoholism, malnutrition, and other toxic conditions and agents which permanently alter brain function

erally, the list of disorders given in Table 1, or something resembling it, is employed.

It is important to note that some of these conditions are progressive and incurable; others are potentially progressive although the progressive process may be arrested; and still others are nonprogressive. The nature of the neuropathology has major consequences for the expected outcome, and thereby for the rehabilitation plan. Thus, some knowledge of neuropathology is needed to make a reasonable assessment of what is possible in regard to rehabilitation.

Neuropsychology

The field of neuropsychology has been and will continue to be extensively discussed in this book. The problem to be discussed here concerns only the basic principles of neuropsychology that should be taught to the rehabilitation staff. First, and perhaps foremost, the behavioral consequences of brain damage are extremely varied as to degree and type of deficit. However, extensive efforts have been made to provide some order to this very vague concept of brain damage. Just as there are ways of classifying various types of neuropathology, ways have been proposed of classifying brain disorders from the point of view of neuropsychology. As an example, we list in Table 2 the organization implied in a book by Heilman and Valenstein (1979). It may be pointed out that the Heilman and Valenstein outline follows a more or less traditional medical model and it might be possible to develop a more dimensionally organized, as opposed to disease organized, taxonomy.

Table 2
Types of Neuropsychological Disorder[a]

Type	Brief description
Aphasia	A group of disorders of language and language-related functions
Alexia	Disorders of reading
Agraphia	Disorders of writing
Acalculias	Disorders of calculation ability
Body schema disturbances	Disorders involving impairment of the body image. This classification generally includes impaired body localization, right-left disorientation, and a loss of ability to identify one's fingers; so-called finger agnosia
Apraxia	Impairment of purposive movement
Visuoperceptive, visuospatial, and visuoconstructive disorders	Impairment of the ability to recognize complex visual stimuli, construct objects in space, and deal with spatial relationships in general
Neglect	Failure to respond or orient to stimuli presented to the side contralateral to a unilateral brain lesion
Amnesic disorders	Disorders of memory
Dementia	Disorders of intellectual function

[a]After Heilman and Valenstein, 1979.

The table represents a way of organizing the major neuropsychological syndromes. It may be pointed out that there are interactions with the neuropathological classification. For example, the usual neuropsychological correlate of a neuronal degenerative disease is a *dementia*. In rehabilitation, however, the goal is not to "cure" the degenerative disease, which is currently impossible in most cases, but to reduce somehow the negative consequences of the dementia. For teaching purposes, this system of classification at the behavioral level clearly illustrates the complexity of brain disorders and the variety of adaptive problems they may produce.

The individual who has not had experience with brain-damaged patients may be confronted with many puzzlements. One finds patients who can speak but who do not understand the speech of others. In a relatively rare condition called *alexia without agraphia*, the patient can write but cannot read what he himself has just written. There are patients who are aware of your presence when you are standing to their right, but not when you are standing to their left. All these behaviors, and a large number of other behaviors like them, are entirely explicable on the basis of structural lesions in various regions of the brain. Indeed, the description of these behaviors and the discovery of their brain mechanisms and localization is one of the major tasks of neuropsychology (Geschwind, 1965). They are not functional psychiatric symptoms and also not behaviors that are under the voluntary control of the patient. Technical information regarding neurological symptoms of this type can go a long way in regard to helping the line staff, friends, and family to deal more productively with the patient. For example, if the patient has a half-field blindness, instruction to all concerned regarding this condition and advice to the effect that the patient should always be approached and spoken to from his sighted side may be extremely helpful.

Misconceptions about Brain Damage

In working with new staff, there is often some need to combine provision of substantive, technical knowledge with some effort to deal with possibly false beliefs and assumptions. Although these beliefs may not be widespread and are frequently not made explicit, they are often influential enough to create significant problems. One set of beliefs that seems to prevail particularly in psychiatric hospitals revolves around the idea that brain-damaged patients present particularly bad problems in management. They are the soilers, the ones who cannot manage their food, and the ones who are most likely to be unpredictably violent. While no one would deny that these beliefs have some basis in truth, they should be put into perspective. First of all, they are not true for most brain-damaged patients. Second, when such behaviors do occur, they are frequently manageable. We will attempt to document this latter claim later on. When dealing with such behaviors as assaultiveness and incontinence in brain-damaged patients, it seems that one must develop a particular attitude that lies somewhere between the belief that these behaviors should be totally tolerated because the patient cannot help himself, and the view that the patient should be dealt with as one would deal with a neurologically intact adult human being. Neither of these extreme approaches is productive. It is generally more useful to view these symptoms as potentially modifiable behaviors through the utilization of methods that are appropriate to the condition of the patient. In other words, do not let the behavior persist on the basis of the belief that it cannot be modified, but at the same time, do not expect to modify it effectively through methods that would normally be effective with neurologically intact people.

Research studies documenting the capability of managing such behaviors as incontinence and violence in adult, brain-damaged patients are lacking, but perhaps certain generalizations can be made from studies done with similar behaviors manifested by individuals with often equal impairment levels. Autistic and mentally retarded children are examples of such individuals. Several studies examined the use of contingent shock for suppressing self-injurious behavior in autistic and retarded individuals (Lovaas & Simmons, 1969; J. Young & Wincze, 1974). Azrin and Foxx (1971) have developed an effective program for toilet training in profoundly retarded adults. The program is a comprehensive package including procedures for detecting voiding, reinforcing voiding that takes place on a commode, and dealing with accidents. The aim of the program is to gradually achieve independent toileting through progressive reduction of the number of prompts provided concerning walking to the commode. The point we are raising here is that it may be possible to apply these programs to brain-damaged adults, since they were originally developed for populations of significantly impaired individuals.

Another common misconception revolves around a complex of issues dealing with the problem of motivation. To a more or less naive observer, many of the symptoms of brain damage are readily attributable to a motivational or volitional etiology. The patient with a dressing apraxia may be seen as being able to dress himself if he really wants to do so. The patient with an amnesic syndrome may be viewed as an individual who "only forgets what he wants to forget." In this case the impression may be based on the fact that memory deficits are often selective, and may involve only certain material (e.g., verbal or nonverbal memory; remote or recent events) (Butters & Cermak, 1980). Beliefs of this type sometimes take on a moral tone, particularly if the patient is known to be receiving compensation for his illness, or if he is seen as escaping some major responsibility through it. In essence, what the psychiatrist Erwin Strauss once charmingly referred to as "high school psychodynamics" often leads to incorrect and nonproductive interpretations of the behavior of brain-damaged patients. Of course, brain-damaged patients are human beings and may on occasion exaggerate their deficits or "repress" or "suppress" undesirable material rather than be amnesic for that material because of structural brain damage. However, it is often possible to discriminate between when the patient cannot remember and when he will not remember.

In our view, the first approach to this problem should be educational in nature. When the new staff member learns that dressing apraxia has to do with a failure to cope with relationships among objects in three-dimensional space, and that the amnesic syndromes are associated with damage to particular structures in the brain that have to do with encoding, storage, or retrieval of specific types of memories, then patients with these disorders may be seen in a new light. The neurobehavior specialist should share this kind of information with the rest of the team and might also recommend pertinent readings from time to time. When a patient with an unfamiliar syndrome appears in the program, the neurobehavior specialist should make efforts to explain the symptomatology to the staff, as in, for example, a regularly scheduled teaching conference.

A third common misconception about brain damage is that it is just one entity manifested in gross and obvious ways. Teuber (1959) was instrumental in demonstrating that the behavioral manifestations of brain damage may not be obvious at all and in fact may require very sophisticated testing to elicit. Some authorities feel that it is unfortunate that the diagnosis of organic brain syndrome was ever invented, since it has had a tendency to engender an oversimplified view of brain-damaged patients. In working with new staff, it is important, through training and example, to illustrate the complexities of structural brain damage in humans. We have already provided a table of the major neuropsychological disorders, but the distinctions suggested there are only the beginning. For example, they do not cover the entire spectrum of what are generally referred to as *hemisphere asymmetries*

(Dimond & Beaumont, 1974). Basically, brain damage to the left-hemisphere may have very different behavioral consequences from right-hemisphere brain damage, although the medical consequences may be essentially the same. The reason for this phenomenon is that in human beings the left hemisphere mediates languages and language-related abilities, while the right hemisphere mediates visual-spatial and other nonverbal abilities. We will deal with the matter of hemisphere asymmetries in great detail later. Here we are using it only as an illustration of an important major categorization of brain damage. Saying that both the patient with a right-hemisphere lesion and a left-hemisphere lesion are brain damaged conveys very little clinically significant information. However, saying that a patient has left- or right-hemisphere brain damage provides an important hint regarding what the associated deficits may be.

Some Basic Rehabilitation Concepts

Thus far, we have been discussing the teaching of facts and principles concerning brain damage. Here, some basic material about rehabilitation will be presented that might be of use in the staff–development process. Rehabilitation, as opposed to immediate treatment of acute illness, is typically a slow and laborious process. The term *burnout* has recently come into common use because it is descriptive of the phenomena associated with extended contact with difficult, seriously ill patients (Edelwich & Brodsky, 1980). The research done regarding burnout has largely involved working with terminal cancer patients, but perhaps some generalization can be made to brain-damaged patients. The phenomenon involves the development of feelings of malice and aversion toward patients with poor prognoses, sometimes resulting in job resignation. While personality factors may play a role in determining who experiences burnout, the work environment may also be important. One important aspect of staff development would appear to be minimization of burnout-related phenomena.

It goes without saying that opportunities should be made available for counseling with staff members regarding job-related personal difficulties. Most large institutions have such a service available, if the staff member prefers not to take up whatever the problem may be with the treatment team. However, there are more straightforward, mechanical ways of diminishing burnout. Some patients seem recalcitrant to any kind of treatment intervention. In some cases, the recalcitrance is related to the progressive nature of certain disease processes, but in others it seems more "functional" in nature. Some patients persist in behaviors that are constant sources of irritation to the staff. Assaultive and incontinent patients clearly represent this category. Some institutions have had good experiences with "patient holidays," in which for varying periods of time the patient is transferred from one ward to another. Thus, the ward he has been on is relieved for some period of time from the burdens imposed by the patient, and the staff can spend more time with other patients. It may be noted that patients of the type being described here typically take up disproportionate amounts of staff time. If adequate communication is maintained among the wards involved, this kind of approach can be carried out in a humane manner, without detriment to the patient in question or to other patients on the various wards. With regard to the matter of assaultiveness, some hospitals have elected to maintain the more chronically assaultive patients on the same ward. While such an approach, in a sense, provides a more or less permanent "holiday" for the remainder of the institution, the effectiveness of programs of this type remains to be evaluated in terms of treatment efficacy and of the impact on the entire institution. In any event, the problem of staff burnout, while only recently labeled, is a longstanding one, and deserves consideration in any program for educating individuals who work with chronic, seriously ill patients.

Another significant aspect of rehabilitation that should be considered during the course of staff development has to do with continuity of care. Often, in the case of brain-damaged patients, considerations regarding returning the patient to his family are crucial. There are, of course, well-understood concepts and practices regarding making family contacts prior to returning the patient home, but in the case of brain-damaged patients, there appear to be certain unique features. For example, we have noted certain interactions between staff or family attitudes toward the patient, and the type of brain disorder the patient has. We have frequently noted that patients with left-hemisphere lesions who are aphasic appear to receive more sympathy from staff and relatives than do patients with right-hemisphere lesions. It is difficult to grasp the nature of the right-hemisphere patient's visual-spatial difficulties even though they may be quite disabling, although a speech difficulty is often immediately apparent. Even within the category of aphasia, Kaplan (1979) has pointed out attitudinal differences in family members depending on the aphasic subtype. Patients with Broca's aphasia or nonfluent aphasia tend to become depressed and are responded to as depressed individuals. Their poverty of language, their efforts to communicate, and their frustration in not being able to do so frequently effectively elicit sympathy and compassion. Conversely, the patient with Wernicke's (or fluent) aphasia may act as if he is not ill and ramble on with speech that makes no apparent sense. Despite his extensive use of speech, which is sometimes seen to the point of "pressured speech," it soon becomes clear that he does not comprehend what others are telling him. In this way, Wernicke's aphasics sometimes seem to engender paranoid attitudes to the effect that they can talk but will not do so in a communicative way, and that they refuse to understand. Furthermore, patients with nonfluent aphasias are more frequently hemiplegic than are patients with Wernicke's aphasia, and thus may engender more sympathy by virtue of the presence of a visible physical handicap.

Rehabilitation and Behavior Therapy

Rehabilitation within the framework of behavior therapy (Lutzker, 1980) requires staff training in the basic concepts and methods of this approach to treatment. First of all, the staff might be encouraged to read an introductory behavior modification text (e.g., Bellack & Hersen, 1977). A series of lectures by a behavior therapist could also be useful. Depending on the nature of the program and the branch of behavior therapy used, there would be much variation in the content of what is taught. However, there appear to be certain essentials, which we will outline briefly.

We would suggest that one very important factor is the history of the field of behavior therapy, particularly in regard to the context out of which it developed. Ogden Lindsley, who is generally given credit for inventing the term *behavior therapy*, was a student of B. F. Skinner at Harvard (Lindsley & Skinner, 1954). Therefore, the field had its origins in the American experimental psychology of the 1940s and 1950s, and more particularly in the operant conditioning or descriptive behaviorism of Skinner (Hilgard, 1956). This history and its later development have numerous implications for clinical practice. First, the theory is learning oriented, and psychopathology is viewed in terms of learned behavior. This view stands in sharp distinction from disease or personality models of psychopathology. In keeping with Skinner's general theoretical position, behavior therapists typically do not seek personality or physiological causes or even correlates of psychopathology. The emphasis is very much on the behavior itself and on the environmental contingencies that maintain or extinguish behavior. It is not necessary to elaborate on these matters here, but for purposes of staff development it is important to point out, particularly to medically trained staff, that behavior therapy stems from a tradition that focuses on behavior. Without belaboring the point, it must be emphasized that behaviorism is not always easy to explain to clinicians, particularly if they were trained in the framework of a medical model or of personality-oriented psychological systems, such as psychoanalytic theory.

The interface between behavior therapy and neuropsychology is of particular interest when one is attempting to develop a rehabilitation program for brain-damaged patients. Perhaps the most obvious point to make is that most psychologists are not as well trained in one field as they are in the other. Behavior therapists may not be familiar with neuropsychology, and many neuropsychologists do not have any particular expertise in behavior therapy. Furthermore, philosophical differences may be present. Neuropsychologists, although interested in behavior, are generally particularly interested in the neurological determinants of behavior. As we have mentioned, behavior therapists tend to be more interested in the environmental determinants of behavior. A reasonable view of this matter is suggested by Bellack and Hersen (1977), who point out that while organic factors can determine behavior in an individual with an organic disorder, not all of that person's behavior is caused by the organic disability. Some of it may be under environmental control. It would seem that a crucial point regarding behavior therapy with brain-damaged patients is that of determining what behaviors are under environmental control, and, therefore, possibly modifiable. With regard to staff development, it would be very important to point out this important distinction. Not all of the behavior of the brain-damaged patient is caused by the brain damage, but some of it is. If the above considerations are accurate, then it is theoretically possible to put some of the behavior of brain-damaged patients under stimulus control.

In order to put these ideas into practice, there would seem to be a few basic points for the staff to learn. We have indicated that the staff should receive some general orientation to behavior therapy, but based on our experience, it seems that there are several points that should be driven home. First, there is the matter of unscheduled reinforcement. It is often the case, in life as well as in clinical settings, that maladaptive behavior is reinforced. In clinical settings in particular, patients are often reinforced for maintaining their pathology. For example, the image of an institutional setting as a "rest home" often encourages nursing staff to reinforce the taking of naps by indolent, apathetic patients. Psychiatrists sometimes have the reputation of at least covertly reinforcing patients to talk about their psychopathology when it might be more socially adaptive to encourage them to talk about other matters when with other people. A related matter has to do with the situation in which the patient is in a behavior–therapy program, but staff members offer reinforcers on occasions when the appropriate behavior is not exhibited. One should not give candy ad lib to someone in a behavior–therapy program when candy is being systematically used as the reward for appropriate behavior. Unscheduled reinforcement that rewards maladaptive behavior or that interferes with ongoing programs should be studiously avoided.

A second, perhaps more general, point involves teaching the staff to look at and describe behavior. In clinical applications, the most pertinent issue is generally that of describing some target behavior that might benefit from behavior therapy. Once such a behavior is designated, the next task is that of determining the conditions under which it occurs. For example, most of us see patients whose symptoms get worse when the doctor appears on the ward, or who become agitated when a visit by a relative is anticipated. These observations provide information concerning what environmental contingencies are maintaining that behavior. When a target behavior is identified, it is then possible to employ some form of intervention while the behavior continues to be monitored. It is also generally possible to employ one of several single-case experimental designs (Hersen & Barlow, 1976) in implementing and evaluating an intervention program. Once the program is formulated, adequate behavior description almost universally involves some form of systematic observation and recording of data. Typically, the recording is graphic in nature, so that one can ascertain the progress of the program at a glance. We will not go into the intricacies of single-subject experimental designs or data recording and analysis here, but would point out that these matters should be taught in some detail to staff working in behavior–therapy-oriented programs.

Rehabilitation and the Rehabilitation Specialties

In Chapter 2, we briefly described the various rehabilitation specialties such as speech therapy, occupational therapy, and so forth. The distinctions among these specialty areas are not always clear and should be clarified during the staff education process. For example, how many of us know the difference between a corrective therapist and a physical therapist, or between an occupational therapist and a manual arts therapist? Particularly in large, complex medical settings, a failure to be aware of these distinctions may provide inconveniences and delays to patients being referred for specialized assessment or treatment. Furthermore, none of these specialty areas work exclusively with brain-damaged patients, and it is important to know what they typically do with their brain-damaged and nonbrain-damaged patients. One very appropriate topic for a staff orientation would consist of a series of presentations by members of these various specialities, so they may explain the nature of their work and how they differ from the other specialities within their institution. The institutional issue is important, because sometimes these specialities do not do the same things in all institutions. For example, corrective therapy as a specialty area exists almost entirely within the VA hospitals. The things done by a corrective therapist in the VA hospital may be done by a physical therapist somewhere else.

With regard to rehabilitation of the brain-damaged patient, it is not the fact that the patient attends a rehabilitation specialty clinic that is important; what is important is what he does there. Thus, to say that a patient is going to occupational therapy as part of his rehabilitation is not sufficiently precise. More desirable would be to say, for example, that the patient is going to occupational therapy to take part in constructional projects that could help in the remediation of a visual-spatial defect. However, to make such a statement it seems clear that there must be a good working relationship between the occupational therapists and the neurobehavioral specialist. The model we are suggesting calls for the programming of the patient's activities in occupational therapy on the basis of the neuropsychological assessment. We are aware that such an objective is somewhat idealistic, and probably exists in practice in very few institutions. However, the nonrational use of rehabilitation specialty facilities in some settings sometimes turns them into little more than time-occupying or recreational facilities rather than clinics that could be of genuine therapeutic value. This type of deficiency is not at all the fault of the occupational therapists and other rehabilitation specialists, but rather is more the responsibility of those who use these facilities, but who do not have an adequate grasp of their potentials.

The Specific Nature of Brain–Damage Rehabilitation

When one consults the rehabilitation literature, it soon becomes clear that the major emphases are on physical rehabilitation and vocational rehabilitation. There is also an emerging field of rehabilitation of psychiatric patients. Hirschberg, Lewis, and Thomas (1964) define the concept of disability in a medical sense as an anatomically defined loss or impairment. Their textbook contains sections on the neurologically impaired patient but stresses such matters as involuntary movements and hemiplegia. Individuals engaged in vocational rehabilitation are very interested in assessment, and particularly in the use of assessment in optimally placing an individual in some job-training facility. Counseling and guidance are characteristically crucial parts of the process. With psychiatric patients, Paul and Lentz (1977) stress adaptive behavior and view resocialization, instrumental role performance, and reduction or elimination of extreme bizarre behaviors as major rehabilitation goals. The term *instrumental role performance* refers to both housekeeping and vocational skills. In a very impaired population it may involve such matters as being on time for activities and paying attention in classes, more than formal vocational training. It would appear, then, that the term rehabilitation is quite vague and has little specific meaning that is separate from the adjective that precedes it; for example, vocational rehabilitation, physical rehabilitation, psychiatric rehabilitation, and so forth.

From the point of view of staff development, it is important to point out that while some staff members may be familiar with rehabilitation concepts and practices as they pertain to the three areas discussed above, there may not be a complete generalization of these concepts and techniques to brain-damaged patients. Thus, it is important to clarify what reasonable rehabilitation goals would be for such patients and to point out that the techniques for attaining those goals are not necessarily the same ones as would be used in other forms of rehabilitation. For example, a major goal of brain-damage rehabilitation, as we are using the term, would be optimal restoration of cognitive and perceptual skills of various types. With the exception of the area of aphasia, there is practically nothing in the traditional rehabilitation literature that describes techniques for such restoration work. One must go to the learning literature, or even to the popular literature in such areas as memory training (e.g., Lorayne & Lucas, 1974). Without this shift in orientation, the clinical problem that often emerges is that the patient is not treated for his major deficit but for more minor difficulties that the rehabilitation specialists know how to treat. For example, it is clear that many brain-damaged patients have poor memory as their major adaptive problem. However, the rehabilitation literature provides little guidance concerning retraining of memory.

Training individuals for working with disabled people often involves some form of role-playing exercise in which the student is artificially disabled and has to cope with a normal environment (Siller, 1979). Thus, students who will be working with blind clients may spend several hours with blindfolds on, or students who will be working with paralyzed individuals might spend some time in a wheelchair; the obvious point being to give the student some experience of what it is like to have a disability. This technique really cannot be used in the case of trying to gain an experiential understanding of the brain-damaged patient. It is not really possible to role play an aphasia, amnesia, or dementia. Therefore, it is not easy to acquire some subjective experience of what it is like to be brain damaged. There are, however, some articles and books by individuals who have sustained brain damage and recovered, such as C. Scott Moss's book about his stroke (Moss, 1973). There are some extensive case studies by Luria, such as *The Man with a Shattered World* (1972) that provide detailed, phenomenological pictures of the experience of brain damage. Gardner (1974) in his book *The Shattered Mind* presents a great deal of descriptive material regarding what happens to people after brain damage. Sometimes aphasic patients can tell you about their subjective experience, which often involves knowing just what they want to say, but being unable to find the words with which to say it. However, most brain-damaged patients cannot provide a description of their condition that is sufficiently detailed to give another person an accurate picture of what they are experiencing. Thus, it is not easy to obtain a subjective impression of what the brain-damaged patient experiences either through some form of simulation exercise or through talking to patients. However, there is a descriptive literature, some of which should be read by trainees. If the opportunity is available to talk to people who have recovered from brain damage, such as a stroke or a head injury, that, of course, would provide an excellent training experience.

Thus far, we have had little to say about the substance of the different techniques that may be used in brain–damage rehabilitation, especially those that are not a part of conventional rehabilitation procedures. Golden (1978) has devoted a chapter of his book on rehabilitation to such techniques, and we will briefly review his outline here. His major headings are *motor disorders*, *verbal functions*, and *spatial skills.* Under motor disorders he has the subheadings of biofeedback techniques for increasing motor control, training motor skill by making automatic motor tasks into complex cognitive tasks, training gross motor function, training fine motor function, and utilization of avoidance conditioning to enhance motivation to move. Under the heading of verbal functions he includes speech comprehension, reading deficits, verbal-spatial deficits, categorization deficits, sequencing deficits, kinesthetic speech deficits, motor speech deficits, memory functions, writing, and mathematical disorders. Under spatial skills he includes basic visual-spatial orientation, orientation in space, and spatial-frontal disorders. The latter category refers to a type of deficit accompanying frontal-lobe lesions in which patients are impaired in the analysis and breakdown of a task into simpler elements, although they can reproduce simple drawings. In his book on restoration of function, Luria (1948/1963) devotes chapters to restoration of motor functions, gnostic functions, speech, and active thinking. Regardless of whether or not one agrees with these categorizations, they clearly outline the realm that we would consider to be within the purview of brain–damage rehabilitation. It is immediately apparent that we are in a different realm from those of vocational, physical, or psychiatric rehabilitation.

Once this basic concept is communicated, the next step involves teaching the various techniques that have been developed for each of the deficit categories. Descriptions of these techniques may be found in the books by Golden and Luria, in the work of the NYU group, and in the neuropsychology and speech therapy journals. In some cases, it is necessary to teach not only the technique but also some of the technical background for the technique's rationale. Thus, for example, in training motor functions, one should know something about the operation of the motor system and how various neurological disorders effect motor abilities. In the case of speech, it is helpful to have some orientation to linguistics so that one is somewhat familiar with such concepts as phonemic analysis, syntactical structures, and semantic relations. For example, a certain form of aphasia may be described linguistically as disorders of acoustic analysis and phonemic discrimination. Hécaen and Albert (1978) point out that the field now known as *neurolinguistics* has contributed substantially to our understanding of aphasia. A brief lecture on basic linguistics might go a long way in at least helping the staff understand the aphasia literature as well as provide a means of organizing what they see in their aphasic patients.

We will not provide an exhaustive description of the various retraining techniques here, but some examples will be given. There is a condition known as *literal visual alexia* in which the patient is unable to identify letters. Instead, on perceiving a particular feature of a letter, he takes a guess as to what it may be. Luria (1948/1963) provides two alternative methods of retraining letter recognition. In both cases, the objective is to replace the haphazard guessing with some form of systematic analysis. In one method the outline of letters is endowed with definite meaning. The goal is achieved by showing pictures of objects paired with words in which the first letter has approximately the same shape as the paired object. For example, a picture of a snake may be paired with the word snake, with the letter *s* drawn to have a particularly snakelike appearance. This pairing encourages a more systematic perceptual analysis of the letters. The second method requires the patient to learn to construct letters out of component parts. The patient is presented with a set of elements out of which letters can be constructed. As the patient learns the principles of construction and how to assemble the units of each letter, the reading defect can be indirectly eliminated or reduced.

Another well-known procedure is the "Block Design Training" of the Diller-NYU group (Diller, Ben-Yishay, Gerstman, Goodkin, Gordon, & Weinberg, 1974). This method utilizes the WAIS Block Design test (Wechsler, 1955a) but as a retraining rather than an assessment method. The patient who makes errors on initial performance is given a series of incremental cues, such as providing an outline of the square the design should be built in, or an outline of the square divided into subdivisions for each block. The authors of this technique view it as a method of training perceptual and psychomotor skills. They were able to demonstrate that successful training generalized to other functional skills.

New techniques of this type continue to be developed, and, in terms of staff training, it would be wise to institute a continuing education program so that staff members may be able to attend workshops and review the literature that describes these methods. However, once the nature of various neuropsychological dysfunctions is understood, it is hoped that the team will be able to develop and implement its own methods. Often one finds that the method follows quite logically from the nature of the deficits, as described by the neuropsychological and behavioral assessment, to the development of a retraining method, and even among the most experienced specialists, it is often a matter of trial and error. Clearly there is no syllabus that provides a comprehensive set of techniques for a variety of deficit patterns. The newness of the field and the need for originality and innovation lead us to conclude that staff development should emphasize the acquisition of a scientific orientation, a matter that will be elaborated further.

Rehabilitation and the Scientific Method

It should seem clear by now that retraining of brain-damaged adults is not an area of standard clinical practice. Since it is so relatively unexplored, it seems wise, at this point, to view our retraining programs as research endeavors, at least to some extent. We are still in the stage of having to be able to demonstrate that these retraining techniques are effective, particularly in view of the persistent belief that the consequences of brain damage are irreversible. We therefore feel that the staff–development process should encourage a research or scientific orientation, rather than a strictly clinical one in which the staff is simply taught to apply well-established methods. Aside from the newness of the field, there are several other reasons for developing scientific standards. The training methods alluded to here and described elsewhere in this book often require an experimental methodology to implement. Furthermore, the point of the whole approach is to apply scientific knowledge acquired from neurobehavioral research to practical clinical problems. It is our view that the unsystematic application of such knowledge may be useless, if not harmful. Finally, in the area of brain–damage rehabilitation in particular, changes produced may be small or subtle in nature. Without organized collection of data, significant changes may not be noted, and certainly cannot be quantified.

The teaching of scientific methodology and basic philosophy of science is straightforward. In most educational settings didactic instruction is often accompanied by some form of practical laboratory experience—as in the case of the traditional experimental psychology course. In clinical settings it is important not only to teach research design and methodology but also to provide some instruction in the conduct of research protocols. Generally, the important point to emphasize is the distinction between clinical practice and research procedures. There are now several texts that may be useful in providing clinicians with an orientation to research procedures (e.g., G. Goldstein, 1980). The aim of such an orientation is certainly not an attempt to make everybody on the staff a practicing scientist but rather to encourage the development of a scientifically oriented treatment program. Perhaps the most appropriate model for what is desired is the so-called metabolic ward or clinical research unit. These facilities are guided by two simple, but often hard to implement, principles: strict adherence to prescribed procedure and systematic data acquisition and recording.

The process of teaching staff to comply with prescribed procedures should begin with the initial orientation. Aside from instruction in conducting protocols, research design, and perhaps some statistics, attitudinal factors should also be considered. It should be pointed out that research protocols need to be followed strictly not for authoritarian but for scientific reasons. When the benefit of a patient becomes an issue, the problem can often be resolved without disruption of the ongoing research. In pharmacological research we have observed that the clinical staff often tends to develop the belief that any adverse effects that emerge always occur in the placebo group, even though the study is being conducted on a completely blind basis. It is good to be able to demonstrate that it is usually not the case, and that most people are generally rather poor judges of who is getting the active drug and who is getting the placebo. It is also important to encourage the development of a nonthreatening atmosphere so that errors are reported to the investigator. Sometimes, errors of omission, in which a prescribed procedure is not accomplished, are committed; sometimes errors of commission are made, in which a prescribed procedure is changed without authorization. In both cases, the staff member should feel free enough to report these matters to the investigator without feeling that punitive action may be taken.

As in most scientific specialties, neuropsychology and behavior therapy have their own methodologies and terminologies. These substantive matters can be taught through a combination of reading and other didactic methods. The idea is that there should be a shared technical vocabulary and a basic understanding of the methods used in each of these areas. Thus, if consultants start using terms like *multiple baseline* or *visual object agnosia*, the staff should know what these terms mean. A related matter has to do with the nature of the data acquired. The acquisition of neuropsychological data is usually reasonably straightforward because they are generally acquired by trained technicians who administer a number of standardized tests and other examinational procedures. However, in the area of behavior therapy, data are often collected by the entire team, and it is generally necessary to provide some instruction in systematic collecting and recording. Perhaps the major point involves making the distinctions between medical record keeping and the kind of record keeping needed in the implementation of behavior–therapy programs. Thus, while a progress note entered in a chart may be anecdotal and unstructured in nature, recording of research data has to be objective and, where possible, quantitative. The information obtained must answer specific research questions. We would strongly recommend that a behavior therapist be asked to teach the staff about the various behavior–therapy designs and about systematic collection and recording of behavioral data.

The scientific atmosphere may also provide a positive morale factor. There is often great gratification in seeing a desirable behavior accelerate on a learning curve, or an undesirable behavior decelerate.

Demonstrations that the theoretical principles underlying the various programs really do work in practice are similarly gratifying. It is nice to be able to specify what the program was for the patient, to be able to show the systematically collected data, and to be able to learn from the experience when planning programs for subsequent patients. We have observed much excitement and interest generated among line-staff members when they felt that they were engaged in a scientific enterprise in which they could participate. In this regard, it is important to listen to suggestions as to procedures that come from line-staff members, but it is perhaps equally important to discourage staff members from implementing these suggestions prior to consultation with the neurobehavioral specialist and the rest of the team. Although good science requires imagination and creativity, it also requires studying phenomena under specified and controlled conditions.

Summary

In this chapter on staff development, we have stressed the content aspects of the training needed to be an effective member of a team devoted to the rehabilitation of brain-damaged patients. It was suggested that the leader of such a team should be what we described as a neurobehavioral specialist: an individual with particular expertise in the problem of normal and abnormal relationships between the brain and behavior. It was pointed out that while the team might consist of individuals from the traditional clinical disciplines, such as nursing and education, special training is required in various neurosciences, methods of behavior therapy, and various rehabilitation techniques. At an attitudinal level, it was felt to be important to correct a number of common misconceptions about brain damage and to foster a scientific-research orientation, since the field is in such an early stage of development and so much more has to be learned.

4

REHABILITATION OF THE BRAIN-DAMAGED ADULT AND BEHAVIOR THERAPY

Introduction

In this chapter, we shall provide a rationale for the use of behavior-therapeutic methods in the assessment and treatment of brain-damaged patients and present some data gathered by ourselves and others to illustrate how various methods derived from behavior therapy can be used effectively in treatment. We will begin with a brief introduction to behavior therapy, go on to provide a rationale for building an interrelationship between behavior therapy and neuropsychology centered around treatment of brain-damaged patients, and conclude with a number of illustrations of how this relationship can work in actual practice.

Some Basic Behavior–Therapy Concepts

We should preface this section by indicating that behavior–therapy research and practice have generated an enormous literature that has been summarized in numerous textbooks at varying levels of sophistication. The person new to the field may wish to read the introductory text of Bellack and Hersen (1977). More advanced treatments may be found in Leitenberg (1976), Hersen and Bellack (1978), Ullmann and Krasner (1965), Wolpe (1973), and H. E. Adams and Unikel (1973). As we indicated earlier, behavior therapy grew out of the school of psychology or theory of learning known as *behaviorism* (Hilgard, 1956), and particularly out of the descriptive behaviorism of Skinner

(1953). There was a point in Skinner's career when he became particularly interested in education, and the applicability of his primarily animal study-based branch of learning theory to educational methods. Some of us will recall the "teaching machines" associated with Skinner and his followers' early efforts in this area. Yates (1970) devoted a chapter of his book on behavior therapy to the historical development of the field up to the time of his writing, tracing it back not only to Skinner but also to work done by the Russian school of psychology founded by Pavlov, and by psychologists in England and South Africa (Wolpe, 1958)—although there is controversy over who started it all. Some authorities say that Ogden Lindsley was the first to use the term *behavior therapy* (Lindsley & Skinner, 1954). He later published (1956) the first report on the application of Skinner's technique, *operant conditioning*, to psychiatric patients.

Skinner separated behavior into two types; respondent and operant. Respondent behaviors are elicited by known stimuli, whereas operant behaviors are emitted responses which need not be associated with known stimuli. Skinner believed that most human behavior is operant in nature. Operant behavior can be conditioned and is viewed as a response that is correlated with reinforcement. Hilgard (1956) provides the example of lever pressing and says that when a subject presses the lever, it is not the sight of the lever that is important but rather the pressing of the lever. The response causes the reinforcement to appear and take place. The significant point is that reinforcement is contingent on response. The strength of the operant increases in relation to the extent to which it is followed by reinforcing stimuli. Ogden Lindsley once gave a lecture that contained what we consider to be a very good, simple example of the application of thinking in terms of operant conditioning to the human situation. When a baby cries at night, the more traditional learning theorists might think of what internal or external stimuli elicited the response of crying. The individual who thinks in operant-conditioning terms would note that when, say, a baby girl cries, the parent goes to the nursery, picks up the baby, and fondles her. Assuming that these events are reinforcing, it can be said that crying as an operant behavior is associated with reinforcing stimulation. In other words, reinforcement (in this case being picked up and fondled) is contingent on a particular operant response (crying). It seems to us that this concept was the conceptual "breakthrough" leading to the widespread application of the operant model in behavior therapy. Therapists using this model started to look less at what the antecedents of behavior were, and more at the consequents. Modification of the behavior could be accomplished by altering these consequents by altering the environmental contingencies. In our example, the baby would theoretically reduce her crying if the parents did not attend to her when she cried, through the process of extinction.

This "breakthrough" and the variety of techniques that were developed as a direct or indirect consequence of it led to a proliferation of behavior–therapy research and practice at least since the mid-1950s. In a recent collection edited by Hersen and Bellack (1978), applications are reported to the following populations: chronic psychiatric patients, alcoholics, children in residential treatment, adults with anxiety and affective disorders, people with marital problems, and children with behavior disorders. Earlier, Yates (1970) reported on applications to enuresis and encopresis, stuttering, phobias, obsessions and compulsions, hysteria, tics, delinquency, psychopathy and criminality, sexual disorders, psychoses, alcoholism and drug addiction, mental deficiency, and "normal" disorders of "normal" people. The latter area includes such subareas as smoking, overeating, compulsive gambling, and social anxiety. Thus, simply by going through the contents of books about behavior therapy, one can see the pervasiveness of the field.

Aside from the applications of behavior therapy, several distinct branches of behavior therapy have emerged. There is the entire area of behavior assessment, concerning which there are several textbooks (Ciminero, Calhoun, & Adams, 1977; Cone & Hawkins, 1977; Hersen & Bellack, 1981) and two journals (*Journal of Behavioral Assessment* and *Behavioral Assessment*). Another major area is social skills training (Bellack & Hersen, 1979; Trower, Bryant, & Argyle, 1978). Still another area includes a variety of techniques for fear reduction (e.g., Wolpe, 1958). One of the more well-known branches of behavior therapy is the token economy (Ayllon & Azrin, 1968; Kazdin, 1977). Although it implies a group approach, many behavior therapists utilize individual approaches, generally by applying some kind of single-subject design. Whereas many behavior therapists use direct trial-and-error learning by the patient, others use an approach called *modeling*. Bandura (1971) is probably the most well-known advocate of this approach, pointing out that a great deal of human learning is achieved by observing social models. Within these branches of behavior therapy there are subbranches such as *in vivo* flooding, desensitization, reciprocal inhibition, covert conditioning, and so forth. We will not define all these terms here, but the interested reader may consult the texts cited above. Rather, we will go on to discuss some of those behavior therapy concepts that might be appropriate for brain damage rehabilitation. In our view, three areas of behavior therapy have particular relevance for rehabilitation of brain-damaged patients: behavioral assessment, skills training using single-subject designs, and token economy.

Behavioral Assessment

Behavioral assessment may be seen as a companion to neurological and neuropsychological assessment, playing the particular role of determining those behaviors that are maintained by environmental contingencies. We have already indicated that whereas brain damage can certainly alter behavior, not all of the behavior of the brain-damaged individual is caused by the disorder. Behavioral assessment in the form of interviewing, naturalistic observation, or analogue techniques can be quite useful in regard to targeting behaviors that might be amenable to treatment, particularly when those behaviors are environmentally maintained. For example, we treated a patient with Huntington's disease, who periodically would go on hunger strikes. Nutrition is particularly important in Huntington's patients because of the large amount of energy they expend on their abnormal movements. Thus, the nursing staff would cajole the patient about eating and spend a great deal of time in trying to get him to eat. It became apparent to us that the hunger strikes were not really caused by his illness but were an attention-gaining device. We reduced this attention, following which the strike shortly ended, and there was no recurrence. Aside from the use of behavioral assessment in identifying maladaptive behavior tied to environmental contingencies, we have found it useful in combination with the neuropsychological tests as an aid in refined specification of potentially treatable deficits that are in fact caused by the brain damage. There is also a borderline situation in which bona fide organic symptoms are increased or reduced in severity depending upon such environmental considerations as who is observing the patient. Thus, for example, the aphasic patient may become more aphasic when the doctor or a relative is present. It is often very important to determine under what conditions the patient is at his best or worst. Behavioral assessment is also a very good method of determining the practical functional consequences of deficits identified by the neuropsychological tests. For example, neuropsychological tests may identify a memory deficit and provide further details concerning its nature, but such tests cannot really evaluate the disability produced by the deficit in regard to activities of daily living. However, systematic observation of amnesic patients, even in a hospital setting, may reveal such behaviors as repeated requests for the same thing, because the previous request was forgotten, or getting lost while going from one part of the hospital to another, possibly because of the patient's lack of memory for where he is going.

The interview of the brain-damaged patient often turns out to be a mental-status examination and, as such, becomes somewhat redundant with the neuropsychological tests. However, the interview may be used to obtain information not necessarily revealed by the neuropsychological tests, such as whether the patient is particularly anxious or depressed, or whether he has social skill difficulties. In the case of neurological patients, the initial interview may immediately indicate the neurological problem, particularly if it is a movement or speech disorder. If the patient is going to be placed in a behavior–therapy program, the interview is a good way of obtaining information regarding appropriate reinforcers and of identifying specific target behaviors that might be treated. The interview could be a good preliminary way of determining what behaviors can be placed under stimulus control and what behaviors are relatively unmodifiable consequences of the organic illness. For example, in an interview we had with a patient with a long history of psychiatric illness and early-life brain damage, it was noted that he never established eye contact. It seemed reasonable that, although we could not alter his basic intellectual deficits, it might be feasible to try social skill training aimed at improving eye contact and other related social behaviors.

It should be appreciated that an interview with a brain-damaged patient may be quite unlike interviews with ordinary people or even with psychiatric patients. The patient may be so aphasic or demented that ordinary conversation is not possible. In some cases, naturalistic observation is the only alternative, although alternative channels of communication can sometimes be established. For example, we sometimes interview patients who are unable to speak or move and who use eyeblinks as a code. Sometimes, the patient can be communicated with in writing. Within the framework of neuropsychology, Luria (cited in Christensen, 1975) speaks of the preliminary conversation, an interview given before administration of the neuropsychological tests. In summary, the preliminary conversation can be useful in establishing several important points: (1) the state of the patient's level of consciousness, (2) the premorbid level of functioning, (3) the patient's attitude toward his illness, and (4) the patient's complaints. While Luria might not have characterized his initial conversation as a form of behavioral assessment, it clearly is different from a diagnostic assessment in that it is not aimed at producing a classificatory label, and there is an apparent relationship to treatment. For example, it is obviously important to know the level at which the patient functioned before acquisition of his brain disorder in order to make reasonable treatment plans. Similarly, it would be important to know whether or not the patient is aware of his illness, and to know about how strongly motivated he is to recover. The entire concept of having an illness from which one wants to recover is not infrequently absent or somehow impaired in brain-damaged patients.

Within the framework of behavioral assessment, one of the major functions of the interview is that of making an initial identification of the target behaviors requiring treatment. In the case of brain-damaged patients, identification of target behaviors should involve the neuropsychological evaluation and the behavioral assessment. Whereas neuropsychologists have been involved with the precise delineation of deficit patterns, it seems that the interview can provide much information regarding how the test-derived deficit patterns are translated into ordinary social behavior. Although rehabilitation specialists such as Diller (1976) call for emphasis on retraining generic abilities, since they are thought to have maximal generalizability, there seems to be an equally good rationale for working with specific target behaviors. Thus, for example, whereas one might retrain judgment in a general way, one might also work with a consequence of poor judgment, such as inappropriate sexual behavior, particularly if it is that behavior that gets the patient into difficulty. The presence of poor judgment can be readily documented by the neuropsychological tests; the problem with sexual behavior can be discovered by interviewing the patient, his family, and other associates.

Following the initial interview, it is often useful to have a period of naturalistic observation. Within the framework of behavioral assessment, observation is not casual watching of the patient, but some systematic method of observing and recording more or less discrete behaviors. Once the behavior is identified, it is often useful to record its frequency, duration, and the circumstances under which it occurs. Sometimes an interval system is used, and the procedure involves recording whether the behavior does or does not occur within some specified time period. Sometimes an observation can be instrumented, such as in the case of putting an activity meter or pedometer on a patient for recording motor activity over a period of time. Generally, some form for recording the raw data and a graphing system is developed. Observations of this sort are generally used to obtain baseline information so that the effectiveness of a treatment program can be evaluated. In general, systematic observation involves some behavior that is going to be treated. The important issues in this kind of procedure involve the related matters of specificity of the target behavior and reliability of measurement. Thus, for example, one would want to record "number of words uttered during a specified time interval" rather than "speech fluency."

We are somewhat unsure of the usefulness of analogue and self-report assessment procedures with brain-damaged patients. Such procedures as role playing, filling out scales, keeping records, and the like often involve abilities that many brain-damaged patients do not possess to a sufficient degree. For example, we once attempted to get an amnesic patient to keep a diary to record events happening during the day. We were unsuccessful in this attempt, because he would forget the whole idea of diary keeping from one day to the next. Furthermore, there is some evidence (Bellack, Hersen, & Turner, 1977) that patients do not behave in "real life" as they do in role-playing situations. We have no hard evidence that analogue and self-report procedures are ineffective assessment methods with brain-damaged patients. However, such patients often have ability deficits that would appear to contraindicate the use of these procedures.

In summary, the interview and the naturalistic observation methods of behavior assessment appear to be quite useful as part of the total initial and ongoing evaluation of brain-damaged patients. It is suggested that these procedures supplement neuropsychological testing in a number of ways. First of all, they may aid in identifying behaviors that could conceivably be placed under stimulus control; they deal with issues, other than the neurological damage itself, that may have strong implications for rehabilitation outcome. Often, behavioral-assessment procedures identify specific behaviors related to the generic behavioral deficits found on the neuropsychological tests. They also have the value of identifying possible reinforcers to be used in treatment. Although we are unsure of the use of various analogue and self-report techniques with brain-damaged patients, their application might merit further evaluation.

Skill Training and Single-Subject Design

The field of behavior therapy may be divided into individual, wardwide, or other group approaches. For example, the token economy system is often a wardwide approach, while such techniques as aversive conditioning for sexual deviation are generally carried out individually. Offering help to the individual is the aim of essentially all clinical practice, but until recently we did not have the methods available for carrying out single-case research. Thus, we had to generalize from group data to the individual, often with unsatisfactory results. Perhaps the roots of the single-case experiment are to be found in Skinner's single-subject research and his well-known disdain for group statistics. However, the approach seems entirely justifiable on clinical grounds, because even though we may treat patients in groups, we are generally interested in helping them individually. Recently, technologies have been developed for doing objective single-case research, many of which are reviewed in Hersen and Barlow (1976). The desirability of single-subject-oriented treatment for brain-damaged patients goes beyond general considerations having to do with the uniqueness of the individual. The actual problem is the great variability of behavioral alterations associated with brain damage. In general medicine, the same drug can be given to a large number of patients with the same illness, allowing for minor variations in dose. This kind of general therapy is nonexistent in the rehabilitation of brain-damaged patients. Deficit patterns are almost exquisitely individualized, to the extent that it is essentially meaningless to speak in terms of any kind of general rehabilitation program. The patient who cannot find words because of a stroke is very different from the patient whose speech is fluent but who cannot remember because of a long history of alcoholism and malnutrition. It would therefore appear that the single-subject design approach is ideally suited to the variety of problems manifested by brain-damaged patients.

The essence of neuropsychologically oriented brain–damage rehabilitation has been skill training. More specifically, the skills tend to be cognitive skills rather than social or study skills, as would be the case in other applications. There is an extensive literature concerning application of various behavior–therapy techniques to the training of cognitive skills, but the majority of the work has been done in the area of rehabilitation of the mentally retarded. For example, Kazdin (1977) has reviewed a series of studies having to do with alteration of verbal behavior and teaching of language to the mentally retarded, utilizing token economy methods. There have been numerous studies, also with the mentally retarded, of how to improve basic self-help skills, such as independent eating and toileting (Drabman, Jarvie, & Hammer, 1978). In the case of psychiatric patients there is a variety of behaviors called *instrumental role performance* by Paul and Lentz (1977). These behaviors include various vocational and housekeeping skills, as well as other skills required for independent living in the community, such as driving or using public transportation. Paul and Lentz taught these skills in their behaviorally oriented program, as do others who use such programs with adult psychiatric patients.

While it is clear that behavior–therapy approaches have been applied to several populations of individuals with various forms of cognitive deficit, the behavioral treatment of brain-damaged individuals with specific cognitive deficits remains relatively unexplored territory. For example, if one were to inquire about the use of behavior therapy in the treatment of aphasia, it would become apparent that little has been done. In his review of aphasia treatment, Kertesz (1979) lists eleven therapeutic approaches, none of which has a behavior–therapy orientation. The only possible exception is the use of teaching machines (Keith & Darley, 1967) which, in any event, turned out to be unsuccessful. The application of various forms of behavior therapy to skill training with brain-damaged patients should be explored, and this venture may be promoted through the use of neuropsychological assessment. It is interesting to note that in a chapter on residential child treatment, Drabman, Jarvie, and Hammer (1978), after rejecting projective techniques and questionnaires as useful assessment methods, mention intelligence and achievement tests, sociometric evaluation, interview, and direct observation as useful assessment methods. However, they do not mention neuropsychological tests, although the population they are addressing most likely contains many children who have what we would describe as *neuropsychological deficits.* In the area of mental retardation, Reitan (1966a) has remarked:

> Neuropsychological examination of individual retarded subjects appears to make a substantial contribution to the understanding of their deficits. On the basis of approximately 400 such examinations, it is apparent that even though intellectual and cognitive abilities are often depressed in comparison with the average level, a good deal of meaningful intraindividual variability is still the rule rather than the exception. (p. 207)

These patterns of intraindividual variability may have strong implications for rehabilitation planning.

In our own research, to be reviewed later, the initially appealing aspect of utilizing some form of behavior therapy with brain-damaged patients was the idea of reinforcement. We were struck by the impression that currently available methods of rehabilitation were more or less didactic in nature, and although there may have been sophisticated neuropsychological analyses done, and often ingenious designing of training methods that took cognizance of this analysis, there was generally little or no rationale presented for why the patient would want to cooperate in these programs. The assumption was implicitly or explicitly made that the desire to restore some lost ability was sufficient motivation to persist in a sometimes arduous and frustrating training program. Nevertheless, we soon found that when we worked with chronic patients in particular, there were often serious motivational problems related not only to restoration of specific functions but to becoming productive in any sense. A popular conception of behavior therapy is that it all hinges on the idea that "reinforcement works" (i.e., people maintain behaviors associated with reinforcing stimuli and extinguish behaviors that are not associated in that manner). While this view, of course, is an oversimplification, it was our idea that the combination of neuropsychological sophistication with a training philosophy that makes use of reinforcement principles could be productive in regard to rehabilitation of adult brain-damaged individuals.

The desirability of utilizing a single-case approach to skill training is based not only on the relative uniqueness of deficit patterns in brain-damaged patients but also with the newness of the field. Hersen and Barlow (1976) suggest that single-case research is an ideal method of determining how a treatment works and point out that it often fulfills a necessary function prior to initiation of group–design studies. They point out that Bergin and Strupp (1972) used the term *technique building* to describe this phase of research. It is not at all unusual in the clinical sciences to begin an investigation with a number of case studies prior to initiation of a full-blown, statistically oriented group study. Neuropsychologically oriented rehabilitation is clearly at a stage in which it has to do more technique building. Working with individual patients in a manner that permits continuous monitoring and alterations of the ongoing program would appear to be a sensible way of proceeding.

The actual content of the skills to be trained and the methods of retraining are discussed in various contexts throughout this book. Such major areas as language, memory, perceptual skills, motor skills, and spatial-relations abilities form the domain from which specific target behaviors are identified. Once these are identified, we believe that behavioral approaches provide, first of all, means of motivating the trainee to continue participation in the program, and second, the kinds of research designs that can generally provide a clear indication of whether or not the program is effective. In general, the approach would involve a number of single-subject case studies, following which it may be possible to construct a set of training modules that may have applicability to groups of patients. For example, it may be possible to ultimately design a memory module or a visual-spatial module.

Having said this, we should emphasize that there are a number of technical problems with utilization of certain design types. For example, skill training would not appear to lend itself to an A-B-A or return-to-baseline design. In this design, baseline behavioral data are acquired and recorded, followed by a treatment intervention period after which the treatment is withdrawn. If return-to-baseline level occurs following the withdrawal of treatment, it would suggest that the treatment intervention was effective in maintaining the modification of behavior that occurred while it was ongoing. For example, if a child persistently speaks out in school but his talking is diminished if the teacher gives him a piece of candy for every 15-minute period during which he does not talk, then when the candy giving is withdrawn the speaking out may or may not return. If it does, the inference can be made that the candy-giving intervention was effective. In the case of skill training, this kind of design is problematic, since once a skill is achieved it is hard to see how a return to baseline can be accomplished. For example, if a brain-damaged patient is retaught the association between letters and sounds through some treatment intervention, it is hard to see how withdrawal of the intervention could produce "unlearning" of these associations. Thus, it would appear that A-B-A designs are not appropriate for evaluating at least certain types of skill training. Perhaps a more promising method is the so-called multiple-baseline design. In this technique, attributed to Baer, Wolf, and Risley (1968), a number of responses are measured, but treatment is only initially applied to one of them. It is hoped that there is little or no change in the others. The same treatment procedure is subsequently applied to all other target behaviors. The therapist can be assured that treatment was effective when changes in the treated behavior are noted without changes in the untreated behavior. In effect, the multiple-baseline method allows for assessment of treatment efficacy without the need for a return to baseline. Gianutsos and Gianutsos (1979) have, in fact, applied a multiple-baseline-across-subjects design to the mnemonic training of verbal recall, utilizing four brain-damaged patients. Initiation of treatment was staggered across subjects so that treatment effects could be distinguished from recovery and practice effects.

Token Economy

The first token economy ward for psychiatric patients was founded by Ayllon and Azrin (1968) at Anna State Hospital in Illinois. The important elements of this program (and those that followed from it) were that it was comprehensive and wardwide, and that it used tokens as generalized, conditioned reinforcers (Kazdin, 1977). Initiation of such programs required reorganization of the ward in a way that allowed for control and continued monitoring of administration of reinforcement. While the results of token economy wards for psychiatric patients have been generally positive, there are no data suggesting that such a program would be effective for a ward of brain-damaged patients. One could object on *a priori* grounds that many brain-damaged patients might be unable to comprehend the meaning of a generalized, conditioned reinforcer and might not see the connection between the token and the primary reinforcers for which it can be traded. Furthermore, it is often not possible to place brain-damaged patients in the kinds of restrictive environments in which psychiatric patients are often placed. It is, therefore, frequently not possible to achieve the necessary level of monitoring of reinforcement and prevention of unscheduled reinforcement.

It is possible to use token reinforcers with individual patients (e.g., Wincze, Leitenberg, & Agras, 1972), and in data to be presented later, we will provide preliminary evidence that this method can be effective with brain-damaged patients. Tokens can be administered for achievement in training programs, for self-care activity, and for a variety of other behaviors. For example, in an informal program, we tried to treat the indolence of a patient by awarding tokens for purposive activity and requiring him to spend them for time-limited periods of napping or sitting in the dayroom. Langer, Rodin, Beck, Weinman, and Spitzer (1979) used tokens to improve memory in elderly individuals. It is our impression, though clearly not yet proven, that while wardwide token-economy programs may not be optimally effective for brain-damaged patients, the use of tokens in individual behavior–therapy programs may be quite useful.

The Wichita Program

The Wichita Program was a federally sponsored research project having to do with rehabilitation of brain-injured young adults. It took place at the Kansas Elks Training Center, a sheltered workshop devoted to prevocational training and job placement for handicapped young adults. The average population of the center is about 130. In general, about half of the clients are mentally retarded, while the other half consists primarily of young people with emotional disorders. There is usually a number of clients with sensory impairment, and many of them have multiple handicaps. One section of the center serves as an extended sheltered-workshop facility primarily for multiply handicapped clients who do not have the potential to function in competitive employment. However, most of the clients receive job evaluation and training with the goals of competitive employment and independent or semi-independent living. The general orientation of the program is toward development of work habits and work skills, rather than toward specific job training. The major training vehicle is an extensive work area in which clients do subcontract work of a varied nature for local industries. The administration maintains an aggressive program for obtaining subcontracts from businesses in the Wichita area.

With the support of a federal grant, a group of investigators was assigned a special area in the center for a research program concerning the vocationally oriented training of brain-damaged young adults. The facility consisted of a large room containing several "stations" for various work-related or prevocational training-related activities. As the program emerged, subjects also worked in other parts of the center, but this room always remained a "headquarters" for the program. The staff for the program consisted of one part-time and three full-time employees. At any particular time, there were seven to twelve subjects enrolled in the program. The staff served primarily as trainers but also assisted in data management and analysis. The program was conducted within the framework of a research design in which a group going through the program was contrasted with a control group that did not receive the program. We will deal with this aspect elsewhere in this book. At this point, we wanted only to describe the general setting as an introduction to the kinds of work done in the area of behavior therapy.

Cognitive Retraining

A common manifestation of many forms of brain damage is impairment of immediate memory: the ability to briefly retain and recite some series of stimuli. Perhaps the most commonly used immediate-memory test is the digit–span procedure, in which the subject is asked to repeat increasingly lengthy series of single digits, usually spoken by the examiner at a rate of one digit per second. A mechanical equivalent of digit span is the Jensen Alternation Board. The subject is asked to retain the order of a series of lights flashed in sequence and to reproduce this order for the examiner. In our version of this procedure, we varied the number of elements from two to eight. Subjects observed the examiner display a particular series of these lights, which were mounted on a board containing a string of five lights. Then they were asked to replicate the series either verbally (visual-verbal response) or by depressing the appropriate switches mounted below the lights (visual-motor response). Figure 2 contains data involving only visual-verbal responses. It is a frequency distribution illustrating the

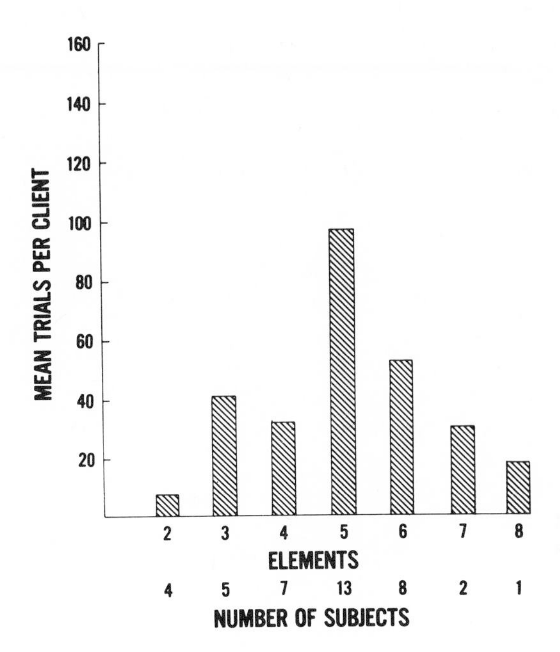

Figure 2. Jensen Board data.

performance of 13 subjects in terms of the number of trials required to reach criterion for each element series. For example, one subject was able to master an 8-element series, and did so in 15 trials. Figure 3 contains learning curves for visual-verbal and visual-motor response conditions for a 5-element series. In both cases there is a gradual reduction in errors as number of trials increases. However, it is apparent that learning seems to be better under the visual-motor condition. Figure 4 contains two curves: one for a 5-element series and the other for a 6-element series. There is a wide disparity in amount of errors made during the intermediate trials, but in the later trials, error rate is relatively low for the 5- and the 6-element conditions. Figure 5 contains data for the one subject who was able to retain an 8-element series. Although he had much difficulty with two elements, once he mastered them, the rest of the task was done well, up to and including 8 elements.

These data are presented mainly to demonstrate that brain-damaged individuals can learn to perform a short-term memory task under conditions of regular instruction, and that they can demonstrate learning phenomena of a type commonly seen in normal subjects. If the

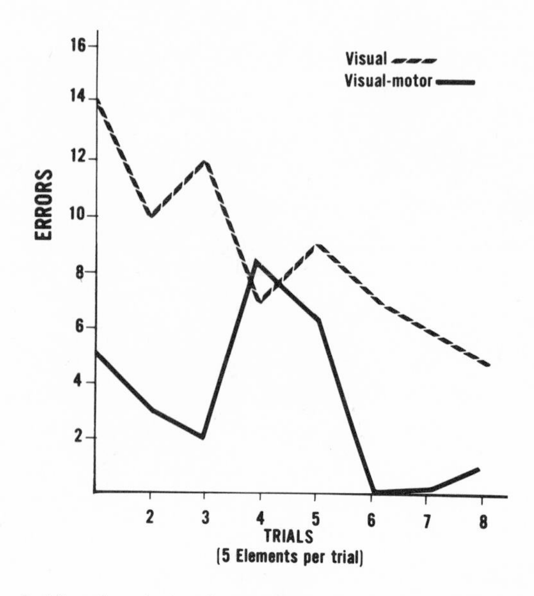

Figure 3. Visual and visual motor learning with the Jensen Board.

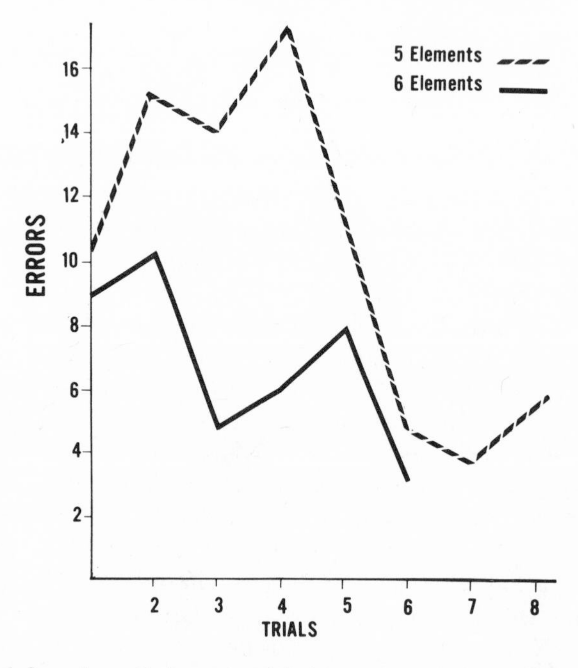

Figure 4. Learning a 5-element and 6-element series on the Jensen Board.

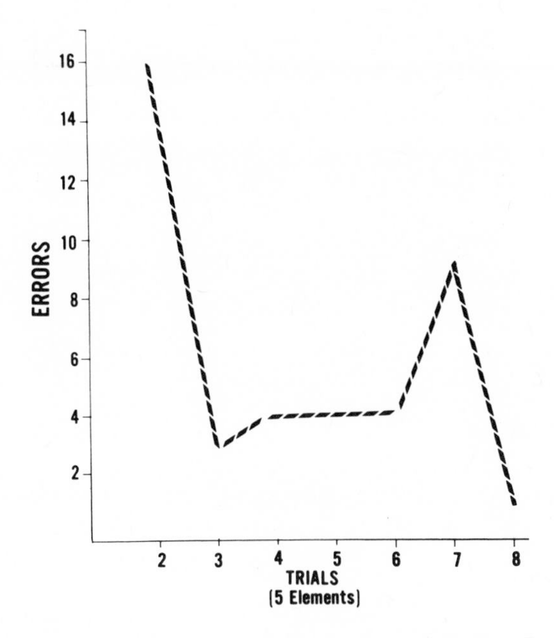

Figure 5. Retention of an 8-element series on the Jensen Board.

brain-damaged individuals had not been able to acquire new learning, then the observed reduction in errors would not have occurred. It may also be noted, particularly in regard to the data in Figure 4, that a strong learning–set phenomenon appears to be present in which once the subject catches on to the requirements of the task and how to go about dealing with it, learning progresses efficiently. In general, it seems that the Jensen board is a reasonably effective device for retraining in short-term memory.

A version of the progressive-matrices test was used to train nonverbal-conceptual skills. The progressive matrices, which is often used as a nonverbal intelligence test, consists of geometric forms embedded in block diagrams. Subjects were asked to study sets of three diagrams, and, for each item, to select from among three choices a fourth diagram that had a logical relationship to the other three. The degree of difficulty could be varied from relatively simple relationships to extremely complex ones. During training, the subject was asked to choose the appropriate block diagram and to verbalize the logical relationship involved. It was hoped that the verbalization requirement would make the subject more aware of his reasoning processes. The progressive matrices are ranked by level of complexity (A through E), and Figure

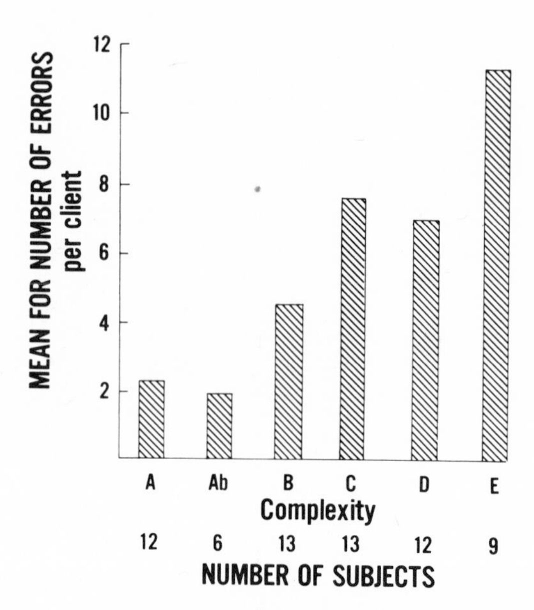

Figure 6. Complexity levels of the progressive matrices.

6 provides information regarding the distribution of subjects able to solve items at each of the complexity levels. Level Ab was only administered to subjects who had difficulty with Level A. Others went directly from A to B. Note that there is the expected gradual increase in errors as complexity level increases. Two subjects were unable to reach a solution at Level A. Nine clients solved Level D and were trained on E. However, only three of the nine subjects trained on Level E solved the problem at this level.

We were particularly interested in evaluating possible differences between right- and left-hemisphere subjects in regard to how they responded to progressive-matrices training. The test appears to measure primarily abstract-reasoning processes (Lezak, 1976), although the abstract concepts to be learned involved complex spatial relationships. Therefore, one would expect that the right-hemisphere subjects would have more difficulty than the left-hemisphere subjects. This belief was generally confirmed. Figure 7 provides learning-curve data for a set of right-hemisphere subjects, and Figure 8 provides comparable data for left-hemisphere subjects. Note that three left-hemisphere subjects were able to solve complexityLevel E, while none of the right-hemisphere subjects could accomplish that task. At complexity Levels B through D,

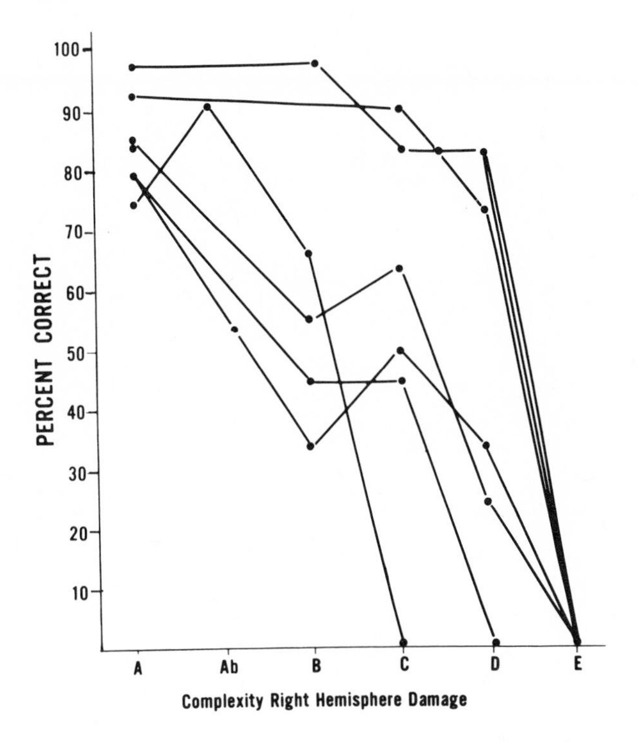

Figure 7. Progressive matrices learning curves for right-hemisphere clients.

it can be seen that the right-hemisphere subjects were less competent than the left-hemisphere subjects.

We suspect that the difference found appears to be related to the specific task requirements and not to differences in overall impairment between the right- and left-hemisphere subjects. To evaluate this matter, we looked at the Halstead-Reitan battery data obtained for these subjects and noted that the right-hemisphere group obtained an average of 74.5% of their scores in the impaired range, whereas the left-hemisphere group had a mean of 75.43% scores in the impaired range. These very similar figures strongly suggest that one group was not more globally impaired than the other.

The progressive matrices "training" itself was a relatively informal procedure in which the subject was encouraged to verbalize his reasoning, with the trainer asking appropriate questions that led the subject to keep working on the problem without giving him cues as to the correct answer. The data presented here only suggest that under

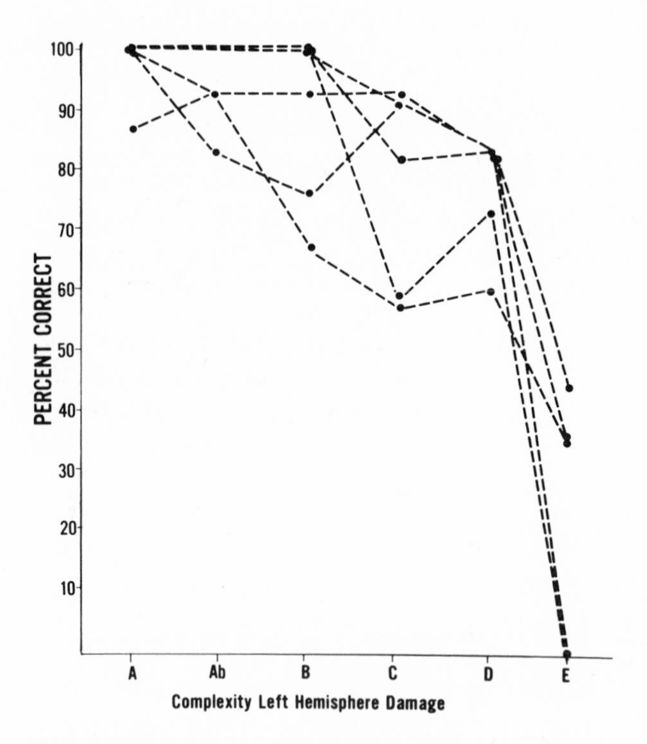

Figure 8. Progressive matrices learning curves for left-hemisphere clients.

some conditions, brain-damaged patients can perform a nonverbal task at a relatively complex level, particularly if the right-hemisphere is not involved. The efficacy of training is only suggested, in that it was demonstrated that several clients, under the prompting conditions used, reached correct solutions at varying complexity levels. Obviously, more detailed studies would have to be done to evaluate specifically the benefits of prompting.

Perceptual-Motor Retraining

For various reasons, many brain-damaged patients have difficulty with hand–eye coordination. Sometimes the difficulty is purely motoric and sometimes it involves central integration of visual- and motor-feedback information. In any event, the result is often slow, awkward motor performance. We used the pursuit-rotor as a hand–eye coordination training device, which, basically, is an instrument consisting of a horizontal revolving disk with a small circular target. The subject holds a stylus which he attempts to keep on the target while the disk is rotating. Time-on-target is the usual performance measure. In our situation, subjects were given immediate auditory feedback when they were off target. When the stylus was on target, a timing device recorded the duration, but when the subject went off target, the timer stopped and a buzzer was activated. Thus, the subjects received both visual and auditory feedback when they were off target. Figure 9 presents time-on-target data for all subjects evaluated. As one would expect to find in normal right-handed individuals, the right hand does better than the left, and there is a decrease in time-on-target as rotation speed increases. Figures 10, 11, and 12 contain individual learning curves for three subjects with right-hemisphere brain damage. Right-hemisphere cases were selected for illustration because, in preliminary studies, we found that they tended to be substantially more impaired on this task than were the left-hemisphere subjects, even though each client showed improvement with practice for most trial sequences. (In this case, a trial sequence was a set of six trials within each rotation speed.) However, there was generally a performance decrement on the final trial in each sequence, probably related to a fatigue effect. Improvement in performance was accomplished for both left and right hands. For early trials, performance level for the left hand was considerably lower than it was for the right hand. With continued

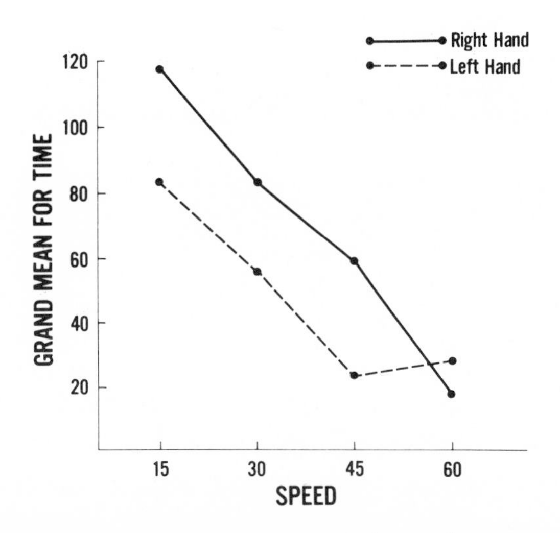

Figure 9. Time on target pursuit rotor data.

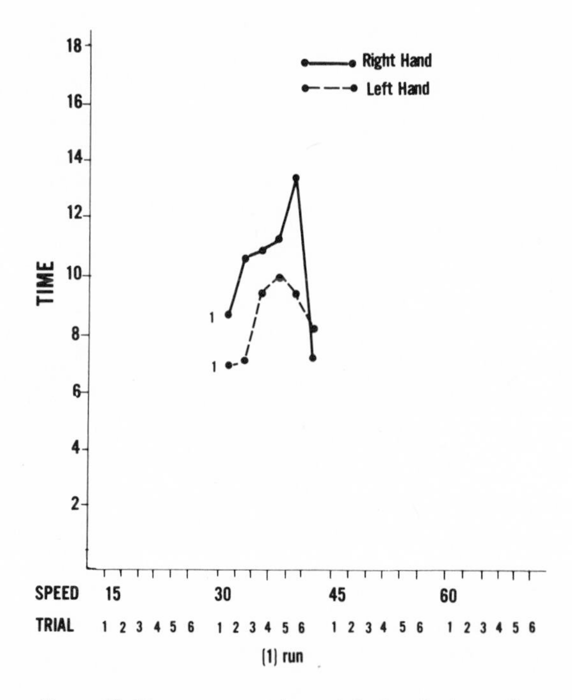

Figure 10. Time on target for a right-hemisphere client.

training, competency with the left hand improved, indicating that psychomotor performance of a limb can be improved even when the brain damage is in the contralateral hemisphere.

A visual-acuity device was designed that required subjects to react quickly to a visual stimulus and to localize a stimulus in space. Subjects were presented with a stimulus–display matrix consisting of 12 lights (4 rows of 3 lights per row). The lights were successively activated in a random sequence. For each trial, when a light was activated, a timer started and the subject was required to point a finger (on which a photocell was attached) at the activated light, in order to turn it off and stop the timing clock. Reaction time in $^1/_{100}$ seconds was recorded for each trial. A session consisted of three runs through random sequences of the 12-light display. Thus, the task required subjects to maintain attention prior to stimulus presentation, localize the stimulus in space, and exercise hand–eye coordination in regard to pointing to the proper light quickly. Figures 13 and 14 summarize the performance of a subject with right-hemisphere brain damage. Figure 13 represents performance with the left hand, and Figure 14, perfor-

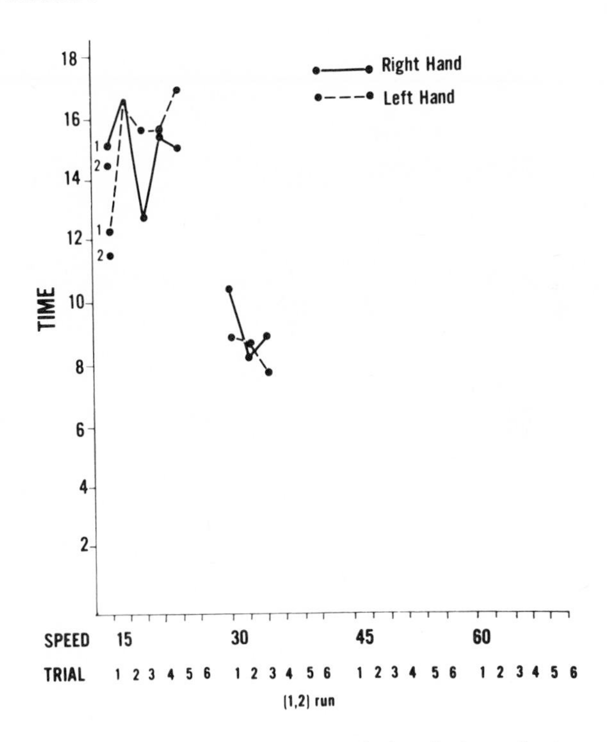

Figure 11. Time on target for a right-hemisphere client.

mance with the right hand. It will be noted that there is, in both cases, a gradual overall reduction in time needed to perform the task. Figures 15 and 16 illustrate the performance of another right-brain-damaged subject, while Figures 17 through 20 contain data for two left-brain-damaged subjects. In all cases, a relatively steady decrease in response time will be noted.

Summary. It would appear that what was accomplished in the cognitive- and perceptual-motor retraining aspects of the Wichita program was a demonstration that brain-damaged subjects could be retrained in a number of skills, utilizing standard instruments borrowed from psychological tests and laboratory procedures. Basically, the findings provide clear evidence that brain-damaged individuals are capable of new learning, and, conversely, the widely held opinion to the contrary has not been supported. We have suggested that these findings are very preliminary. We know very little about generalization from these laboratory tasks to real-life situations, but later on we will be able to show that this retraining, at least in a general way, was

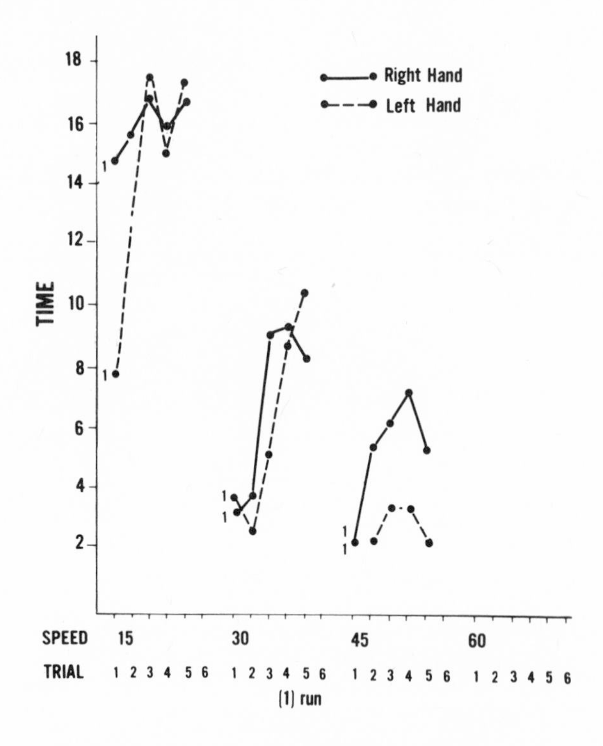

Figure 12. Time on target for a right-hemisphere client.

associated with improvement in neuropsychological test performance. It was also shown that in certain instances, notably the progressive-matrices training, there were differences between right- and left-hemisphere cases in regard to learning efficiency. However, it is important to note that both right- and left-hemisphere cases did learn. Therefore, one might conclude that if visual-spatial skills are important to the general adaptive functioning of the patient, retraining in that area should not be avoided even if it is a defective area for the patient.

The Neuropsychology Center

In Chapter 9 we will consider the case of the patient with long-standing, chronic brain damage and describe an institutional program devised for such patients. Here, we will be concerned only with reviewing a number of cases in which behavior–therapy-oriented programs were conducted. The difference between the Wichita and the

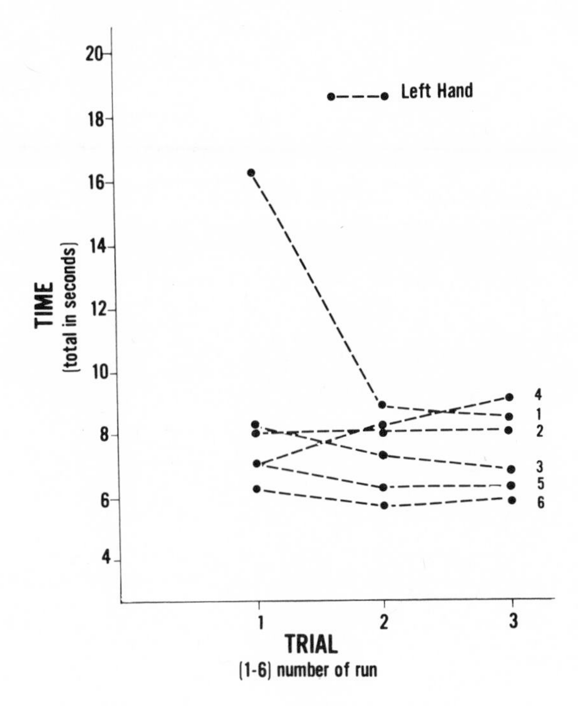

Figure 13. Visual data for a right-hemisphere client—left hand.

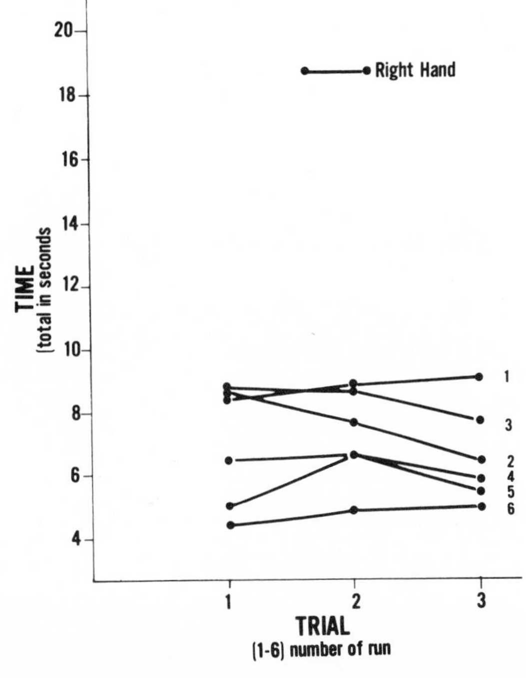

Figure 14. Visual acuity data for a right-hemisphere client—right hand.

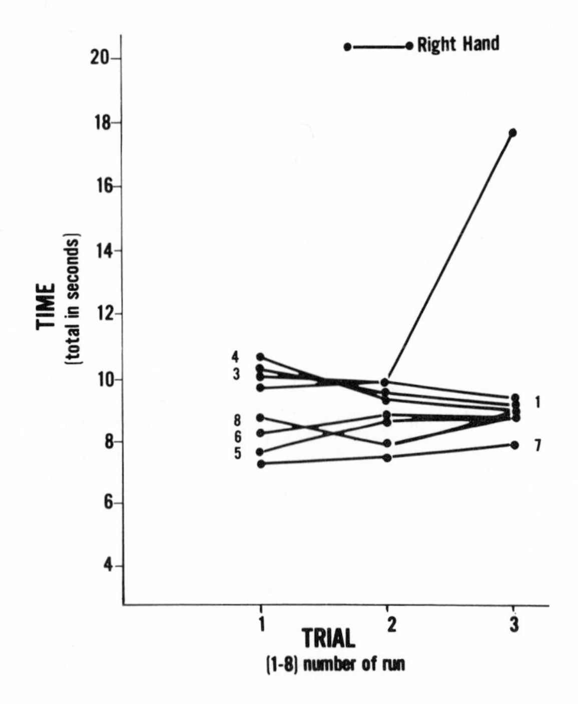

Figure 15. Visual acuity data for a right-hemisphere client.

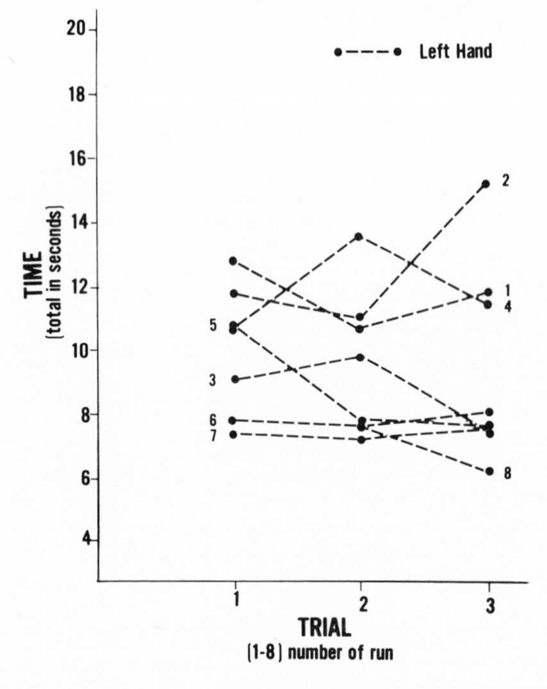

Figure 16. Visual acuity data for a right-hemisphere client.

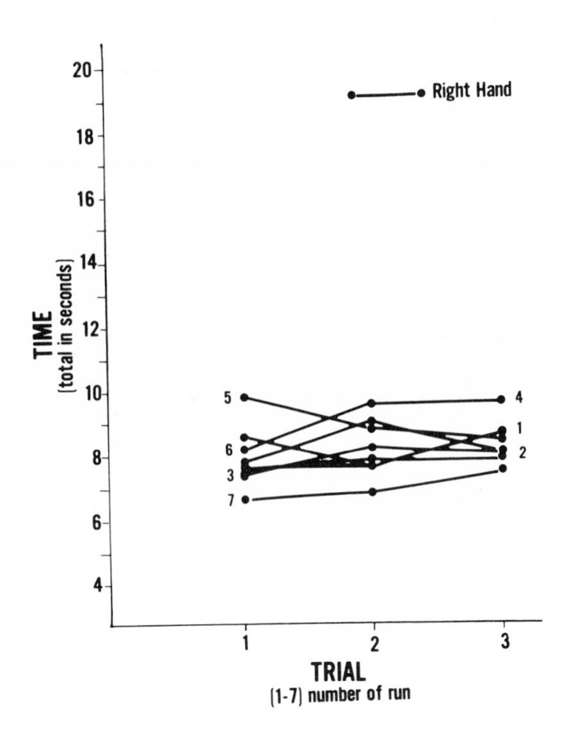

Figure 17. Visual acuity data for a left-hemisphere client.

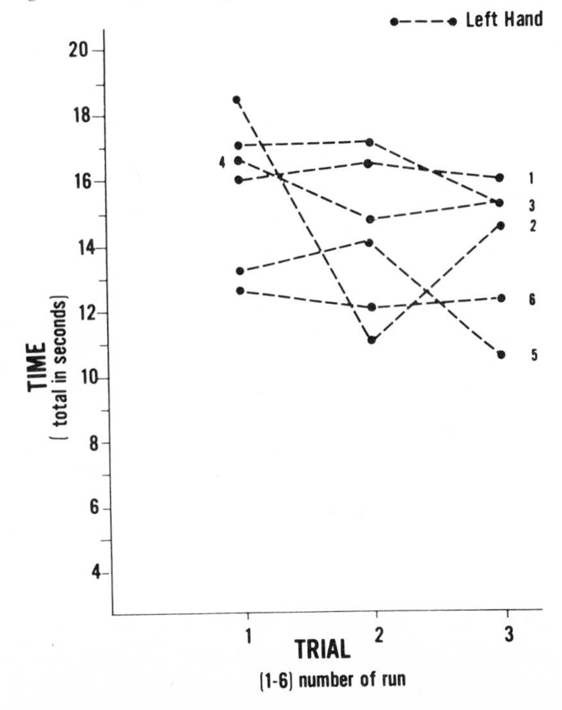

Figure 18. Visual acuity data for a left-hemisphere client.

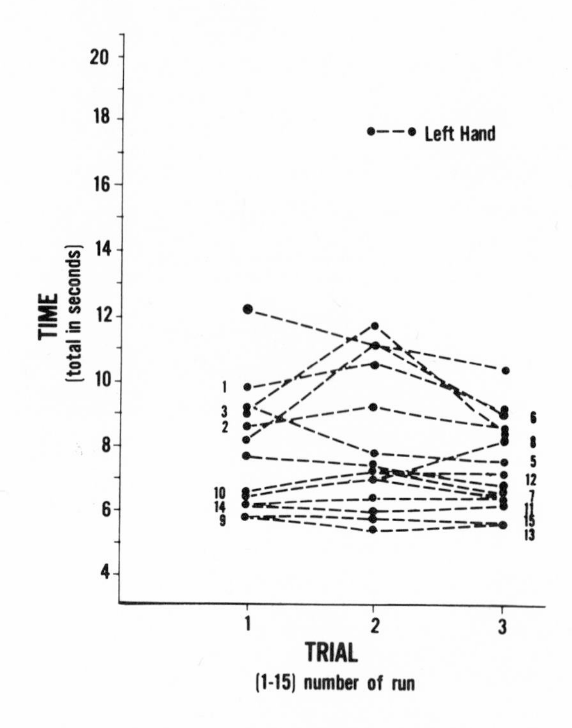

Figure 19. Visual acuity data for a left-hemisphere client.

Neuropsychology Center programs involves both setting and population. The setting for the Wichita program has been described. The setting for the Neuropsychology Center is an inpatient ward of a neuropsychiatric hospital. The Wichita subjects were primarily young men who had recently sustained head injury. The Neuropsychology Center subjects were middle-aged to elderly, and suffered from a variety of brain disorders.

A Case of Inappropriate Verbalization

The patient was a middle-aged man with the diagnosis of Korsakoff's syndrome. Although he exhibited the dementia and amnesia characteristic of this condition, he also made numerous requests for candy and chewing gum, often to inappropriate people at inappropriate times. These requests, made to staff, other patients, and strangers became essentially this patient's entire social behavior repertoire. The quantity of the requests proved to be an irritant to other patients and

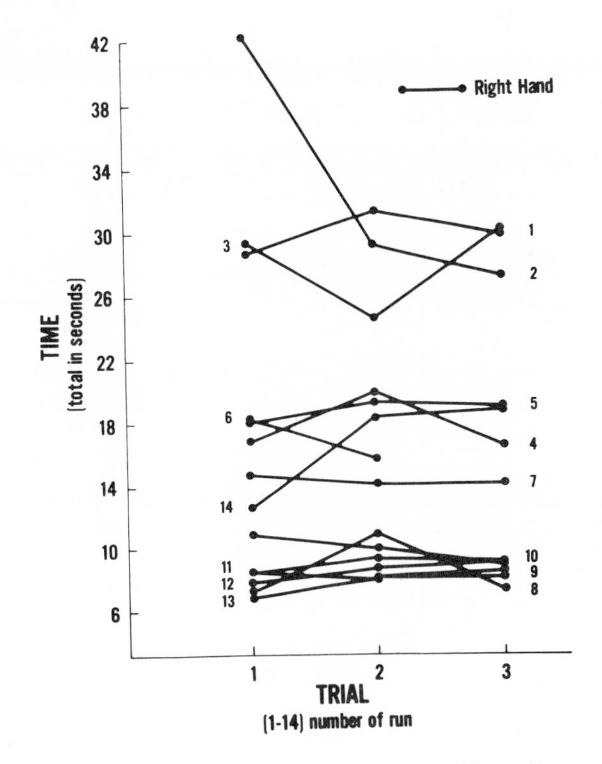

Figure 20. Visual acuity data for a left-hemisphere client.

prevented more productive social behavior from emerging. "You got any candy or chewing gum?" became the patient's major item of spontaneous conversation. It turned out that the patient did enjoy candy and chewing gum, so that these foods seemed to be appropriate reinforcers. The following program was established. The number of requests per hour were recorded as baseline data over a period of several days. In this case, a token system utilizing a *response cost method* seemed appropriate and was utilized. The response cost method is a procedure in which the subject is deprived of a positive reinforcer, or some penalty is invoked for exhibiting some inappropriate behavior. Kazdin (1977) indicates that the response cost method has been used as a special variation of the token economy. In this case, the patient was given a supply of tokens every morning, which could be redeemed in the evening for candy or chewing gum. However, every time a request was made during the day, a token was taken away. Under this system, the number of requests was gradually reduced over a period of several weeks, and essentially disappeared from the pa-

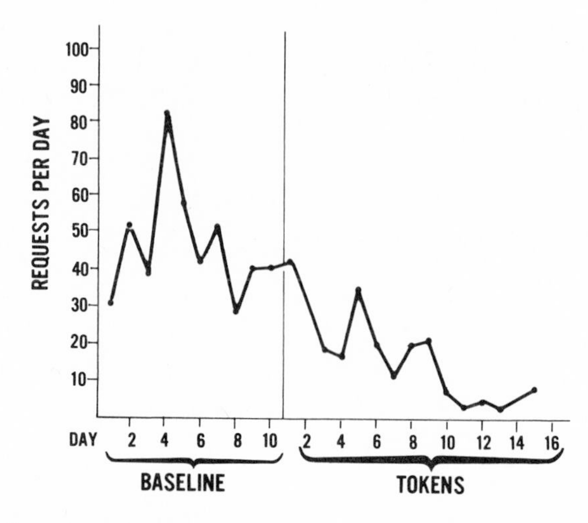

Figure 21. Token treatment for inappropriate verbal requests.

tient's repertoire. The token system was subsequently removed, and the behavior did not return. Relevant data are presented in Figure 21.

A Case of Fecal Incontinence

The patient was a severely demented young man who had sustained massive head trauma primarily involving the frontal lobes. Although the patient exhibited numerous behavioral difficulties, the one dealt with initially was frequent fecal incontinence. The neurological evaluation indicated that the incontinence was not a function of decreased sphincter control, and judging from the patient's physical condition, he was able to control his bowels. With regard to a choice of reinforcer, a difficult decision had to be made. The patient's only obvious source of enjoyment was cigarette smoking, and he would be given individual cigarettes whenever he requested them. In the judgment of the ward physician, it would not have been appropriate to attempt to limit the patient's smoking despite the potential health hazard. In view of the importance of continence for this patient's future disposition in combination with the severity of his dementia, the decision was made by the treatment team to use cigarettes as a reinforcer, despite the potential health hazard. Based on the same considerations, it was also felt that some degree of cigarette deprivation was justified on the basis of the potential for improving the patient's general life situation.

The program established for this patient involved making his supply of cigarettes contingent on appropriate toileting. Cigarettes were withdrawn and were issued only after a request to be taken to the toilet. After he left the toilet, he was immediately given a cigarette. The results of the program are presented in Figure 22. Even though the effectiveness of the program was not complete, episodes of incontinence were reduced from a high of six in one day to a fairly stable level of not more than two per day. In this case, an inadvertent return to baseline was achieved in connection with a visit by the patient's mother. The mother, not fully appreciating the contingency program, gave the patient cigarettes on request, and as can be seen in Figure 15, there were six episodes on the second day of her visit. The high level of episodes continued throughout the visit, and their frequency never really stabilized thereafter. There were great fluctuations, ranging from no episodes to eight of them in a single day.

Two Cases of Memory Training

The first patient, in this group of two, had severe Korsakoff syndrome and was densely amnesic for recent events. In his case, memory training was approached by asking him what items of information he felt were important for him to remember. He indicated that he would like to be able to recall the name of the ward (23 West), the times cigarettes are distributed (9A.M. and 3P.M.), and the ward visiting hours

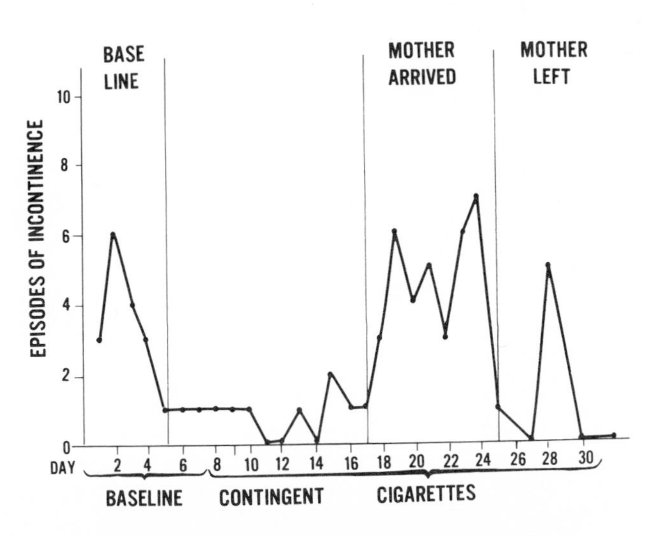

Figure 22. Contingency management for incontinence.

(1P.M. to 8P.M.). We added to the list the name of the technician who would be doing the training (Marilyn). It was established that the patient could not recall these items spontaneously. The training was based on the *Premack Recency Principle* (Homme, 1965; Premack, 1965) which states that learning can be enhanced by associating new, to-be-learned material with well-consolidated and overlearned old material. Initially, the patient was interviewed on several occasions with the aim of building up a list of old associations. As is characteristic of patients with Korsakoff's syndrome, this man had relatively well-preserved old memories. He could recall the names of his wife and children, his career in the military, his home town, former occupation, and related events. Thus, it was possible to make up a list of items available in his long-term, remote memory. These items were then paired with new items, specifically the things he wanted to learn. For example, we paired the name of one of his daughters (Jane), which he recalled spontaneously, with the name of the ward (23 West). We worked with one pair at a time.

At the initial training session the patient was told that we were going to try to help him with his memory by connecting something that he did remember with something that he did not. He was then told that the first connection would be between his daughter's name and the name of the ward; Jane and 23 West. He was asked to say "Jane-23 West" five times. Following this procedure there was a 5-minute break filled by conversation, after which he was asked to give five more repetitions. This procedure was repeated for a total of three times. The entire procedure was repeated for three days. In addition, the patient was probed by the ward staff three times a day for approximately a week. Thus, for example, a staff member would ask the patient in the morning, in the afternoon, and in the evening, "What is your ward?" and record the response. This procedure was followed for each new pairing. If the patient gave the wrong response, he would be given the stimulus word and was asked what it reminded him of; for example, "What does Jane remind you of?" The data for the four pairs are presented in Table 3.

It was noted that on incorrect trials, giving the stimulus word was generally ineffective in eliciting the correct response. Nevertheless, the procedure did work and the patient was able to retain the information over at least the length of the probe period, about a week. The limitations of the retention time are, of course, unknown, and it is not clear that the application of the recency principle was actually what produced the effect. Further research is needed to clarify these issues, but on a pragmatic basis, the procedure appeared to have been effective.

Table 3
Frequencies of Correct and Incorrect Recalls

Item	Pair	Correct	Incorrect	Total probes
Ward name	Jane-23 West	3	0	3
Cigarette time	Pittsburgh-9 and 3	13	1	14
Visiting hours	Dorothy-1 to 8	15	2	17
Name of teacher	New Castle-Marilyn	10	0	10

Neuropsychological literature (e.g., Butters & Cermak, 1980) and clinical observation attest to the fact that amnesic disorders associated with Korsakoff's syndrome or damage to the temporal-limbic system tend to be severe and generally irreversible. Based on the findings with the preceding patient, and more casual observation of other patients, we believe that although it may not be possible to improve the memories of these patients in a general way, it may well be possible to teach them specific items of information that are important to them and/or important for adaptive functioning in general. The next patient to be presented was administered memory training with this hypothesis in mind. The task was to determine whether or not we could improve the patient's performance on the basis of regular repetition of a standard set of memory items. In this case, we used the memory scale from the Luria-Nebraska neuropsychological test battery (Golden, Hammeke, & Purisch, 1980) as the standard stimulus material. This scale includes a variety of memory–test items such as word–list learning, recall following interference, recall of gestures, and paired-associate learning. The patient was a middle-aged man who suffered from an episode of cardiac arrest leading to substantial cerebral anoxia. The patient, who had formerly been a successful businessman, came out of the episode with a severe memory disability. He had retained this disability for about five months prior to training, and so it was unlikely that any training effects could be attributed to spontaneous recovery.

The training in this case simply consisted of administering this scale in the standard manner for a period of 21 days. The scores achieved each day are plotted in Figure 23. On the Luria-Nebraska battery, a *t* score higher than 60 is considered to be abnormal. This patient's initial score was 75, suggesting a moderate degree of memory disability. Note in Figure 23 that by the fourteenth day his score was down to 50, the mean for the normal population. Actually, a score of 50 represents a practically error-free performance on this scale. Thus, this patient who was severely disabled in regard to recalling even very

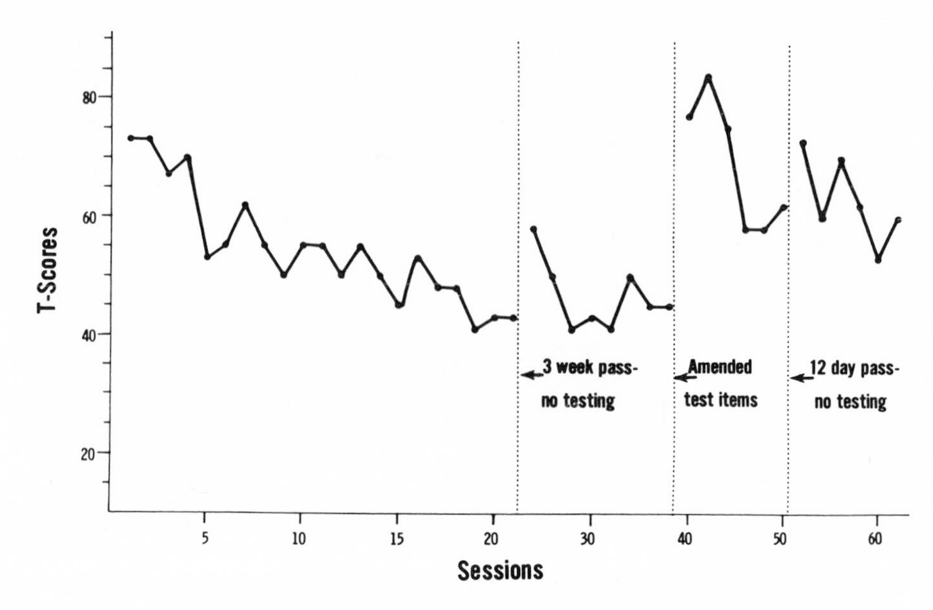

Figure 23. Luria-Nebraska Memory Scale *t*-scores across sessions.

basic facts, such as the name of his wife, was able to learn the Luria-Nebraska memory items.

In order to ascertain whether this specific learning generalized, we constructed a tentative alternate form of the scale, primarily through equating items for imagery, frequency, and associability. Note also in Figure 23 that when the alternate form was first administered, performance became a great deal worse; even worse than his initial performance on the original scale. However, it improved rapidly over the next several days. When, however, the patient could not be tested for several days because of a leave of absence from the hospital, performance deteriorated; but, at the point at which testing was continued, performance seemed to recover again. The preliminary conclusion one might want to draw from this case, and the one presented previously, is that severely amnesic patients can be taught specific contents that may be important for adaptive function. When training is discontinued, however, there may be a rapid deterioration in performance level.

Summary. Whereas the Wichita program dealt mainly with generic abilities, the Neuropsychology Center deals more or less with specifics. Perhaps this distinction is based largely on population differences. The center patients were clearly more impaired than the Wichita clients, and there seemed to be more need to deal immediately with concrete situations. In any event, we have shown that chronic, severely disabled patients could be helped if the goals were modest and relatively delimited. Even with the few cases presented, it is clear that the content of the behavior therapy may vary greatly from such matters as social behavior and maintenance of continence to rather specific memory training. A great deal of the treatment that goes on in the center involves more informal training, with no systematic attempt to collect data or employ some research design. For example, the program mentioned elsewhere of using tokens with an individual patient to discourage indolence was done on an informal basis. This matter of informal programs within the context of an appropriate ward milieu will be elaborated on in Chapter 9.

The Work of Others

Any review of the work of others requires the drawing of a fine line between what is and is not behavior therapy for treatment of neuropsychological defects in adult patients with acquired brain damage. Golden (1978) alludes to the use of such behavior–therapy-oriented techniques as multiple-baseline design, but there is no indication that investigators have really applied this design to the treatment of brain-damaged patients, with the exception of the Gianutsos and Gianutsos (1979) study. Again, we are not including the large amount of work done with the mentally retarded and with other patient groups with cognitive deficits, such as autistic children and schizophrenic patients. Similarly, we are not including here the many approaches to such areas as speech rehabilitation (Holland, 1982; Kertesz, 1979) that are not behavior-therapy oriented. The work of the NYU group, discussed elsewhere in this volume, is also not a behavior–therapy orientation. Indeed, Diller (1970) has taken issue with the application of behavior–therapy concepts to brain–damage rehabilitation. Correspondingly, when one reviews the behavior–therapy texts, our experience indicates that any search done for studies of behavioral treatment of adult brain-damaged patients will not yield much. Although there is an extensive literature regarding the behavioral problems (such as incontinence or memory difficulties) that brain-damaged patients may have in common with other groups, there is very little literature concerning the remediation of these difficulties *in brain-damaged patients.*

It is probably fair to say that the literature that is available has placed heavy emphasis on the areas of memory and speech. In recent years, and particularly since the publication of an important article by

Lewinsohn, Danaher, and Kikel (1977), there has been a growth of interest in memory training for elderly and brain-damaged individuals. A major question has been whether or not the various memory training methods used with normal people who want to improve their memories can be applied to clinical populations. A major class of these methods involves the use of visual imagery. A visual image associating two items is either generated by the learner or provided by the teacher. It is generally thought to help matters if the image is made as ridiculous or absurd as possible. For example, if the word pair is "tree-ear" the learner might think of a tree growing out of someone's ear. Lewinsohn, Danaher, and Kikel, working with a group of brain-damaged individuals of various types, were able to show that imagery training was effective in improving memory performance over a brief period of time, but the effect was not maintained over a one-week interval. The effect obtained was significant for a verbal paired-associate learning task, but not for a task involving the pairing of names with faces. Poon, Fozard, Cermak, Arenberg, and Thompson (1980) reviewed the literature on visual-imagery training with the elderly and reported that 14 of the 17 studies reviewed obtained positive results. It should be pointed out, however, that caution should be exercised in generalizing from studies done with the normal elderly to clinical populations of brain-damaged adults.

Again, using elderly, normal people, Langer, Rodin, Beck, Weinman, and Spitzer (1979) found that the use of a token system, in which subjects were given redeemable poker chips as reinforcing stimuli, was helpful in assisting subjects in the recall of names and activities. Subjects given tokens for accurate recall of information requested by the experimenter improved significantly more than subjects who were given redeemable chips, but not in a response-contingent manner. It was also found that the token-contingent group took less time than the noncontingent group to find needed information, and that the token-contingent group did better than the noncontingent group on tests of short-term and remote memory administered after training. The procedure used in this study did not utilize visual imagery or any other mnemonic techniques. The subjects were simply asked questions, the answers to which would be made available through information-seeking in the environment. For example, one question was "How many male residents are there on this floor?" Thus, we have another example of the efficacy of memory training in the elderly. However, subjects were ambulatory nursing home residents who were judged by the staff as being alert. Again, we do not know how this procedure would work with brain-damaged patients.

There is a body of literature that has strong implications for memory training with brain-damaged, amnesic patients, but that does not provide direct evidence for the efficacy of any particular training method. We are referring to the basic research into the neuropsychology of memory, and specifically the research of Butters and Cermak (1980) with alcoholic Korsakoff's patients. A basic problem with the memory function of the Korsakoff's patient is an information-processing deficit in which incoming information is not adequately encoded. However, extensive research has suggested that under conditions of appropriate cueing and instruction, Korsakoff patients can encode in a relatively normal manner. The problem is that when left to their own devices, they encode in an inefficient manner; one that makes newly learned material highly susceptible to interference. Based on a series of experiments, Butters and Cermak (1980) make these remarks:

> Under the appropriate instructional conditions, alcoholic Korsakoff patients seem to be capable of semantic encoding. However, when they were not so instructed they seemed to prefer to rely on their encoding of the acoustic dimensions of words. If this is true, then it would not be surprising that the alcoholic Korsakoff patients' STM (short term memory) decays rapidly, or that it is highly susceptible to interference, as information encoded solely on the basis of its acoustic features decays more rapidly than materials encoded semantically. (pp. 76–77)

Instruction in semantic encoding is generally accomplished through a cued recall procedure in which the subject is given the name of the category for which the to-be-recalled word is an exemplar. It would appear, then, that the functional memory of patients with amnesic disorders may be improved through providing more efficient encoding strategies than the patient would ordinarily use.

A direct test of the efficacy of a reinforcement-oriented technique in regard to improving the memory of adult brain-damaged patients was made by Dolan and Norton (1977). The task involved teaching the subjects the names of the staff through pairing the names with photographs, and also teaching the answers to a series of questions about the ward environment. Subjects were divided into three groups. One group received verbal and material reinforcement, the latter consisting of redeemable canteen coupons; the second group received only verbal reinforcement; and the third group received no training at all. Verbal reinforcement consisted of praise for a correct response. It was found that both trained groups improved on posttesting significantly more than the control group. However, there was no difference between the material-reinforcement and verbal-reinforcement groups. In other words, addition of material reinforcement to praise did not enhance learning significantly. The authors discussed this finding and suggested that the high level of motivation that these patients initially had did not necessitate use of material reinforcement. It may be mentioned that we had the same experience in the case of the Wichita program, although we did not study this phenomenon as systematically as these investigators.

In summary, application of behavior–therapy-oriented procedures to memory training shows some promise. However, work of the Lewinsohn group suggests short-term effects, lasting perhaps less than a week. It was also pointed out that memory training that is efficacious with nonneurologically impaired individuals, both young and old, may not be as effective with brain-damaged patients. However, detailed studies of memory in patients with various kinds of amnesic disorders, notably work of Butters and Cermak (1980), may provide valuable information concerning the kinds of cueing that might be helpful in enhancing memory. In other words, knowledge of the type of memory disorder the patient has (e.g., an encoding deficit, a storage deficit) may be important in regard to planning an appropriate treatment program.

It is probably still true that the most active efforts to rehabilitate brain-damaged patients are in the language area. As indicated elsewhere in this book, there are several schools of language therapy based on varying theoretical frameworks. We have also pointed out that behavior therapists have done language training, although primarily with the mentally retarded (e.g., Guess, Sailor, Rutherford, & Baer, 1968; Lutzker & Sherman, 1974). Literature in this area involving the use of token economy is reviewed by Kazdin (1977). The question is whether or not these procedures are applicable to adult brain-damaged individuals, notably aphasics. For example, Kazdin (1973) was able to use token withdrawal to reduce speech dysfluencies in adult retardates. Many forms of adult-acquired aphasia, notably pure motor and Broca's aphasia, involve impairment of speech fluency. In the past, language therapists working with aphasic patients have reported on efforts to treat these patients with behavior–therapy-oriented methods (see Lahey, 1973; Sloane & Macauley, 1968). Some time ago, a major effort was made involving programmed instruction, often using an operant model. The names most often associated with this approach are those of Martha Taylor Sarno and Audrey Holland. Sarno, Silverman, and Sands (1970) and Taylor (1964) describe this approach in terms of a progression from preverbal behavior to matters as complex as syntax. Holland and various collaborators have provided case and descriptive material involving the use of programmed instruction (Holland, 1969,

1970; Holland & Harris, 1968; Holland & Levy, 1969, 1971; Holland & Sonderman, 1974; Sparks & Holland 1976). The essence of programmed instruction involves targeting the desired language behavior and devising a structured, stepwise program to achieve it. Such methods as melodic-intonation therapy, in which the patient is taught to make simple statements by intoning them as melodies, also involve programmed instruction (Sparks, Helm, & Albert, 1974; Sparks & Holland, 1976).

An explicit, behaviorally oriented hypothesis involving language rehabilitation has been offered by Basso, Capitani, and Vignolo (1979) in the following remarks:

> The hypothesis underlying rehabilitation is that all intentional verbal behavior, if appropriately reinforced through training, leaves a trace in the cerebral structures that are potentially capable to carry out linguistic activity, and this trace increases the probability of occurrence of subsequent correct verbal behaviors. (p. 192)

Although Basso *et al.* did not test this hypothesis directly, they did provide one of the very few controlled studies of the efficacy of speech therapy. They used the so-called stimulation method (Taylor, 1964) with aphasic patients, in which the patient is given a number of exercises and assignments with the aim of facilitating increasingly intentional, voluntarily controlled language. It was shown that a large sample of aphasics undergoing language rehabilitation improved in various aspects of language behavior more than an untreated control group. Although this study was criticized by E. Miller (1980) on the basis of the adequacy of the control group, it stands as perhaps the most impressive demonstration of the efficacy of speech therapy with aphasics that is available at present.

When one looks in areas other than language and memory, there is very little literature. Most reviews of the area of rehabilitation of neuropsychological deficits either stress the language and memory areas (Kertesz, 1979; E. Miller, 1980) or emphasize the early work of Luria's group (Luria, 1948/1963) and the work of the NYU group, reviewed elsewhere in this book. E. Miller (1980) calls our attention to an impressive case study by Saan and Schoonbeck (1973) in which a patient was treated for dyscalculia within the framework of a multiple-baseline design. Nevertheless, when one considers the variety of neuropsychological deficits in such areas as motor function, visual-spatial abilities, various gnostic or perceptual functions, and higher cognitive abilities, there is surprisingly little literature in the behavior–therapy area. In the area of such specific behaviors as eating, toileting, various aspects of social behavior, and self-care, one might wish to borrow techniques reported in the mental retardation and psychiatric patient literature, but the actual application of these techniques to adult patients with acquired brain damage has apparently not been reported. In his recent review, E. Miller (1980) also points out that neuropsychological deficit is the only area in which clinical psychologists work that has not received some impact from behaviorally oriented treatment and management methods.

Summary

In this chapter, we have suggested that there may well be some links between theory and methods of behavior therapy and care and treatment of adult brain-damaged patients. Following a very brief introduction to the field of behavior therapy, we pointed out that certain of its methods could be applied to the assessment of brain-damaged patients and to the devising of novel treatment methods. It was emphasized that despite widespread application of behavior therapeutic approaches to various clinical populations, adult brain-damaged patients still constitute a group that has received very little impact from behavior therapy. We suggested that single-subject design methodologies as well as techniques, such as skill training and token economy, might provide effective treatments for brain-damaged patients, not only in the more widely studied areas of language and memory, but also in regard to specific social and adaptive behaviors, as well as certain of the less well-studied neuropsychological defects such as visual-spatial or gnostic disorders.

By reviewing the work of others, as well as our own work in two different settings, it is possible to draw some preliminary conclusions. Most investigators who have worked in this area tend to describe their findings as modest but promising. Structured behavior–therapy programs usually accomplish what they specifically set out to do. Thus, for example, in the Dolan and Norton (1977) study, subjects did, in fact, learn the names of staff members through pairing the names with photographs. Data from the programmed instruction literature in the language area indicate that aphasic patients do learn the material that is taught. Although some might not be overly impressed with this general finding, it seems important to us in view of the commonly held belief that brain-damaged individuals are incapable of new learning. It has been well demonstrated that this belief is clearly not the case. Nevertheless, it is still appropriate to raise questions regarding generalizability of the results obtained to ordinary adaptive function. In the language area, Holland (1980) suggests that although it has been demonstrated that abnormal behavior can be modified in controlled clinical or laboratory situations, we should go on to the next step of looking at how language functions in more natural contexts. The assessment of the success of many of the language and memory procedures was accomplished with formal testing, rather than with observation of behavior in natural settings.

A related problem area has to do with permanency of obtained effects. In regard to memory training, it seems as though initially obtained improvements may not last beyond a week. However, a follow-up of a case reported some time ago by Holland and Harris (1968) indicated that the patient continued to be nonaphasic for more than ten years since the case was reported (personal communication). It should be recognized that the matter of *training effects* interacts with *recovery processes* in general, and, except in the case of controlled studies, it is not really possible to say whether some permanent effect is a result of the training or of recovery from the underlying neurological disorder. It is important to note that recovery from aphasia is quite common, whereas recovery from such amnesic disorders as Korsakoff's syndrome is essentially unheard of. For example, the famous case H.M. (Scoville & Milner, 1957), who was followed for many years, showed no sign of recovery of memory. Thus, the question of the extent to which retraining is helped along by nature is crucial when one is trying to evaluate the efficacy of retraining.

The role of reinforcement in many of the studies is of particular interest. For example, it will be recalled that in the Dolan and Norton (1977) study, material reinforcement did not add to what was accomplished with praise. On the other hand, Langer, Rodin, Beck, Weinman, and Spitzer (1979) found that contingent token reinforcement was superior to noncontingent issuing of tokens in regard to eliciting improvement on a probe-recall test of immediate and remote memory. In the case of the Wichita program, early attempts at token economy were not successful. Perhaps one difference between the Langer *et al.* (1979) studies and the other studies was that subjects were elderly but were not brain-damaged individuals with debilitating deficits. In the case of the brain-damaged patient, motivation for recovery may well be sufficient to sustain learning. It may not be necessary to add material reinforcement. Luria (1962/1966) supported this view:

> In all the cases described above an injury to the cerebrum has caused a disturbance of some necessary factor for the performance of a mental activity, making a certain operation impossible. The patient has, however, been

> left with a clear awareness of his defect, and has developed a persistent need to compensate it. (p. 232)

The one exception reported by Luria is the patient with a massive frontal-lobe lesion. Luria then went on to present evidence in support of the view that large frontal-lobe lesions impair the mental motivation of activity. Thus, one might wish to use external, material reinforcers with frontal-lobe patients, but they may be superfluous in the case of patients with lesions elsewhere in the brain. In the case of the patient we presented earlier with the problem of fecal incontinence, he did have a massive frontal-lobe lesion, and it was clear that he was not at all motivated to modify this behavior without external pressure to do so.

Perhaps the most pertinent summary statement that one could make about application of behavior therapy to brain-damaged patients is the same as the comment made by most investigators of this matter—the results are modest but promising. The great need now is for more data, preferably in the form of innovative and well-designed studies. It seems necessary to develop an educational technology based on current knowledge concerning the nature of various neuropsychological deficits. The most efficient method of obtaining this information from individual patients is through neuropsychological assessment, the topic to which we will now turn our attention.

5

NEUROPSYCHOLOGICAL TESTING AS A MEANS OF PLANNING REHABILITATION PROGRAMS

Clinical neuropsychology has traditionally been concerned with the assessment of individuals who have brain lesions and the state of their cognitive, perceptual, and motor skills. Such assessments have been conducted at varying levels, ranging from prediction of presence or absence of brain damage based on a single indicator, to prediction of type and locus of lesion from the results of an extensive series of tests. The former kind of procedure, although generally rapidly and efficiently accomplished, has received much criticism because of its failure to elicit underlying brain mechanisms as well as its limited capacity to describe the pattern of preserved and impaired abilities found in the brain-damaged individual. Thus, the neuropsychological test-battery approach has become increasingly popular in recent years as a means of assessing functioning on an individual basis (Filskov & Boll, 1981). Correspondingly, neuropsychological testing has tended to become an area of specialization within the broader field of psychological testing.

The development of neuropsychological test batteries has been accomplished for various purposes by a number of investigators. Some of them are quite specialized, such as the series of tests developed by K. Goldstein and Scheerer (1941) which are all devoted to the study of varying forms of abstract behavior. Batteries for right-left discrimination and finger localization have been described by A. L. Benton (1959). Semmes, Weinstein, Ghent, and Teuber (1960) utilized a battery of tests for evaluation of somatic sensation. Bentin and Gordon (1979) have developed a specialized battery for the purpose of assessing so-called cognitive asymmetries, or differences in information processing between the right- and left-cerebral hemispheres. A battery, quite different from these, was developed in Ward Halstead's laboratory during the 1930s and 1940s. Its first comprehensive description can be found in Halstead's book, *Brain and Intelligence* (1947). While this battery had a specific purpose, that of assessing functioning of the frontal lobes, its content was very broad ranging, covering numerous areas of psychological functioning and containing different kinds of procedures. This battery of tests, greatly expanded and explored by Halstead's student, Ralph M. Reitan, has provided the major thrust for the growth of the battery approach in clinical neuropsychology.

As a consequence of some 25 years of research by various investigators, the Halstead Battery and associated procedures has shown itself to be a highly effective instrument in the elicitation of brain–behavior relationships (Boll, 1981; Reitan, 1966a; Russell, Neuringer, & Goldstein, 1970). Utilizing the various findings, clinicians have been able to predict, with reasonable accuracy, site and type of lesion in individual cases. Thus, the Halstead Battery has become an important tool in the diagnosis of brain damage, because with it, the clinician is no longer limited to a probability statement to the effect that the patient is or is not brain damaged. Rather, he can offer a description of what may be the localization of the lesion, the underlying neurological process determining the patient's symptoms, and the behavioral evidence for the postulation of such a process. This behavioral evidence consists of the pattern and level of test performance, the tests themselves evaluating a number of cognitive, perceptual, and motor abilities.

This tradition of a general neuropsychological battery has been followed by Golden and his various collaborators (Golden, 1981) in the development of the Luria-Nebraska neuropsychological battery. This battery is different in content from the Halstead-Reitan battery in that it consists of a long series of small tests developed by Luria over many years of practice. These tests were compiled by Christensen (1975) and divided into a number of scales (e.g., memory, expressive speech, etc.). Golden and his collaborators provided a scoring system for these scales and standardized the battery, using various samples from brain-damaged, psychiatric, and normal populations. Whereas the Luria-Nebraska would appear to be a good battery for rehabilitation planning, its potential in that regard has not yet been realized. Therefore, although we will be saying little about it in the present work, the reader should be aware of its existence and its promising potential.

In recent years, there has been a trend toward going beyond the diagnostic aspects of neuropsychological testing, because it seemed that its potential as an assessment instrument had been well-demonstrated, and the time had come to explore its possible role in the treatment planning that usually follows diagnosis. In its initial phase, neuropsychological testing was often of value in the treatment of the acutely ill person who was a candidate for surgical intervention. The neuropsychologist was frequently called upon to aid the neurosurgeon in localizing the lesion prior to operation. However, this situation generally represented the exception rather than the rule, with the major problem being that of rehabilitation of individuals with chronic lesions. This trend was accompanied by increased optimism regarding recovery of function in brain-damaged individuals, and the realization that more could be done after surgical or other treatment than during the acute phase of the illness. Thus, the usefulness of neuropsychological testing in the planning of rehabilitation programs has become an area in which substantial research efforts could be fruitfully initiated.

To a clinical neuropsychologist, the formulation of rehabilitation programs rather than of diagnoses represents something of a challenge. While the description of the pattern of abilities and the relating of this pattern to possible underlying brain dysfunction is well within his repertoire, the neuropsychologist is not accustomed to taking the next step of reccmmending a treatment program based on his diagnosis. The use to which test results are put is frequently not determined by him, but by the person for whom he is providing information. Programming thus puts him in a new role and induces him to look at his test data in a different way. However, it may be a way that is consistent with his model of the human brain and of the way he feels that brain function is related to behavior.

A Neuropsychological Model of Cerebral Organization and Function

The current state of neuropsychological research allows only for a relatively simple model of cerebral organization. However, it is an empirically derived model based on studies of individuals with brain lesions of various types and in various regions of the cerebral hemispheres. In the past, there has been a tendency to postulate certain connections between brain structures and behavior on the basis of insufficient information in regard to a number of considerations, including the methods employed, the extensiveness of the evaluation, and the number of subjects in the sample. Thus, during the nineteenth and early twentieth centuries, neurology was heavily influenced by a naive localization theory postulating the existence of certain highly discrete structures as mediators of highly specific behaviors. The reaction to this view in the form of Gestalt or organismic models (K. Goldstein, 1939), in which the brain was thought to always function in a holistic manner, is perhaps now seen as something of an overreaction. Each view certainly seems to hold some truth, but neither one seems sufficiently sophisticated in the light of current evidence.

Naive localization theory makes the unfortunate assumption that the brain is organized in accordance with certain categories of behavior. If we reflect on this matter, it becomes clear that these categories are constructs created for delineating types of behavior, and need not correspond with anatomical regions in any consistent way. Simply because we delineate handwriting as a category of behavior, it does not necessarily follow that there is a handwriting center in the brain. Certainly, the fact that human beings can write must mean that the brain mediates this behavior in some way. Our point is that it need not perform such mediation through the action of a circumscribed anatomic structure. Other alternatives are possible, such as Luria's (1962/1966) concept of functional systems. Nevertheless, continually accumulating evidence leads to the firm belief that the brain is a differentiated organ, and although it may always function as a whole in some respects, there is certainly strong evidence for some degree of localization of function.

A type of localization of function is seen in control of sensation and motility in bilaterally symmetrical organisms. In organisms with crossed nervous systems, the right half of the brain, for the most part, controls the left half of the body and vice versa. Thus, we have the principle of contralateral control as a means of regional localization. Along the anterior-posterior axis, there is currently little doubt that the frontal half of the brain mediates motility, while the posterior half mediates various forms of sensation and perception. With regard to the major lobes, the frontal lobes seem to be related to motility, the temporal lobes to hearing, the parietal lobes to somatic sensation, and the occipital lobes to vision.

Contemporary critics of localization theory do not appear to deny that the primary sensory and motor functions are localized as described above. Their view appears to be that the higher cortical activities, such as memory, reasoning, and learning, are not anatomically localized but are mediated by integrative activity of the whole brain. Teuber and Liebert (1958) may have contributed significantly toward a resolution to this controversy through postulating that there are both specific and general effects of brain damage. In essence, the theory supported may be a function of the methods of investigation used. Tests of simple functions tend to support localization theory, while more complex tests support holistic theory. It is likely that most contemporary neuropsychologists accept some version of the Teuber-Liebert formulation (e.g., Satz, 1966). The brain is capable of highly discrete, localized activity, and at the same time it can perform highly complex, integrative activity utilizing vast networks of neurons.

Another aspect of regional localization seems to apply only to the human brain. The two cerebral hemispheres, although essentially anatomically identical, appear to mediate differing kinds of complex behavior. Here we are referring to the principle of functional asymmetry of the cerebral hemispheres, or *cerebral dominance:* the left-cerebral hemisphere mediates language function, whereas the right-hemisphere has to do with a number of non-verbal skills that can generally be described as spatial relations abilities. When damage occurs to the left-hemisphere, the individual often acquires a condition known as *aphasia*—a general term denoting some kind of impairment of communication ability owing to focal structural brain damage. The syndrome picture most frequently accompanying right-hemisphere brain damage is called *constructional apraxia*—an impairment of the ability to deal with spatial relationships.

In summary, a neuropsychological brain model has been developed with regard to localization of function and can be recapitulated in these points:

1. As behaviors become increasingly complex, correspondence between those behaviors and some local anatomic region in the brain becomes less precise.
2. Regional localization appears to exist for primary sensory and motor functions. In this regard, the principle of contralateral control as well as assignment of specific functions to the major lobes are useful concepts.
3. There is a functional asymmetry of the human cerebral hemispheres, in which the left-hemisphere mediates language abilities, while the right hemisphere has a corresponding role for spatial relations skills.

In proposing this model, we do not wish to imply that more precise localization of function has been ruled out. Indeed, there are numerous examples in the current literature supporting more detailed localization. We may mention, as a few examples, the interesting relationships between the hippocampus and memory (Milner, 1964), the question of specific localization for varying types of aphasia (Goodglass & Kaplan, 1972), and Teuber's "corollary discharge" theory of frontal-lobe function (Teuber, 1964; Teuber & Mishkin, 1954). Ideally, a complete neuropsychological assessment should incorporate these new developments as they occur. Such incorporation, however, involves the step of going from experimental procedures as used in research studies to psychometric procedures with established validity, and with the other attributes usually expected of standardized psychological tests. These new developments have not reached that stage as yet.

The model being proposed has another feature that provides additional difficulties, particularly for naive localization theory. It has to do with the type of brain damage rather than the locus, and with the individual in whom the brain damage occurs. The general principle is that localization of any brain lesion does not provide all the necessary conditions for some particular behavioral outcome. Rather, the locus interacts with a number of other variables, all of which have some role in determining the resultant behavior. Thus, in attempting to understand the consequences of brain damage, it is necessary to introduce a number of independent variables that are often not considered in neuropsychological research. They include, but are not limited to, the process that produced the lesion, the age of the person, the age of the person at the time the lesion was acquired, and the age of the lesion.

We often tend to think of brain damage in terms of trauma. The neuropsychologist working with animals generally produces his lesions by extirpation of various areas. The extensive studies done by Teuber and his group have primarily involved individuals with penetrating missile wounds. The difficulty with the trauma model occurs when we try to generalize from traumatic brain damage to a particular region of the brain to lesions to the same area produced by other processes. When this is done, we often find that the behaviors we expect to find do not appear. As an example, we may take the case of aphasia. Earlier, we said that aphasia occurs when brain damage involves the left-cerebral hemisphere. However, if the process happens to be a demyelinating disease, particularly multiple sclerosis, we are very unlikely to see aphasia. Correspondingly, if the process is a stroke that occurred many years prior to the examination, it is also likely that there will be no apparent aphasia, despite the fact that the lesion might be in the speech areas of the left hemisphere. In reality, aphasia is usually seen in cases in which there has been localized destruction of cortical tissue that occurred rapidly and recently. Thus, we are likely to see it in cases of recent, open-head injury or strokes involving the left-hemisphere but less likely to see it in slowly progressive illness or residual brain damage. The variables making the difference here appear to be recency of lesion acquisition and the acuteness or chronicity of the lesion.

Other considerations include the age of the person at the time of lesion acquisition. An elderly man with impaired brain function related to his age will be less adaptable to, say, a traumatic injury than will a younger man with more healthy brain tissue. A lesion produced in a child before he has acquired certain skills will have different consequences from the same lesion acquired by an adult. These examples suggest that there is always a type-locus interaction, and that the behavioral outcomes of brain damage cannot be specified unless the nature of these interactions is considered. The clinician knows the significance of this point, because most of the brain-damaged patients he sees probably do not have well-localized traumatic lesions. Indeed, in the general population most brain damage is not produced by trauma but by various general systemic illnesses, such as atherosclerosis, infection, and various degenerative processes that are not fully understood (e.g., Alzheimer's disease).

In summary, we have presented a model of brain function and brain damage that utilizes some well-established principles of cerebral localization. In addition to these topographic principles, it also considers the underlying processes producing the brain lesion in relation to the topography. An attempt at applying this model in its most simple form was made by Russell, Neuringer, and Goldstein (1970), who attempted in the case of localization to discriminate among subjects with left-hemisphere, right-hemisphere, and diffuse lesions. With regard to the underlying process, they attempted to distinguish among subjects with acute, static, and congenital lesions. They were successful in that predictions made based on the principles forming the model were accurate at high levels of statistical significance. It is hoped that future research will allow for an evaluation of the model in a more sophisticated form.

The Neuropsychological Test Battery as a Tool for Testing the Model

The material presented above was not developed prior to empirical investigation but rather evolved as a result of a large number of studies of individuals with brain lesions. Comprehensive reviews of these studies are published elsewhere (Boll, 1981; Reitan, 1966a; Russell, Neuringer, & Goldstein, 1970) and need not be repeated here. The general effect of this body of research has been that the battery of tests devised by Halstead (1947) with supplements provided by Reitan and various associates (Reitan, 1966b) is currently probably our best instrument for description of various parameters of brain dysfunction. It can not only detect the presence or absence of brain damage but also lateralized lesions and provide accurate predictions of the underlying process. It provides sufficiently rich material to allow for description on an independent basis of various parameters of brain dysfunction in individual cases. The experienced neuropsychologist, using test data alone, can frequently accurately predict the site of the lesion and the specific underlying process. In addition, he can provide a profile of the subject's deficits and preserved abilities. It has thus become possible to produce accurate descriptions of brain lesions solely on the basis of behavioral indices, when such indices are integrated into a viable model of brain function.

Before describing the promised next step of using these test data in the planning of rehabilitation programs, it is necessary to describe the tests, and what they do individually and in combination. Without such description, the relationships between test patterns obtained and the rehabilitation recommendations would remain obscure.

The Halstead Battery has gone through many versions over the years. However, the following tests from the original battery are the ones that remain in common use: (1) the Category Test, (2) the Tactual Performance Test, (3) the Speech Sounds Perception Test, (4) the Seashore Rhythm Test, and (5) the Finger Tapping Test. The Time Sense and Flicker Fusion Tests that were part of the original battery are not described here because in several studies they have been shown to be of limited sensitivity. Reitan supplemented the Halstead Battery with a number of other procedures, including the Wechsler-Bellevue Intelligence Scale, the Trail Making Test, the Reitan-Heimburger Aphasia Test, and a series of tests of elementary perceptual skills, including fingertip number writing, finger discrimination, and tests of tactile, auditory, and visual imperception. Many users of the battery now substitute the Wechsler Adult Intelligence Scale (WAIS) or WAIS-R for the Wechsler-Bellevue. Reitan also added the procedure of doing the Finger Tapping Test with the nonpreferred hand following completion of the test with the preferred hand.

We will provide brief descriptions of each of these tests, along with some indications of what they are thought to measure and of how they are used in neuropsychological interpretation.

The Category Test

This test measures the ability to identify the concept or principle that governs a series of geometric forms, or, in some cases, verbal or numerical material. It consists of seven groups of stimuli. A specially built apparatus is used, consisting of a display screen, four horizontally arranged numbered switches placed beneath the screen, and a control console. The subject's task is to press the switch that the picture on the screen suggests to him. If he presses the correct switch, he hears a pleasant chime. Wrong answers are associated with a rasping buzzer. The goal is to always get the chime, which can be achieved through identifying the concept that unites all the figures in the series. The conventional score on this test is total number of errors for the seven series combined.

The Category Test is quite complex, and taps a variety of functions and abilities. However, for the most part, it appears to be a test of the capacity to abstract, and thus is similar in its essential purpose to the tests of K. Goldstein and Scheerer (1941). One of the more frequent effects of brain lesions, regardless of locus or type, is impairment of the ability to conceptualize or to generalize from individual instances to some rule or principle. It appears to be a good measure of the individual's capacity to cope in a general way with the complexities of a normal environment. Some terms appropriate in describing performance on this test would include capacity to organize, planning ability, ability to transcend the immediate stimulus, judgmental capacity, and ability to solve complex problems.

From a diagnostic standpoint, the Category Test is probably the best test in the battery for discriminating normal from brain-damaged individuals. Although there has been much research done concerning the relation between abstraction ability and frontal-lobe function (Halstead, 1940, 1947; Rylander, 1939) nothing definitive has emerged. In general, the Category Test does not appear to have any particular localizing or lateralizing value (Doehring & Reitan, 1962; Shure & Halstead, 1958). It appears to be assessing a general or nonspecific effect of brain damage, and performance may be impaired with a lesion anywhere in the cerebral hemispheres. Individuals with extremely poor Category Test scores often suffer from extensive brain dysfunction that often is recently acquired or rapidly progressive in nature. Such individuals may have substantial difficulties in managing their affairs in an independent manner and generally require some kind of structured or sheltered setting. Individuals with relatively good Category Test scores, but who do have brain lesions, usually have relatively discrete lesions that are not rapidly progressive. These individuals may have specific cognitive, perceptual, or motor deficits but can often function independently in the community. Improvement on the Category Test on retesting indicates, as would improvement on the other tests, that from a neurological standpoint, the individual probably does not have a progressive lesion. Behaviorally, it means that he has recovered in a general way, is probably better organized, and can deal with his environment in a more effective, independent manner.

The Tactual Performance Test

The material for this test consists of a modified version of the Seguin-Goddard formboard. The test procedure is divided into two parts. The first part consists of having the subject place the ten blocks into their proper locations on the board. He must do this blindfolded, without having seen the board previously. The procedure is repeated three times, once with the preferred hand only, once with the nonpreferred hand only, and once with both hands. Time-to-completion for each trial and total time for the three trials constitute the conventional scores for this part of the test. However, some investigators prefer using minutes-per-block as the score, since some subjects do not complete the test and thus cannot get an accurate total time score. The second part of the procedure involves drawing the formboard from memory. The board is removed, the blindfold taken off, and the subject is asked to draw a picture of the board containing as many blocks as he can remember. He is also asked to draw the blocks in their proper location on the board. There is no warning at the beginning of the test that he is going to have to do this. Two scores are derived: the Memory score—or number of blocks correctly recalled—and the Location score—or number of blocks drawn in their approximately correct position on the board.

Much information can be derived from the Tactual Performance Test. There is a measure of the subject's capacity to use each hand individually, and thus an opportunity to compare functioning of the two sides. The third performance trial provides an opportunity to examine the subject's ability to coordinate one hand with the other. The drawing of the board provides material to evaluate the subject's perceptual-motor coordination and spatial relations abilities as well as to assess some aspects of memory. There is more that can be learned based upon additional information about this test. In general, normal individuals show improvement from trial to trial. If they do the first trial with their right hand, they should do better on the second trial, even though it is done with the nonpreferred hand. Additional improvement is expected on the third trial, which is done with both hands. Brain-damaged individuals frequently do not show trial-to-trial improvement. Although the research evidence (Reitan, 1959a) suggests that brain-damaged individuals generally do show improvement that does not differ on an absolute or proportional basis from what is found in normals, clinical analysis of individual cases often reveals that such improvement does not occur, or occurs to a lesser degree than is generally found in normals. It is likely that there is a difference between those that do and do not improve in regard to capacity to learn from prior experience. At times, performance on the third (both hands) trial does not improve over the second trial, and indeed sometimes gets worse. In these cases, it can be inferred that the subject has difficulty in coordinating the operation of his two hands, sometimes to the extent that one hand actually interferes with the functioning of the other, thereby detracting from efficiency of performance. Prognostically, those brain-damaged subjects who show trial-to-trial improvement may respond more favorably to relearning-type rehabilitative procedures than do those who do not.

The right–hand versus left–hand comparison of Tactual Performance Test performance can be useful in lateralization of brain lesions. Reitan (1959c) reported successful results in this regard, but Goldstein and Shelly (1973), utilizing a slightly different method of data analysis, were able to classify only 50% of their cases, all of whom had lateralized lesions, into the correct group on the basis of the discrepancy in performance between hands. This result is exactly equal to chance. Clinically, the right-left performance discrepancy can be viewed as a weak lateralization indicator that should not be used to predict a lateralization without supportive evidence from other test results.

Like the Category Test, the Tactual Performance Test tends to be done poorly by individuals with lesions anywhere in the cerebral hemispheres. The reason for this is somewhat puzzling since one might reason that a tactile task would be more sensitive to dysfunction of the parietal lobes than it would be to equal degrees of dysfunction elsewhere in the brain. Although there is some evidence for this inference (Teuber & Weinstein, 1954), it is nevertheless true that Teuber and Weinstein also found that all their brain-injured groups (divided into subjects with anterior, intermediate, and posterior focal lesions) did significantly more poorly than control subjects. It is possible that the nonspecific relationship between Tactual Performance Test performance and brain localization has to do with the more general aspects of the task. The subject is confronted with a problem of a type he probably never had to solve before. It is rare to have to solve a complex problem by sense of touch alone. Thus, the novel problem-solving aspect of the task may be more pertinent to an explanation of its discriminatory power than the fact that it involves the tactile modality. It is clear that many brain-damaged individuals with intact tactile sensation and normal motor function do exceptionally poorly on this test. For this reason, it seems necessary to look at the Tactual Performance Test as a measure of complex problem-solving ability rather than as a test of tactile perception or tactual-motor coordination.

The Memory and Location components of the test are also good discriminators of brain damage, but the general ability they are tapping is unclear. On the basis of factor analytic studies, no clear picture emerges, except to say that these measures either load on their own factor (G. Goldstein & Shelly, 1971) or on a factor with other complex problem-solving tests (G. Goldstein & Shelly, 1971, 1973). There is some evidence that individuals with right-hemisphere lesions do more poorly on the Location measure than do individuals with left-hemisphere lesions, but this discrimination cannot be made with the Memory measure (G. Goldstein & Shelly, 1973). It would not seem appropriate to simply treat these measures as indices of memory ability, since on the face of it, they seem to be tapping a highly specialized form of memory. Further specification may be made by calling it intermediate-range, nonverbal, incidental memory, but even this description may not capture the essence of the task. Perhaps the most pertinent consideration is that the normal brain can incorporate and store material presented in one modality (tactile) and output it in a different modality (visual-motor).

Unlike the Category Test and the performance component of the Tactual Performance Test, we can offer no clear behavioral correlates of good and poor performance on the Memory and Location components. Poor performance does not necessarily mean that the person will have a poor memory in the sense that we usually use this expression. The Location component in particular probably bears some relationship to spatial imagery, and may be important in regard to performing certain analytic and synthetic tasks such as making assemblies, comprehending how the parts of a complex object fit into the entire configuration, and, perhaps to some extent, mathematical ability. Many clinicians and researchers have noted a relationship between spatial imagery and ability to manipulate numbers mentally, but we know of no systematic neuropsychological research in this area. In essence, the

Memory and Location components are sensitive indicators of brain dysfunction, and improvement on retesting probably denotes some generalized recovery as well as some improvement in the ability to store and recall spatial configurations. However, the adaptive significance of this area of functioning has not as yet been clarified by appropriate investigations.

The Speech Sounds Perception Test

The stimulus material for this test consists of a series of 60 multiple-choice items, each of which contains four nonsense words. A tape recording is played to the subject that presents, for each item, one of those four words. The subject's task is to underline the word he hears. The nonsense words themselves are all constructed of a consonant prefix, a double *e* digraph, and a consonant suffix (e.g., ween). The prefixes or suffixes may contain one letter, as above, or two letters, as in "geend." For each item, aside from the correct word, there is a word in which the prefix is correct but not the suffix, a word in which the suffix is correct but not the prefix, and a word in which both prefix and suffix are incorrect. The subject must listen to each word presented on the tape, read the alternative answers, decide on the one he feels he heard, and underline that answer on his multiple-choice form. Therefore, while the test seems to be mainly a measure of the ability to discriminate among sounds, it also requires reading ability and ability to remain attentive throughout the length of the task (13 minutes). Since the test is not a measure of auditory thresholds, the volume at which the tape is played may be increased or lowered to suit the subject. Whereas severe deafness may rule out use of the test, valid performances can be achieved by individuals with substantial hearing losses.

Assuming that the subject can read the items and that he can maintain attention, the aim of the Speech Sounds Perception Test can readily be achieved. It is that of assessing auditory discrimination: the ability to detect the difference among sounds. Referring to our model, it will be recalled that although the temporal lobes mediate hearing, the speech area is generally in the left-cerebral hemisphere. Therefore, this test would appear to be a measure of the intactness of the left-temporal lobe. Although we know of no formal research study that demonstrated this relationship, there is a good deal of clinical evidence for it. In its most extreme form, difficulty in auditory discrimination is seen in individuals with receptive or Wernicke's aphasia. Goodglass and Kaplan (1972) state that this type of aphasia is associated with a lesion in the posterior portion of the first temporal gyrus of the left-hemisphere. The primary symptoms are impaired auditory comprehension and paraphasia, but these authors point out that the paraphasia may be secondary to the auditory comprehension defect. Individuals with this disturbance may be impaired in the ability to monitor their own auditory output. In many individuals with left-hemisphere lesions, we do not see a full-blown aphasia but more subtle difficulties in interpreting the speech of others. These individuals are not deaf in the usual sense, but cannot integrate the incoming auditory signals in a normally precise and rapid manner. Frequently, the Speech Sounds Perception Test is sensitive to these subtle difficulties in instances in which a standard aphasia test may not be.

The Seashore Rhythm Test

This test was borrowed from Seashore's tests of musical talent. It is generally administered on tape immediately after the Speech Sounds Perception Test. The stimulus material consists of 30 pairs of rhythmic patterns. The two members of the pair may be the same or different, and the subject's task is to write an *S* on his answer sheet when the members of the pair are the same, and a *D* when they are different. The total task involves listening to each pair, deciding whether the two parts are the same or different from each other, recording the decision in writing, and waiting for the next pair. The pairs are given in relatively rapid succession, and the tape may not be stopped once it has been started. It is necessary for the subject to both keep up and keep the pairs separate. Unlike the Speech Sounds Perception Test, the items are not numbered and so the subject must keep track of where he is. The answer sheet is divided into three columns, and the letters of these columns are announced at the beginning of each one. Thus, there is some possibility of reorientation but it is a limited possibility.

The Rhythm Test discriminates well between individuals with and without brain damage (Reitan, 1955); why it does so is far from clear. Reitan (1966a) describes it as follows: "This test appears to require alertness, sustained attention to the task, and the ability to perceive differing rhythmic sequences" (p. 167). It is likely that the former two points are more significant than the last one. Perhaps the subject's sense of musical rhythm is the least important factor in regard to the question of identifying brain damage. We might add to the above description the possible role of short-term memory, since the subject must remember the first pattern in order to compare it with the second. In general, we tend to use the Rhythm Test as a measure of the capacity to sustain task-oriented attention. It differs from the kind of function assessed by recollection of digits in that it requires a lengthy, sustained effort rather than an immediate response to a single-stimulus set.

To our knowledge, no investigator has implicated any area of the cerebral hemispheres as crucial for performance on the Rhythm Test. It thus plays a very different role in the battery from the Speech Sounds Perception Test, despite some superficial similarities. It is more of a test of brain damage in general, and in ths way resembles the Category Test.

Subjects who do poorly on the Rhythm Test may be characterized as readily distractible, unable to concentrate, and as having a limited attention span. Improvement on the test may indicate that the subject adopted more of a task-oriented attitude, and is now capable of more complex learning and retraining since he is better able to maintain concentration.

The Finger Tapping Test

The apparatus for this test consists of a mechanical counter to which a typewriter key is attached. The subject is asked to tap as rapidly as possible for a series of 10-second trials. Utilizing the standard Halstead apparatus after proper calibration, the normal person should be able to tap about 50 times with his preferred hand, and perhaps five to seven fewer times with his nonpreferred hand, during a 10-second trial. The original standard administration of this procedure calls for the administration of five trials with the preferred hand. Reitan added an additional five trials for the nonpreferred hand. In some laboratories, trials are continued with a maximum of 10, until there are five consecutive trials all of which are within five taps of each other. If this is not achieved by the 10th trial, the five consecutive trials that come closest to this standard are used. The important points about this test are that the subject must use only his index finger (i.e., not his entire hand or arm) and that he is encouraged to tap as rapidly as possible. Attempts to eliminate the effects of fatigue are made through providing adequate rest periods between trials.

Clearly, the Finger Tapping Test is a measure of pure motor speed. Reduction in speed of many components of the motor system frequently accompanies brain damage of many kinds. Therefore, the Finger Tapping Test is a good indicator of brain dysfunction in general. However, it serves a number of other useful diagnostic functions in the battery. It provides an opportunity for comparison of the right and the left hemispheres, since it involves measurements for both the right and left hand. Unlike the Tactual Performance Test, G. Goldstein and Shelly (1973) found that the Finger Tapping Test discriminated between individuals with right- and left-hemisphere lesions at a statistically significant level. In clinical interpretation, reduced motor speed is often used as a sign of "long tract" pathology. That is, reduced tapping speed is seen more often in individuals with cerebral vascular or demyelinating diseases and less often in cases of cortical atrophic and degenerative diseases.

From the point of view of rehabilitation, tapping speed is very important as an index of recovery from hemiplegia or hemiparesis. It provides a quantitative measure of the return of function to the involved limb. In a more general way, it provides information with regard to how much faster or slower the patient can move.

The tests discussed thus far constitute the part of the Halstead-Reitan neuropsychological battery taken from Halstead's original tests. The remaining tests are those added to the battery by Reitan.

The Wechsler-Bellevue or Wechsler Adult Intelligence Scale

It is not necessary to describe these tests here as they are well known and described elsewhere (Wechsler, 1944, 1955b).[1] We will therefore provide some information about how the Wechsler Scales are used in neuropsychological interpretation.

The first point to be made is that the Wechsler is not used as the indicator of presence or absence of brain damage. The Halstead tests do a better job of making this discrimination (Reitan, 1959b). The Wechsler is used to provide additional information that the Halstead tests do not elicit. The Wechsler provides a great deal of data concerning verbal intelligence; it also provides a good test of psychomotor speed—Digit Symbol—and an excellent test of spatial relations ability—Block Design. Since Reitan's original publication on the subject, the relationship between the verbal and performance scores and cerebral lateralization has become a much discussed issue (Reitan, 1955). The general principle is that in individuals with brain lesions, if the verbal IQ or verbal-weighted score is lower than the performance IQ or performance-weighted score, the lesion will be lateralized in the

[1]We will use the term "Wechsler" or "Wechsler Scales" to denote studies done with either the Wechsler-Bellevue or the WAIS.

left-cerebral hemisphere. In individuals with right-hemisphere lesions, the reverse relationship is said to hold. Individuals with diffuse lesions have little difference between their verbal and performance scores. The research emanating from this finding is reviewed elsewhere (G. Goldstein, 1974; Reitan, 1966a; Russell, Neuringer, & Goldstein, 1970), and will not be elaborated upon here. The current conclusion appears to be that these principles work for individuals with acute lesions, such as recent strokes and rapidly progressive cerebral neoplasms, but not for static lesions, such as old head injuries and old strokes. For our purposes here, it may be said that the verbal-performance relationship should be looked at as one of the many indicators of lateralized dysfunction to be found in the neuropsychological test battery.

Some other features of the Wechsler that are interesting from a neuropsychological standpoint are that the Digit Symbol subtest is particularly sensitive to many kinds of dysfunction, and is of special importance when the score on it is lower than on Block Design, Picture Arrangement, and Picture Completion. Rennick (cited in Russell, Neuringer, & Goldstein, 1970) developed a system for scoring the Digit Symbol subtest in relation to Block Design, Picture Arrangement, and Picture Completion. This score is converted to a rating and included as one of the scores making up a version of Halstead's Impairment Index. There is clinical evidence for relationships between certain of the subtests and certain of the lobes. Block Design is thought to be a parietal-lobe indicator; Picture Arrangement, a right temporal lobe indicator; and Similarities, a left-temporal lobe indicator.

Clinically, the IQ scores are often useful in assessing premorbid level of functioning. Unless the subject is aphasic, the verbal IQ often gives a good indication of the level of functioning prior to acquisition of the brain damage. There is also some degree of correlation between IQ and some of Halstead's tests. Most clinicians, in interpreting neuropsychological test material, take into consideration the IQ level. A score of 100 errors on the Category Test (52 errors is the upper limit of normal) takes on different significances depending on whether the subject has a 120 or an 85 Full-Scale IQ. In general, one might say that the IQ level has a tempering effect on interpretation of the other tests. From the standpoint of rehabilitation prognosis, this consideration is of extreme importance. We may be optimistic about recovery from brain damage, but we cannot reasonably expect the subject to surpass his premorbid level. If the Verbal IQ in particular is low, it is often the case that it was also low prior to acquisition of the brain lesion. In these cases, we cannot reasonably make demands on the subject to do better than would be expected of an individual with his premorbid level.

The Trail Making Test

This test consists of two parts, A and B. Part A consists of a sheet of paper containing circled numbers scattered randomly over it. After going over a sample containing the same type of material, the subject is asked to connect the numbers in order as rapidly as possible. Performance is timed. Part B is similar to Part A, but more complex. The sheet of paper contains both numbers and letters. The task is to go from a number to a letter to a number, for example 1 to A to 2 to B, and so forth. Again, time-to-completion is the score. The usual scores for this test are time-in-seconds for Part A and for Part B. Some investigators use the total of the two.

Research indicates that the Trail Making Test is a highly valid indicator of brain damage (Reitan, 1955, 1958a). In this regard, the score on Part B is a far superior discriminator than is the score on Part A. Reitan and Tarshes (1959) also found that individuals with left-hemisphere lesions did more poorly on Part B relative to Part A than did individuals with right-hemisphere lesions. While subsequent clinical investigation did not strongly support this finding, it still may be useful as a lateralization indicator within the context of other test results.

While the Trail Making Test is an apparently simple procedure, analysis of the kinds of performance required leads to the conclusion that it is actually rather complex. At the most basic level, on Part A, the subject must know the sequence of numbers. Aside from this, he must perform the task rapidly, and so the element of psychomotor speed enters. Part B requires more than A in regard to symbolic functioning and also involves psychomotor speed. Perhaps more significantly, Part B requires an ability that K. Goldstein and Scheerer (1941) have referred to as "simultaneous function" or the ability to carry on two activities concurrently. In the case of the Trail Making Test, Part B, the subject must keep the number and letter sequences in mind at the same time. Still another feature of the task is that it involves thinking while involved in action. Decisions must be made as the task is proceeding, and perhaps we are seeing some model of what we would describe in a natural situation as "thinking on your feet."

Improvement on the Trail Making Test is a good sign of general recovery and has implications similar to improvement on the Category Test. Generally, it does not simply mean that speed has been increased, as we might conclude from improvement on Finger Tapping, but rather that the subject is more capable of organizing his behavior and of engaging in effective goal-directed activity. It is interesting to note that in a factor-analytic study (G. Goldstein & Shelly, 1972), Trail Making, Part B, and the Category Test had very similar patterns of factor loadings. There was also a +.58 correlation between the two measures.

The Reitan-Heimburger Aphasia Screening Test

This test provides a rapid survey of the major language functions. It is a modified and shortened version of a test developed by Halstead and Wepman (1949). The term *screening* is used advisedly, and neither the Halstead-Wepman nor the Reitan-Heimburger procedures should be viewed as substitutes for the many more extensive aphasia examinations currently available (e.g., Goodglass & Kaplan, 1972; Holland, 1980). The test material consists of a small booklet and a recording sheet, and a blank sheet of paper and a pencil to be used by the subject. The booklet contains stimulus material for the various test items which evaluate such functions as word finding, spelling, handwriting, right-left discrimination, calculation, reading, and other language-related behaviors. Generally, the test is not formally scored, but is reviewed by the neuropsychologist who typically lists the various aphasic symptoms found. However, Russell, Neuringer, and Goldstein (1970) have constructed a preliminary scoring system that may be of some assistance to individuals unfamiliar with the major aphasic syndromes. The researcher has the choice of using either this system or the number of symptoms as his quantitative measure.

There is little question here concerning what the test measures. It measures language and language-related abilities, and thus may be viewed as an examination of the left-hemisphere. It is not an adequate examination for localization within the hemisphere based on the pattern of aphasic symptoms, and it should not be used for diagnosis of aphasia, since it does not clearly differentiate among the various aphasic syndromes. Although certain inferences can be drawn on the basis of clinical correlation of various items, there is no objective, systematic means of reaching a definitive diagnosis. For example, one section of the test includes having the subject read the sentence, "Place your left hand to your right ear," and then performing this action. If the patient fails, it may be because he has alexia (inability to read), a right-left confusion, an ideomotor apraxia (inability to engage in purposeful movements), or a body agnosia (inability to identify body parts). Clinical observation and correlation with other test items are needed to decide among these alternatives.

With regard to the rehabilitation question, language is often a crucial factor. Fortunately, aphasia is frequently not a permanent residual of brain damage, and recovery may often be rapid and almost complete. Indeed, if aphasia does not disappear perhaps within a year of the time of acquisition of the lesion, one might become concerned about the possibility of a progressive process, particularly if it can be shown that the person did not have language difficulties premorbidly. Recovery can be expedited by starting the individual in speech therapy as soon as possible after the acquisition of the lesion. While there is some degree of natural recovery, there is little question that an active speech therapy program can hasten the recovery and enhance the possibility that it will be complete.

Perceptual Disorders

This portion of the battery consists of a number of brief tests of perceptual skills in the tactile, auditory, and visual modalities. These tests all involve comparisons of the two sides, and therefore provide additional lateralization information. Certain test procedures that follow have become a part of the standard battery.

The Fingertip Number Writing Test. The subject is asked to close his eyes and to identify by touch numbers written with a pencil on his fingertips. There are 20 trials administered for the right hand and 20 for the left, in a prescribed order. The score is the number of errors made by each hand. This test provides a simple measure of tactile recognition ability. When this ability is found to be impaired to a substantial degree, the term applied to the condition is *astereognosis* or *tactile agnosia.* In the case of the present test, the term tactile agnosia is more precise, since astereognosis technically refers only to the loss of ability to recognize three-dimensional objects by touch. Although the parietal lobes are the primary mediators of tactile discrimination, clinically, a deficit on this test is seen in individuals with lesions in a variety of sites.

The Finger Discrimination Test. The subject is taught a numbering system for his fingers; thus, the thumb is "1," the forefinger "2," and so forth. He is then asked to close his eyes and the examiner randomly touches individual fingers as the subject calls out the number of the finger touched. Twenty trials are administered for the right hand, and then the procedure is repeated for the left hand. Again the score is the number of errors made by each hand. The purpose of this test is that of detecting a condition known as *finger agnosia.* The nature of finger agnosia is well described by A. L. Benton (1959). It is defined in detail in Benton's book, but the most basic definition is, "relative or absolute inability to show or name the individual fingers" (p.2). In its classic form, bilateral finger agnosia is thought to reflect a disturbance of the body schema, but it may simply represent a tactile discrimination deficit, particularly if it is unilateral.

Tactile, Auditory, and Visual Imperception Tests. The basic phenomenon evaluated by this test is that in some cases of brain damage, single stimulation—be it tactile, auditory, or visual—is normally appreciated, but a stimulus to the same area may not be perceived when it is applied simultaneously with a similar stimulus to another area. This effect has been known for some time, but was popularized and carefully studied by Bender (1952). In the case of the tactile modality, the subject is lightly touched on either his hand or his cheek. Intensity of touching may be increased until the subject reliably reports feeling the stimulus. Then, double stimulation, which may consist of touching a hand and a cheek or both hands simultaneously is intermingled with single stimulation trials. The question to the subject is always, "Where did I touch you?" A number of trials are run, involving various combinations such as right hand-left hand; right hand-left cheek, and so forth. The score consists of the number of times the subject fails to perceive one of the stimuli on those trials in which double stimulation is administered. Misses on the single-stimulation trials are also recorded but are not counted as instances of imperception.

Similar procedures are used for the auditory and visual modalities. In the case of the auditory imperception test, the tester stands behind the subject and rustles his fingers with his right hand, his left hand, or both hands simultaneously. Again, the subject may hear the sounds normally when each ear is stimulated individually, but he may not report hearing the rustle in one ear or another under conditions of double stimulation. In the case of the visual test, the subject sits facing the tester, who extends his arms into the edges of the subject's visual fields. The stimulus is a small movement of the thumb and forefinger, and the subject is asked to report whether the movement took place in his right or left field. Trials in which only one hand is moved are intermingled with trials in which both hands are moved simultaneously. The procedure is repeated for the upper, middle, and lower visual fields, with the number of occasions on which the subject reports seeing movement of one hand only, when there was movement of both, constituting the score.

In general, individuals without brain lesions can perform the imperception tasks without any errors, or perhaps with a few initial errors. On the other hand, consistent production of errors often can denote severe pathology of the cerebral hemispheres. The comparison of right and left sides can often help in identifying the lateralization of the pathology. On retesting, reduction in amount of imperception may reflect recovery from an acute process. The tests of imperception do not appear to be measuring primary sensory functions in that the subject performs normally on single stimulation. The defect is more subtle, and possibly exists at the level of attention. That is, the subject who demonstrates imperception may be impaired in his ability to deploy his attention to more than one area at a time.

General Indices of Impairment

In order to obtain a global index of level of impairment, a summary score called the Impairment Index or the Average Impairment Rating is computed. The former term is the one introduced by Halstead (1947); the latter was introduced by Rennick (cited in Russell, Neuringer, & Goldstein, 1970) and is computed in a slightly different manner from Halstead's method. In the case of both indices, the logic used is that the more tests performed in the brain-damaged range, the greater the likelihood that there is a lesion present. That is, the normal individual may perform on one or two of the tests at a level that would put him past the cut-off score for brain damage. However, as the number of tests scored in this range increases, so does the probability that some kind of lesion is present. Since these indices are based on all the tests, they provide the most sensitive indicators of presence or absence of brain dysfunction. However, they do not provide information concerning the type or locus of damage.

In the case of the Average Impairment Rating, computation must be preceded by a preliminary step in which all the tests are converted to a 5-point rating scale, the points corresponding to excellent, normal, mildly impaired, moderately impaired, and severely impaired performance. In most cases, the ratings are based on level of quantitative test scores, but in a few cases they reflect judgments made by the neuropsychologist on the basis of his review of the test material. Ratings are made for the following tests:

1. Category Test
2. Tactual Performance Test—Total Time
3. Tactual Performance Test—Memory
4. Tactual Performance Test—Location
5. Speech Perception
6. Rhythm
7. Finger Tapping

8. Trail Making, Part B
9. WAIS Digit Symbol
10. Aphasia Screening
11. Spatial Relations
12. Perceptual Disorders

The first nine measures are rated on the basis of quantitative cutoff scores; the last three are based on judgments made by the neuropsychologist (cf. Russell, Neuringer, & Goldstein, 1970). The Average Impairment Rating is computed by simply summing the ratings and dividing by twelve. Average Impairment Ratings of 1.35 or below are considered as normal. Anything between 1.36 and 1.55 is borderline, and values of 1.56 and above are considered to be clearly within the brain-damaged range. Change for the better on retesting suggests that the subject has improved in a general way, but the Average Impairment Rating does not provide information concerning the nature of that impairment. It could be based on substantial improvement on one or two tests, or on small amounts of improvement on many of the tests. In order to assess the pattern of improvement, it is best to look at the profile of ratings.

Neuropsychological Tests and Rehabilitation Planning

With this brief introduction to a neuropsychological test battery, it is now possible to describe how level and pattern of performance may be used in the formulation of training programs. It may be reiterated that in many clinical settings, psychological testing is used primarily for the descriptive and diagnostic information it provides. Those who formulate and implement the treatment plan tend to be different individuals from those who administer and interpret the psychological tests. For purposes of presentation, it may be of value to separate the diagnostic material needed for formulation of a training plan into three areas: the premorbid level, the current level of performance, and the performance pattern. In other words, it is important to know what the person was like before he acquired his brain lesion, how much impairment he has in a global way, and what his current status is regarding the pattern of preserved and impaired abilities. We will try to describe how neuropsychological tests are used to determine each of these factors.

Estimating the Premorbid Level

When one assesses level of functioning in a subject, there is always some question as to the relative contributions of his functioning prior to acquisition of the identified brain lesion, and his functional level after its acquisition. On the surface, it would appear that this question can be readily resolved, not through the neuropsychological assessment but through inspection of records and social history data. Indeed, Wilson, Rosenbaum, and Brown (1979) have demonstrated that a demographic estimate of premorbid IQ based on age, sex, race, and education is superior to an estimate based on IQ test indices thought to reflect premorbid level. This may be the ideal approach but there are a number of practical difficulties. First of all, there may simply be unavailability or nonexistence of medical records or other important documents. An informant might not be available. The subject may not be in a condition that allows him to report accurately on his past life. There is also a whole complex of considerations that revolve around the reliability of the informant, who for various reasons will not or cannot provide accurate, unbiased information. For these reasons, attempts have been made by students of psychological-assessment methods to develop techniques based on contemporary test data for predicting the premorbid level. An early effort at accomplishing this goal revolved around the so-called Babcock Hypothesis (Babcock, 1930), which, simply stated, suggests that vocabulary level is a stable, nondeteriorating aspect of mental functioning. Therefore, it may be used as a contemporary standard against which degree of deterioration, as measured by other kinds of tests, can be estimated. Babcock developed a test based on this hypothesis, in which the subject's scores are compared with those obtained by normals of the same "vocabulary age". Wechsler (1944) also developed a differential test score method of measuring mental deterioration. Using the Wechsler-Bellevue Scales, he distinguished between tests which hold up with age and those that do not: the so-called Hold and Don't Hold subbatteries. The Hold tests are Information, Comprehension, Object Assembly, Picture Completion, and Vocabulary. The Don't Hold tests are Digit Span, Arithmetic, Digit Symbol, Block Design, Similarities, and Picture Arrangement.

The problems with these methods of estimating degree of deterioration revolve around both theoretical and empirical issues. In regard to the Babcock Hypothesis, Yacorzynski (1941) pointed out that the Vocabulary test accepts as correct word definitions at several levels of abstraction. Thus, the deteriorated subject can continue to do well on the test and at the same time give much more primitive responses than he was capable of in the past. Wechsler, in a later edition of his book (1958), admitted that the Hold-Don't Hold distinction was not supported by a number of studies. Matarazzo (1972) pointed out that this situation had not changed during the intervening years. An additional difficulty with using verbal tests as an estimate of premorbid level of functioning is that doing so completely ignores the area of aphasia. One could go very far astray if vocabulary test scores are taken as estimates of premorbid ability in aphasics.

Despite these discouraging research findings, clinicians are often successful in distinguishing between optimal and deteriorated performance. It can readily be understood that it may be quite important to distinguish between the normally functioning individual who experienced rapid deterioration as a consequence of a brain lesion acquired during adulthood, from the individual with a history of brain dysfunction antedating the acquisition of an additional lesion during adulthood. In the latter type case, the lesion acquired during adulthood may often turn out to be a "red herring," with further evaluation revealing that the major adaptive difficulties found were also present prior to the lesion acquired in adulthood. For example, we treated a patient who was referred because he sustained a depressed skull fracture in an auto accident. However, it was later found that he had a history of mental retardation. It is clear that the rehabilitation planning for such an individual must be different from plans and expectations made for the premorbidly normal individual who sustained a lesion during adulthood.

In analyzing the way in which premorbid level is assessed clinically, these three points are relevant: (1) Despite the discouraging research findings, there are instances in which the vocabulary level or Verbal IQ do provide adequate estimates of premorbid level of functioning. (2) The full neuropsychological test battery provides more information for making judgments than does the WAIS given as an individual test. (3) Neuropsychological research and clinical practice have provided patterns of performance that characterize various diagnostic groups within the global category of "brain damage."

In our experience, estimating the premorbid level very frequently reduces to determining whether the level of performance seen at present can be attributed to recent brain damage or to factors that have been present since birth. In more neuropsychologically oriented language, we try to distinguish between acute and static lesions on the one hand, and congenital lesions on the other. In the case of the individual with a congenital lesion, language development is frequently severely impaired, and so we would naturally expect to find poor performance on verbal tests. Individuals with other types of lesions also do poorly on such tests. Earlier we referred to the relationship between lesions of the left-hemisphere and language dysfunction. The question then becomes one of whether the individual has language deficiencies on a developmental basis or an aphasia acquired during adulthood. Here is where the remainder of the neuropsychological battery comes into play. We can look at the aphasia test and see whether or not there is more direct evidence for adult-acquired aphasia than can be provided by verbal intelligence tests. We can also look at the tests for perceptual and motor dysfunction in order to determine whether or not there is evidence for left-hemisphere brain damage. In the case of acquired aphasia we would expect to find it; we would not in the case of congenital brain damage. The adult aphasic may also show exceptionally good abilities in some nonlanguage-related areas. The individual with congenital brain damage is less likely to do so. Russell, Neuringer, and Goldstein (1970) developed a set of rules for identifying congenital brain damage on the basis of neuropsychological test data. Essentially, these rules call for a level and pattern of test performance that is in the mildly to moderately brain-damaged range (i.e., the Average Impairment Rating must be more than 1.55 but less than 3.00), exhibiting the following alternative conditions: either the WAIS Full Scale IQ is less than or equal to 98, and mixed hand or hand-eye dominance is present, or the WAIS Full Scale IQ is less than or equal to 80, and either perceptual and motor functions are relatively normal or there is a less than 6-point discrepancy between Verbal and Performance IQs. In less complex language, congenital brain damage is characterized by mild to moderate dysfunction on the neuropsychological tests, no more than minimal signs of lateralization, and a relatively low IQ. Many individuals with congenital brain damage have mixed hand, or crossed hand-eye, dominance, and the presence of either can be viewed as a diagnostic sign of congenital damage if the IQ is no more than 98.

In estimating the premorbid level, the clinician can also piece together bits of information on the basis of general knowledge of brain–behavior relationships. For example, the aphasia examination can provide numerous clues that can help the clinician to recognize the language of the individual who became aphasic during adulthood. If the subject cannot read following acquisition of a brain lesion, it is likely that he could not read before it either, unless a rarely occurring pattern of symptoms involving other signs of neurological dysfunction is also present. Loss of the ability to read, or acquired dyslexia, as an *isolated* symptom, is a rarely occurring consequence of brain damage sustained during adulthood. These considerations naturally bring up the matter of the relation between neuropsychological test performance and education. Boll (1981) has reviewed the literature in this area and has shown that the findings are rather complicated. However, in two studies (Finlayson, Johnson, & Reitan, 1977; Prigatano & Parsons, 1976) it was shown that education was not significantly related to neuropsychological test performance in brain-damaged individuals, although there are significant correlations in the case of nonbrain-damaged individuals. Thus, deficits exhibited on neuropsychological tests by brain-damaged patients are generally not readily attributable to limited education.

The significance of estimating premorbid level in regard to rehabilitation planning cannot be overemphasized. There is a great deal of difference between teaching someone something he never knew, and reteaching some previously learned skill or knowledge. Perhaps of greater significance is the level of achievement expected of the subject. The individual with congenital brain damage often has a lifelong history of failure in competition with his peers. If he is placed in a rehabilitation program along with individuals who became brain damaged during adulthood, he is again at a disadvantage if the goals set for him are the same as for the others.

Assessing the Amount of Impairment

Although we are using the term *brain damage* in a rather general way, we should remain aware that the term refers to a wide variety of conditions and an enormous range of severity. Certain investigators entirely reject the use of the term, preferring to refer only to the specific conditions in which they are interested at the time. Brain damage is viewed as a "wastebasket" term. One could defend the use of the term as defining a group of individuals who share certain common problems, but in general, specificity is needed to arrive at meaningful diagnoses and treatment plans. As an obvious illustration of this point, we might compare the case of the individual with "brain damage" resulting from a rapidly progressive malignant neoplasm, with the individual who sustained a head injury at some time in the distant past. In both cases "brain damage" is present, but in the first instance the individual is seriously ill, possibly terminal and in need of some radical form of treatment such as neurosurgery. In the second case, the person may be in good health, can be expected to live for a normal life span, and may not require any active form of medical treatment. One could raise some serious question concerning the meaningfulness of placing these two individuals in the same category—for whatever purpose.

In rehabilitation planning, one of the more significant parameters is extent of impairment. Certain kinds of brain lesions give rise to quite discrete, mild forms of deficit, whereas other kinds produce pervasive, severe impairments. It is important to know where the individual subject lies along this continuum. Generally, this task can be accomplished by constructing some global measure of level of performance. In the case of the neuropsychology battery, we have the Average Impairment Rating and the three IQs as measures of this type. Knowing these values, it is also important to know how they were produced. A high Average Impairment Rating can be produced by a pattern of mild to moderate deficit on a large number of the contributing tests, but it can also be the result of very severe impairment on a limited number of tests with relatively normal function elsewhere. The same principle holds for the IQ measures. This problem of the patterning of the test results will be discussed in the following section of this chapter. For the moment, let us ignore how the global scores were produced and concentrate on the scores themselves.

As a general principle, the global index is helpful in determining the mode of approach to the subject and in setting goals. Again resorting to an extreme illustration, most clinicians are not likely to approach the elderly, debilitated patient with an "organic brain syndrome" in the same manner that they would a younger patient with a discrete, focal brain wound. The level at which one carries on conversations with these individuals, the manner of giving instructions, the closeness of supervision, and many other aspects of the treatment relationship are carried on quite differently in these two cases. In general, the approach to the more impaired individual should aim at simplification of instruction, division of the task into small components, repetition as needed, and avoidance of abstract or symbolic conversation. The trainer should be a particularly active participant, and generally does best by not modeling his orientation on a nondirective basis. These standards can be increasingly relaxed in correspondence with the degree of intactness of the subject. In the case of the more impaired subject, it is important to avoid creating a situation that is overwhelming in its complexity. In such cases, it is possible that the subject will react with what K. Goldstein (1939) termed a "catastrophic reaction," something like a massive anxiety attack.

It is hoped that neuropsychological test level-of-performance data can prepare the trainer and aid him in modulating his initial approach. This modulation should occur eventually, but it is hoped that the tests will expedite the process. In some cases, the subject has built a strong facade to conceal his deficits, and it may take some time to discover their nature and extent. The tests tend to make an immediate penetration of this facade. Level-of-performance data also appear to bear some relationship to outcome. Recovery in the area of brain dysfunction is rarely complete, and so the level obtained at the beginning of training will have a strong influence on the ultimate result. Recovery to a normal level is usually an unrealistic expectation in the case of the severely impaired subject. It is more realistic for mildly impaired individuals. Another aspect of the problem has to do with the subject who has some specific deficit such as aphasia. Valuable prognostic information can be obtained from knowledge of the global level of performance. The aphasic individual who is otherwise relatively intact has a much better prognosis than does the more severely impaired individual. Prognosis thus need not be based solely on the severity of the specific symptom, in this case aphasia, but can be sharpened through obtaining an estimate of the subject's more general capacity to relearn.

In summary, level-of-performance data, such as the Average Impairment Rating or IQ scores, provide global indices of the extensiveness of deficit. Although they are combined summary scores, and so cannot lead to specific treatment recommendations, they are useful in planning the general approach to the subject and in setting realistic goals.

Determining the Pattern of Assets and Liabilities

Performance pattern constitutes the most complex level of diagnostic material needed for the planning of training programs. The general assumption is that retraining cannot be the same for all brain-damaged individuals but should be appropriate in some way to the individual's profile of abilities. These profiles can vary extensively, and it would not be feasible to map out a treatment plan for every possible variation. For presentation purposes, we will first distinguish among three classes of profiles. First, there is the type in which there is no apparent specific defect. Second, there is the type in which there is a specific defect in the area of language, and third, there is the type in which there is a specific defect in the area of spatial relations abilities. We might recall from the brain model described earlier that, from a neuroanatomic standpoint, we are referring to individuals with diffuse, left-hemisphere and right-hemisphere lesions, respectively.

The individual who is generally impaired, but who does not have a specific deficit, is probably the hardest one to plan for. In these cases, the neuropsychological test profile approximates a straight line, but level of performance can vary from mildly to severely impaired. These are the cases that we are hard put to say anything substantial about, since there is no specific deficit that can be pinpointed, nor is there an identifiable asset that can be exploited. In cases of this type, perhaps the best general plan is that of attempting to upgrade general intellectual functioning while exploring a number of areas that might appeal to the subject in regard to a vocational choice or further education. Areas that might be tried include abstraction ability, memory, psychomotor speed, perceptual-motor coordination, and other similar areas of general ability that are frequently more or less impaired in the individual with diffuse brain damage.

It is important to note that these are the people who tend to "fall between the cracks" in many clinical settings, since they are not typically prime candidates for treatment by the established disciplines of speech therapy, physical therapy, perceptual retraining, and so forth. The usual experience is that the more severely impaired individuals cannot benefit significantly from the specialized modalities, whereas the more mildly impaired may not have any problems with such specific areas as speech or motor function. More general treatment techniques such as Folsom's "Reality Orientation" (Folsom, 1968; Mitchell, 1966; Stephens, 1969; Taulbee & Folsom, 1966) would seem to be appropriate for the individual with severe diffuse brain damage. However, this method is designed primarily for hospital inpatients who are substantially confused, disoriented, and forgetful. It is not appropriate for individuals with mild to moderate generalized brain dysfunction, especially younger people. As suggested, training for such individuals should be organized around some vocational or educational goal and should emphasize the more general types of abilities. In the case of our own work, we did in fact initiate this type of training in such areas as memory, attention, and psychomotor speed. Details of these programs are described elsewhere in this volume.

Rehabilitation planning for the individual with an acquired language defect (as opposed to a developmental one) is more straightforward. In some cases, a course of speech therapy can constitute the major treatment. Such cases, however, occur relatively infrequently for a number of reasons. First of all, aphasia rarely appears as an isolated symptom. Particularly in the case of the stroke patient, it is often accompanied by paralysis or paresis of the right side of the body. Less frequent, but not at all rare, is loss of vision in the right visual field. In most cases there is also generalized loss of intellectual functioning, as would be measured by such instruments as the Category or Trail Making Test. Therefore, the individual who has acquired aphasia as an adult frequently does not only present his language problem, but motor, perceptual, and general intellectual deficits as well. To further complicate the matter, these individuals often have accompanying physical health problems. In the young stroke patient, there is often concern about the stability of the cerebral blood vessels, particularly if he is found to have a vascular malformation. Such individuals are high-risk cases for having additional strokes. In the middle-aged and older stroke patients, there is often concern about the extent of generalized atherosclerosis, and about cardiovascular functioning in general.

Individuals with aphasia resulting from head trauma may not have these general health problems, although they frequently have seizures. If this is the case, then they are subject not only to the distress associated with the seizures themselves, but also the limitations society places on epileptics in regard to vocational opportunities.

Traditional rehabilitation programs for individuals with left-hemisphere brain lesions emphasize speech therapy and physical therapy if there is hemiparesis. This approach is certainly a rational one, in that following the immediate medical intervention that occurred at the time of acquisition of the lesion, the major concern logically becomes restoring as much speech and motility as possible. During this phase of treatment, it is unlikely that the neuropsychological test results would lead to recommending any other course. The question remains, however, of what happens to the patient after speech and physical therapy have accomplished as much as they can. He may be discharged from the hospital (or from regular outpatient care) in order to go back out into the world. Many patients reach this point but do not go on to live productive lives. The neuropsychological tests, when given at this stage, almost always reveal substantial residual deficit in these individuals. In the cases of those with language problems, as in other cases, cognitive deficits not explainable on the basis of the residual aphasia are often noted, as are various elementary perceptual and motor abnormalities. It is at this point that the neuropsychological tests can be used as a means of documenting these deficits, and of developing recommendations concerning what more can be done.

In the case of the subject with a language problem, rehabilitation recommendations can often be based on three considerations: (1) the severity of the residual aphasia, (2) the presence or absence of a clearly definable compensating asset, and (3) the over-all level of performance. In those cases in which the degree of aphasia is substantial, a recommendation for continued speech therapy is usually made. In regard to vocational possibilities, one could reason that certain jobs require primarily verbal skills, while others require mainly spatial- and perceptual-motor abilities. This distinction is not always clear-cut, as in the case of typing, but it is useful. The subject with a substantial degree of residual aphasia is generally not a very good candidate for training in a verbal-type job. For such individuals, even the most apparently simple clerical work can be overtaxing, and the training effort would be doomed to failure. However, if this substantially aphasic subject had a clear asset, as might be revealed by, for example, an exceptionally high Block Design or Tactual Performance Test-Location score, he would be considered a good candidate for training in some reasonably high-level spatial-perceptual-motor-oriented occupation.

In regard to this analysis we would like to make two parenthetical points. First, the problem in vocational planning for brain-damaged individuals is often not one of finding tasks that are sufficiently simple, but rather one of finding tasks having the appropriate content. What we may think of as an exceptionally simple task, for example, placing cards in alphabetical order, may completely overwhelm the aphasic. Yet he might be able to perform what would be considered a highly complex task, such as constructing an elaborate device out of component parts. Second, the nature of aphasia is frequently misunderstood, since we frequently equate fluency and extensiveness of speech with level of intelligence. In the case of the aphasic this equation might be misleading. Many aphasics retain highly adequate intellectual functioning despite their language difficulties. Therefore, there is no reason to treat the aphasic as though he were functioning at a lower level than other brain-damaged individuals. There is, however, the problem of establishing a channel of communication with him, so that he can be worked with at an appropriate level.

A final major consideration is the global level of performance. We would look initially at the Average Impairment Rating in order to make this assessment, but we might turn to some of the individual tests as well. The point here is that of determining the level of the to-be-trained-for task. Spatial and perceptual-motor tasks cover a wide range from simple assembly line construction to high-level technical and artistic activities. If the subject has a clearly identifiable asset in this area, and a good Average Impairment Rating, we might recommend high-level training or no training at all, if we feel that the individual is ready simply to go out and work. To the extent that the reverse is true, we would recommend increasingly simple, less demanding vocational preparation. In individuals with gross perceptual or motor impairment, the Category Test may provide a more accurate estimate of general level than would the Average Impairment Rating. In this case, the Average Impairment Rating may be overinfluenced by the subject's physical handicaps, and fail to reflect his true level of intellectual functioning.

The Patient with a Spatial Relations Deficit

Many of the considerations mentioned concerning the individual with language difficulties also obtain for individuals with spatial relations problems. The major problem in this area is that there is no refined and sophisticated treatment procedure that can be viewed as equivalent to speech therapy. Therefore, training in spatial relations skills has to be more improvisatory and informal than training in speech. However, numerous methods suggest themselves once a clear understanding of the nature of the problem is reached. Spatial relations difficulty, or *constructional apraxia* as it is more technically called, means that the individual is impaired in his ability to organize material in three-dimensional space. The space referred to is external to his body, and so the term does not refer to right-left confusions and other disturbances of body spatiality. The term *spatial orientation* is sometimes used in describing this area of functioning, but may be somewhat confusing. Typically, the individual with constructional apraxia is oriented in that he knows where he is. His difficulty emerges when he must move around in his spatial environment, and so what we are describing is really a disturbance of action. Therefore, the essence of retraining involves getting the individual to do things; to engage in goal-directed behavior. Except in unusually severe cases, the individual can find his way around familiar locations, but has excessive difficulty in assembling parts into a desired entity. The defect is generally demonstrated in such tasks as block-building, puzzles, and copying from a model. In one patient we treated, the defect was observed in a difficulty in producing the proper orientation to a voice coming from behind. The difficulty expresses itself dramatically on the Halstead-Reitan neuropsychology battery in the subject's inability to copy a Greek Cross with one continuous line. In individuals with constructional apraxia, the finished version often reveals significant deviation from the proper spatial configuration of a cross, despite the fact that there is typically no difficulty with those components of the motor system needed to execute the drawing. That is, the difficulty is clearly not one of motor awkwardness, slowness, or unsteadiness.

Numerous possibilities present themselves as training devices:

jigsaw puzzles, maps, and games of various sorts. If one particular goal of the training is that of specifically enhancing spatial relations abilities, the most important criterion for choosing training materials is the absence of cues to problem solution that are verbal in nature. Many apparently "nonverbal" tasks are highly subject to solution by verbal mediation. For example, the Category Test is an apparently nonverbal task, in that the stimulus material consists mainly of geometric forms, and a verbal response by the subject is not needed. However, the subject may work on the task through the formulation of a number of verbal hypotheses concerning the correct solution; for example, "The correct answer is always the missing quadrant." Thus, concept identification tasks do not make for very good spatial relations training material. Nor do tasks that involve geometric forms, if these forms can be named for example, square, triangle, and so forth. On the neuropsychology battery, the tasks that come closest to being nonverbal are the copying of geometric forms and Block Design.

The distinction between verbal and nonverbal also has implications for vocational training and placement. While producing subassemblies may be an appropriate job for someone with aphasic difficulties, it is somewhat less appropriate if he must read or orally comprehend complex instructions. Correspondingly, if a verbally oriented job is desired, and typing is chosen, then some concern must be given to the subject's capacity to master the spatial organization of the letters on the keyboard.

In summary, retraining of the individual with a specific spatial relations defect can follow the same general model as that proposed for the individual with a language defect. The subject may be trained with a variety of spatially oriented tasks including puzzles, games, and particular kinds of work assignments. The most important considerations in selection of these tasks are that (1) they deal with problems in organizing external space, and that (2) they are resistive to solution through verbal mediation.

We selected these three patterns because they are relatively representative and commonly occurring ones. In principle, however, a similar approach can be taken to individuals with difficulties in a number of other areas, including memory, psychomotor speed, concept formation ability, attentional ability, and numerous other areas of mental functioning. As a rule, training programs in these areas designed for individuals with brain lesions have not been developed. The neuropsychology battery can be used to identify deficits and to provide other significant data, such as global level of performance, but the training procedures often have to be improvised in accordance with the performance pattern of the individual. One group of investigators (Ben-Yishay, Diller, Gerstman, & Gordon, 1970) used actual tests as initial training devices. We might add that an inspection of a catalog of instruments used in experimental psychology may provide a wealth of ideas. Devices such as memory drums, pegboards, reaction time apparatuses, and display panels for concept–identification studies all have possibilities of being converted from their original research and teaching aims to training devices.

On Going from Test Results to the Rehabilitation Plan: Direct Remediation versus Detouring

The guidelines offered above clearly do not tell the clinician how to review a set of neuropsychological test results and use them in planning and implementing a specific rehabilitation program. Although the state-of-the-art in this regard is still at the stage of improvisation, perhaps an admittedly small number of specifics can be suggested. One major issue has to do with the matter of whether one should directly attack the deficit or detour around it. In initial rehabilitation efforts, it is customary and proper to attempt to restore to the individual his most significant and prominent losses. Immediately after a patient has a left-hemisphere stroke, there is no question that every effort should be made to restore his speech, rather than to ignore it as a problem and divert rehabilitation efforts to some other area. However, as the condition becomes more chronic, and as the level of recovery appears to be stabilizing, the direction of further rehabilitation efforts becomes more open to question. If it is clear that further recovery from the residual symptoms does not appear likely, but the person seems capable of going back to work with some retraining, there is a real question as to the nature and direction of this retraining. Should one continue to work on the residual deficits or develop compensatory skills in some relatively unaffected area? This matter often interrelates with the nature of the premorbid occupation. Should we attempt to return an aphasic lawyer to his practice, or a structural engineer with constructional apraxia to his work, or should we encourage and train them in the direction of finding some other vocation? Answering this question in any individual case is a highly complicated matter, but in making retraining recommendations based on neuropsychological test performance, it is one that must be addressed. On the basis of the test performance, should we exploit the strengths or try to shore up the weaknesses? Ultimately this is a matter of subjective judgment; nevertheless we will try to present some of the criteria used in making this decision. Obviously, the premorbid status of the patient and the degree of progressiveness of his condition are crucial considerations.

In the case of the individual with a specific deficit, one important consideration would be the severity of the deficit. For example, if an aphasic subject received a rating of 3 or 4 on the aphasia test (moderately or severely impaired), the recommendation would be that vocationally oriented retraining should avoid the language area. If he held a job that required extensive language skills (e.g., schoolteacher or clerk) the recommendation would be that the subject should be encouraged to set his goals toward finding some other kind of occupation. Continued speech therapy would also be encouraged, but for the purpose of preserving residual language ability, rather than for vocational training purposes. Global level of performance would also be an important consideration. If the subject performed at a high level vocationally (e.g., high school or college-level teacher), training toward a new occupation would be recommended even if aphasia were absent. In this case the difficulty would be the general intellectual demands of the work, rather than the more specific language difficulties.

In the case of the individual with a mild specific deficit and mild global impairment, our tendency would be to recommend that attempts be made to return him to his premorbid occupation, even if it involves the skill in which he is deficient. In this case, the prospect of returning to his old work would appear to be more easily accomplished than training in an entirely new occupation. In some cases, we might recommend a return to the same area but not at the same level. A former lawyer might become a legal clerk; a former engineer might serve in some capacity as an assistant to other engineers. We would guess that there would be a strong motivational difference between subjects with severe and mild deficits. The individual with a mild deficit would probably see less of a need for changing occupations when he can still do some of the things he used to in his former position. The individual with severe deficit might be able to see the need for changing much more clearly.

The recommendation for the patient without a specific deficit would be based largely on level of performance. In this case, the subject has not lost some specific skill that prevents him from doing what he used to do. Rather, he has lost some degree of general adaptive capacity. In some little-known but potentially significant studies done in Halstead's laboratory (Halstead & Rennick, 1962) it was found that high-level agreement could be reached between level of performance on the Halstead tests and merit ratings given independently by supervisors. However, on the basis of longitudinal testing, it was found that whereas merit ratings and deteriorative changes corresponded well in the case of high-level executives, deterioration might be found on the tests but not in work performance in the case of lower level employees, such as blue-collar workers. In other words, not all jobs demand the same level of adaptive capacity, and whereas some slippage is tolerable in some cases, it is intolerable in others.

In some cases, a recommendation for prevocational training may be made. This recommendation is particularly appropriate for patients who are so severely impaired that any kind of vocationally oriented training seems unrealistic. Such patients might benefit more from direct training in such basic skills as speech, memory, psychomotor speed, spatial relations, and so forth. In some cases vocational training might have more of a chance in succeeding if it is accompanied by some form of basic skill training. For example, a report might recommend that the subject be placed into some form of manipulative, nonverbal occupational training, but that speech therapy should be continued simultaneously. In this way, the patient could learn a job that does not directly require language abilities, but he would receive enough language training to preserve, and perhaps enhance, his capacity to communicate in and out of his work environment.

Program Formulation for Working with Specific Deficits

The best way of describing how to use neuropsychological tests in the planning of rehabilitation programs for patients with specific deficits is probably by means of a number of case illustrations. Such detailed illustrations will be provided elsewhere in this book. Here, we will deal with more general considerations with regard to a number of areas, including memory, visual-spatial functions, language disorders, conceptual-reasoning difficulties, and perceptual disorders. In general, we will deal with these areas as though they reflected specific deficits. Thus, for example, when we discuss memory training, we will discuss it in terms of a memory disorder being the patient's only neuropsychological deficit. Furthermore, we will attempt to limit the discussion to those inferences that can be derived from neuropsychological tests alone. Thus, we will describe how the tests can be used to identify and detail the deficit, assess resources available for coping with it, and aid in formulation of a specific training program. We have already discussed more general matters, such as estimating premorbid level, assessing global degree of impairment, looking at the patterns of assets and deficits, and determining whether to attempt to directly remediate the deficit or provide appropriate training in an unaffected area. Now we will get on to some of the specific contents as applied to the situation in which we want to deal directly with the deficits.

Memory

The identification of a specific memory deficit or amnesic disorder by means of neuropsychological tests can generally be accomplished in a straightforward manner. However, the Halstead-Reitan battery does not include formal memory tests, and if a memory disorder is suspected, the clinician would do well to supplement the battery with additional tests such as the Wechsler Memory Scale or the Memory Scale of the Luria-Nebraska battery. The concept of specific memory disorder implies that the patient maintains average intelligence, does not have a language disorder that could confound memory testing, and has a severe difficulty in regard to recollection of past events. A large (20-point or more) discrepancy between the memory quotient (MQ) derived from the Wechsler Memory Scale and the full-scale Wechsler IQ is often accepted as a "neuropsychological definition" of an amnesic disorder, especially if the IQ is in the average range (Butters & Cermak, 1980). Once the presence of the disorder is established, the next task is generally that of determining the type of amnesia the patient has. First, it is important to determine that the patient does have a memory disorder and not an attentional disturbance. For example, if his major difficulties are with Digit Span or the Rhythm Test, rather than with recall of stories, pictures, or word lists, the likelihood is that the patient has an attentional rather than a memory disorder. Remediation for retraining in the case of attentional disorders is different from memory training. It often involves such methods as reaction time and/or vigilance tasks (Ben-Yishay, Rattok, & Diller, 1979). Once an attentional disorder is ruled out, one might look for material specificity of the memory disorder, determining whether the memory problem involves verbal material more than nonverbal material, or vice versa. As with other abilities, patients with specific verbal memory defects tend to have left-hemisphere brain damage, while right-hemisphere brain damage is more common with patients who have nonverbal memory difficulties (e.g., memory for faces). Another dimension relates to the temporal span of the amnesia. Some patients have difficulties only with recent memory while others have more global disorders involving both short-term and long-term memory. We might also want to do an information processing analysis in order to determine whether the memory difficulty is at the encoding, storage, or retrieval phase of the process. Finally, it is generally desirable to make some kind of assessment of whether the patient's performance improves with cueing, and if it does, it is often useful to know what kinds of cues are of most benefit and when the best time in the training process occurs for provision of cues. All this information can be derived from careful neuropsychological assessment. Once it is obtained, the next step is that of devising an educational technology for remediating the observed deficit. One important decision to make in regard to memory training relates to whether one should attempt strategy training and teach the patient some mnemonic device, such as visual imagery, or teach him specific contents that are important for adaptive function. The neuropsychological assessment should provide relevant data in regard to the patient's capacity to benefit from strategy training. Pertinent considerations include the general level of intellectual functioning and the severity of the amnesia. Simply put, patients with very dense amnesias tend to forget the strategies they are taught or cannot learn the strategy in the first place. However, we have been able to show that with extensive repetition of small units of specific material (e.g., the nurse's name) learning does take place. We have documented a case of this type elsewhere in this book. In a study done with left-hemisphere stroke patients, Gasparrini and Satz (1979) demonstrated that memory training using visual-imagery techniques was superior to training using verbal mediation. Theoretically, these patients' relatively intact right-hemispheres could encode visual images, while their impaired left-hemispheres were deficient in regard to processing verbal information through the usual linguistic channels. Thus, through information based on neuropsychological assessment, suggesting that the patient has impaired verbal-processing skills but intact visual-perceptual skills, a strategy involving using the spared ability to aid in restoring an impaired ability can be evolved. This process is the essence of how neuropsychological assessment can be used in planning remediation programs.

Visual-Spatial Disorders

Visual-spatial disorders are made manifest on tests primarily through measures involving copying geometric forms or solving nonverbal problems requiring analysis of complex spatial configurations. Block design and progressive matrix-type tasks are the ones most commonly used. With respect to both diagnosis and rehabilitation planning, it is important to know whether the deficit is expressive or receptive. Expressive deficits involve the ability of the patient to produce complex constructions, and so are often referred to as constructional deficits or as constructional apraxia. Receptive deficits are perceptual problems in which complex visual patterns do not make sense to the patient. These patients have difficulties with ambiguous and embedded figures and tend not to be able to recognize what they are in a normal manner. In this case, no construction is required, but simply an identification of the pattern. Although we have indicated that patients with specific visual-spatial disorders tend to have their major brain damage in the posterior portion of the right hemisphere, it has recently been demonstrated that both the right and the left hemispheres play a role in constructional tasks (Kaplan, 1979). The important point here is not so much the matter of localization but the question of identifying the specific nature of the visual-spatial disturbances. Like the amnesias, there are several types of this class of deficit. The major distinction, however, appears to be between the constructional deficits, which may involve a problem with synthesis of spatially complex material, and the perceptual deficits that have more to do with the analysis of complex visual configurations. Thus, for example, Kaplan (1979) has demonstrated that some patients do poorly on the WAIS Block Design Test because they break the rectangular pattern of the model in their construction, while others never break the configuration but make errors in copying the internal pattern of the design. Aside from this important distinction, it is also useful to distinguish between a visual-spatial disorder and a right-left confusion. Right-left orientation is actually a language-related ability having to do with the patient's learning and retention of a verbal symbolic system for indicating directions. Retraining to resolve right-left confusion is a different matter from visual-spatial retraining.

Once the neuropsychological tests have identified the visual spatial disturbance and have made the distinction between its expressive and receptive components, retraining planning can begin. In the case of constructional difficulties, we have available the block–design training devised by the NYU group. Through a process of cueing, the patient is taught to construct increasingly complex designs. Luria (1948/1963) has also described a method in which the patient practices copying geometric forms, and is required to analyze his mistakes in terms of spatial coordinates (up, down, right, and left). In Chapter 4, we showed how the progressive matrices test can be used as a visual-spatial retraining task. One of the more debilitating forms of perceptual-type visual-spatial disturbance is the inability to recognize familiar faces (prosopagnosia). We know of no established training programs for this condition, but if it appears to be relatively chronic, one might think in terms of teaching the patient to identify people through their clothing, voice, or characteristics other than the appearance of their faces. Visual-spatial disturbances also sometimes have implications for reading and calculation difficulties. In these cases, the patient has difficulty with identifying the spatial characteristics of letters or with the type of visual or spatial imagery that is often helpful in working with relationships among numbers. A frequent problem with attempted remediation of these cases is that the underlying deficit is not properly assessed, and sometimes strenuous but unsuccessful attempts at remediation are made through the use of traditional verbal-symbolic methods. Perhaps individuals with these disorders would benefit from preliminary training in visual-spatial, perceptual, and imagery skills. It must be admitted that methods of this type have not been well worked out, but it is possible that such techniques as block–design training or copying with correction could conceivably be of benefit to patients with these difficulties. With regard to calculation, Levin (1979) describes the spatial type of acalculia as "impaired spatial organization of numbers frequently reflected by misalignment of digits, visual neglect, inversion (e.g., 12 and 6), and reversal errors (e.g., 12 interpreted as 21), and inability to maintain the decimal place" (p. 129). It would appear that this description in itself suggests a retraining program based on practice with correction of the various types of error.

Language Disorders

Because the treatment of language disorders is a specialized area, it is usually accomplished by speech pathologists or therapists who have an interface with human neuropsychology in the area of aphasia and related language disorders. Often neuropsychologists and speech pathologists collaborate with each other in aphasia-related research or clinical practice. The interested reader is referred to review chapters by Kertesz (1979) and Holland (1977) for overviews of the various approaches to speech therapy. The assessments used for treatment programming are generally different from the kinds of neuropsychological tests we have been describing in that they tend to be rather specialized methods for evaluation of many aspects of language. They are commonly referred to as *aphasia tests*, and there are several of them (e.g., Eisenson, 1954; Goodglass & Kaplan, 1972; Holland, 1980). We will not review the various tests and treatment methods here except to comment briefly on the usefulness of neuropsychological tests, as distinct from aphasia tests, in rehabilitation planning for aphasic patients. Neuropsychological test batteries, such as the Halstead-Reitan battery, screen for aphasia but do not provide detailed aphasia examinations. In this regard, they often alert the clinician that the patient could benefit from a more extensive language evaluation. However, perhaps the more important matter is that the neuropsychologist can tell the speech therapist things about nonlinguistic aspects of the patient's functioning that could aid in treatment. First, information can be provided about the patient's general level of impairment utilizing such indices as some version of the impairment index or the IQ values. Second, the neuropsychologist can provide data concerning perceptual difficulties the patient may have, such as hemifield inattention. Such information can be of great value in regard to dealing with the patient physically. For example, the therapist might do best by working with the patient within the range of his normal visual field. Perhaps most interesting, however, is that the neuropsychologist can tell the speech pathologist about the right hemisphere. The traditional thinking made the assumption that with recovery and treatment, the right hemisphere somehow learns to mediate at least some of the language functions originally controlled by the left hemisphere. Recent neuropsychological research, however, has shown, first of all, that the right hemisphere may independently possess many language abilities (Zaidel, 1977). Furthermore, it seems clear that the right hemisphere and left hemisphere utilize different information processing strategies. However, right-hemisphere strategies can be applied to tasks normally approached with left-hemisphere strategies and vice versa. Thus, it is possible that a patient who can no longer utilize cognitive processes usually mediated by the left hemisphere to solve a linguistic problem can be taught to use right-hemisphere processing. A very clear example of this method is given by Carmon, Gordon, Bental, and Harness (1977), who reported the case of a patient with an acquired alexia resulting from left-hemisphere brain damage who was retrained to read by using a right-hemisphere perceptual strategy. More specifically, it is thought that right-hemisphere function is characterized by processing information in a holistic manner and has the function of synthesizing information rather than analyzing it. The left hemisphere processes information in a sequential, analytic manner. Thus, for example, the right hemisphere would be better than the left at a pattern recognition task, while the left hemisphere would be better than the right at a task requiring phonetic analysis. Thus, in the case of dyslexia, the patient who cannot read by breaking words down into phonemes may be able to learn to read through learning to recognize whole words as perceptual patterns.

With regard to the planning of such programs on the basis of Halstead-Reitan battery data, the patient with a language disorder who does poorly on the aphasia screening test and the verbal Wechsler tests, but who does well at Block Design and Object Assembly, may be a good candidate for right-hemisphere strategy training. It is assumed on the basis of this test pattern that the patient has impaired left-hemisphere function but has more or less intact right-hemisphere function. Therefore, rather than attempting language therapy requiring information-processing abilities the patient no longer has available, the idea is to substitute processes he does have available in the service of restoring language.

It should be mentioned that Luria (1970) and various co-workers were actively engaged in language therapy over a period of many years. The treatment philosophy was heavily oriented toward neuropsychology, and many innovative treatment programs were worked out for individual patients. The interested reader is referred to the traumatic aphasia book (Luria, 1970) and the book on restoration of function (Luria, 1948/1963) for detailed descriptions of these programs. Here, we will provide only an example. Luria has described a condition called *dynamic aphasia,* in which the patient is essentially incapable of producing spontaneous speech. The neuropsychological basis for this condition is thought to be a disturbance in formation of what is described as the linear scheme of the sentence. The localization is thought to be in the inferior portion of the left-frontal lobe. Patients with this condition cannot express thoughts or produce even simple verbal expressions. The treatment centers around having the patient express himself by writing down on separate pieces of paper fragments of the theme he wants to express. The fragments can be written down in any order, but at the end, the pieces of paper have to be ordered so as to form a coherent narrative. Thus, the neuropsychological concept involving the inability to sequence elements of a spontaneous expression is directly translated into a treatment program through aiding the patient in mastering this kind of sequencing, thereby restoring his capability of processing from the plan to speak to the execution of the narrative.

Conceptual Reasoning Difficulties

Perhaps one of the most universal and most disabling features of brain damage is the impairment of the ability to think in conceptual terms. This disability was described in detail by K. Goldstein (1959b) and K. Goldstein and Scheerer (1941), in their extensive considerations concerning the relationship between brain damage and impairment of the abstract attitude. Despite the pervasiveness and high frequency of occurrence of this symptom, little is known about its remediation. There is even a question as to whether abstract reasoning abilities can be restored in brain-damaged patients. There has been an extensive literature controversy regarding whether the abstract attitude is a qualitative or quantitative phenomenon (Goldstein, Neuringer, & Olson, 1968; Reitan, 1958c, 1959a). K. Goldstein argued that abstraction is a qualitatively distinct level of function and does not form a continuum with concrete thinking. Reitan (1958c, 1959a) argued that the reverse is true, and that the brain-damaged patient is only more or less relatively impaired in abstraction ability when compared with the normal individual. However, K. Goldstein and Scheerer (1941) and Scheerer (1946) pointed out that in the process of adjustment, brain-damaged patients develop the ability to find concrete solutions to abstract problems. Clearly, this tendency should provide an important lead regarding treatment. In many circumstances, while the formation of an abstract concept is the most efficient and definitive solution to a problem, it may also be solved to some extent with concrete thinking. There are numerous clinical examples of this phenomenon. We can offer the one of the patient who is able to reproduce an abstract geometric form from memory (The Stick Test) by giving the form a name. Thus, although he may be unable to reproduce two lines converging toward each other at 45-degree angles, he may be able to reproduce a "roof." It is thought that whereas the most efficient way to do well on the Category Test is by means of abstract reasoning, partial solutions to several of the problems can be solved with concrete thinking (Simmel & Counts, 1957). Indeed, it has been something of a challenge to construct tests that can only be solved by abstract reasoning.

One might suppose that direct training with categorization or sorting-type tests in which the patient is given cues concerning formation of the appropriate concepts might be an obvious way to proceed in this area. Thus, for example, the patient might be taught to "pass" the Category Test or the Wisconsin Card Sorting Test. We know of no systematic attempt to do this in a treatment context, although Reitan (1959a) showed that brain-damaged patients, despite making more errors than normals, improved as much as normals on Subtest VI of the Category Test relative to Subtest V. Since Subtests V and VI are based on the same organizing principle, the supposition is that what was learned while taking Subtest V generalizes to some extent to Subtest VI. Thus, there is some evidence that brain-damaged patients can improve their performance on an abstraction task on the basis of previous experience with a related task. Luria (1948/1963) described a treatment program for patients with a difficulty related to impairment of abstract thinking which he described as a disturbance of active thinking. These patients have difficulties in various aspects of spontaneous behavior because they cannot maintain a flow of thought. They cannot produce spontaneous narratives, write essays, or engage in other behaviors that require a sequential flow of thought. The treatment involves providing the patient with what are described as "transition formulae" or cues as to what comes next. The patient is asked to produce a narrative, but is given a card to look at on which are written such words as "However," "Since," "Although," and so forth, and is encouraged to consult this card to find the appropriate connecting word when he has difficulty in passing on to a subsequent part of the narrative. A similar procedure is to encourage the patient to write down components of the narrative in whatever order they occur to him. He is then asked to make up a plan by arranging these components in their proper order. Finally, the patient is asked to use the plan in producing the narrative. Luria reported success with these and related methods in getting patients who could not carry on active, spontaneous thinking to do so. In essence, by providing "pegs" on which to organize the structure of a narrative or other sequence requiring active thinking, such thinking can be restored.

It is unclear whether this disturbance is the same thing as the more widely reported disturbances of conceptual reasoning. Luria (1948/1963) indicates that this particular problem is generally associated with lesions of the convex surface of the frontal lobe, and is quite specific. However, the nature of the disorder is such that it can be mistaken for a dementia or generalized impairment of intellectual function. Luria claims that in these cases the logical structure of abstract thinking is unaffected, the patient's difficulty only involving the flow of thought. Neuropsychological assessment of such patients should reveal normal language, except when the requirement involves producing a narrative, normal perceptual abilities, and preserved ability to solve simple intellectual problems. Thus, it would appear that we have a treatment method for what may appear as a conceptual deficit but really is not, and it is unclear as to what benefit this treatment may have for patients with significant impairments of abstraction ability. Such patients very frequently have the same difficulty in sequential flow of thought but may not benefit from the use of "transition formulae," because the basis of the disorder is different from those patients who have an active thinking disturbance as a specific, isolated symptom.

Although the Halstead-Reitan battery, particularly the Category

Test, and similar procedures provide excellent means of identifying and quantifying impairment of conceptual ability, no systematic attempt to remediate this frequently observed deficit has been accomplished. Perhaps that is because this particular deficit is essentially irremediable, a view that would be compatible with K. Goldstein's theory. We need only repeat this oft-quoted dictum, "Even in its simplest form, however, abstraction is separate in principle from concrete behavior. There is no gradual transition from one to the other" (K. Goldstein, 1951, p. 60). Elsewhere, K. Goldstein (1959b) makes the point even more bluntly: "loss of abstract capacity cannot be regained by retraining. Only improvement of the brain damage may more or less restore the impaired capacity" (p. 775). If this situation is the case, and no one has determined that it is not, then the only alternative would be that of providing the patient with problem-solving strategies that do not involve abstraction. We often do this informally, as when we provide the patient with a schedule for his daily activities rather than require him to plan out his own day, but we do not yet have formal procedures for this type of rehabilitation.

Perceptual Disorders

Neuropsychological tests readily identify various types of perceptual disorders on the basis of which retraining procedures can be readily devised. A perceptual disorder from the neuropsychological standpoint involves those situations in which there is a failure of perception in the presence of normal sensory function of the modality in question. Thus, what we commonly describe as deafness or blindness are not neuropsychological perceptual disorders. This is not to say that neuropsychology in general cannot assist the deaf or blind individual. However, that is another matter which we will not be discussing in this section. As we have defined them, the major perceptual disorders are the so-called gnostic disturbances and disturbances of attention. The classical gnostic disturbances are relatively rare, and will not be dealt with in detail here. They are generally called *agnosias,* and may affect the visual, auditory, or tactile modalities. Thus, visual agnosia would be the inability of a person who is not blind to identify objects visually, although they may be identifiable by touch or by their characteristic sounds. Sometimes the gnostic disorders blend with the language disorders, since these recognition disturbances may involve only symbolic stimuli. For example, visual-verbal agnosia is the inability to recognize letters, a severe form of dyslexia. The point of retraining in these cases is that of teaching the patient, through various means, to recognize objects. However, the teaching method used has a great deal to do with the modality involved, and with whether the disorder involves only verbal stimuli, only nonverbal stimuli, or both. For example, if the patient has a global disturbance of visual recognition, sometimes retraining methods involving the extensive use of eye and head movements traced around figures that cannot be recognized as a whole are effective. The patient may be encouraged to refrain from guessing what the object may be on the basis of an inadequate search, or he may be taught to search more systematically, perhaps by a series of left-to-right transverse sweeps, as in reading.

Neuropsychological tests are not only useful in regard to diagnosing gnostic disorders, but qualitative observations of the patient's test performance can provide important cues as to how the disorder is coped with. For example, observation of excessive head movements while trying to identify an object should give the examiner the cue that the patient is tracing around the object that he knows is there but that he cannot identify. This naturally occurring method of adaptation can be built on as described. Similarly, patients with disturbances of tactile recognition can be observed to move their fingers while attempting to identify an object placed in their hands. They are trying to use joint cues as a supplement to the information provided by tactile sensitivity alone. Again, this tendency can form the basis for a program.

It should be pointed out that with the exception of a small number of settings, most readers of this book will rarely if ever see a patient with a pure, modality-specific gnostic disorder. The more commonly seen perceptual disorder is inattention or neglect. The most widely studied aspect of this phenomenon is visual neglect, in which attention is not paid to one or the other of the visual fields. The phenomenon is reported to occur most frequently in the case of right-hemisphere brain damage (Heilman, 1979), and so, more often than not, we deal with neglect of the left visual field. Heilman (1979) points out that treatment may be informal or formal in nature. With regard to the informal methods, the bedfast patient should be positioned such that his good side faces the area in which most social interaction takes place. Efforts should also be made to discourage the ambulatory patient from driving or working around dangerous equipment if he has a visual neglect syndrome. Diller and Weinberg (1970) have reported the existence of high levels of accident proneness in hemiplegic patients with neglect syndrome. The NYU group (Diller & Gordon, 1981; Diller & Weinberg, 1971) has developed formal retraining methods for patients with visual neglect. Basically, the treatment involves assisting patients to scan their entire visual field through an anchoring or cueing technique. In the case of reading, these patients frequently start in the middle of a line, neglecting the left side of the page. The strategy used was to anchor the left side of the page by some visual marker, such as heavily drawn vertical lines. As the patient learns to turn his head to the left so that the left side of the page appears in his right visual field, the markers are gradually faded. Another method used by this group involves the use of a scanning machine (Diller, Ben-Yishay, Gerstman, Goodkin, Gordon, & Weinberg, 1974). The device consists of a board with a target that can be moved at variable speeds, containing 10 colored lights, spaced 7 inches apart. One part of the training involves having the patient follow the target around the edges of the board. As proficiency increases, the target speed is increased, and the patient is asked to sit closer to the board. The other part involves flashing the lights and asking the patient to identify by color which lights are on. As proficiency increases, the patient is required to go from searching for lights in the center of the array to lights presented to his impaired hemifield. Finally, he must search for lights simultaneously presented to the right and left fields.

Neglect phenomena provide a good example of how a simple neuropsychological test (in this case the visual imperception test) can identify a symptom that might have gone unnoticed on casual clinical observation, and lead to the implementation of an established treatment program. Although such examples may not be as common as we would like, they may provide a model for what is possible in other areas. Neglect is a relatively subtle but significantly disabling symptom because of its relation to a propensity for having accidents, the difficulty it may provide in regard to reading and other cognitive activities requiring visual perception, and the difficulty it provides with regard to social interaction. Although the symptom may gradually disappear over time in some cases, or the patient may eventually adapt to it on his own initiative, it would appear that a relatively simple treatment program can significantly expedite alleviation of the symptom.

The Physically Handicapped Patient

In many instances, brain damage produces a combination of physical and cognitive deficits. The patient with Huntington's disease and the stroke patient typically have significant impairment of motor function along with some form of intellectual or perceptual disability. Sometimes a brain lesion can produce true blindness for a portion of the visual field as opposed to the neglect phenomenon. Numerous texts have been written concerning the treatment and rehabilitation of the physical difficulties frequently experienced by brain-damaged patients (e.g., Hirschberg, Lewis, & Thomas, 1964) and we will not deal with that area here. We will comment briefly on only two areas: the problem of the patient with a hemiplegic or hemiparetic preferred hand and staff orientation to physical disability in brain-damaged patients. With regard to the matter of loss of function of the preferred hand, it is important that special attempts be made to train the individual in the use of his nonpreferred hand so that he can write with it and perform other skills ordinarily requiring the use of the preferred hand, such as using tools or eating. With regard to the matter of writing, the neuropsychological assessment can be quite useful in regard to distinguishing between *agraphia,* a verbal-symbolic difficulty with handwriting related to the aphasias, and simple inability to write because of impaired motor function. In the latter case, teaching the nonpreferred hand to write may be quite straightforward; but if the patient has agraphia, much more complex retraining is required involving such matters as restoration of grapheme–phoneme associations. The matter of staff orientation relates to the need for teaching by the neurobehavior specialist discussed elsewhere in this book. In our experience, certain of the physical difficulties associated with brain damage appear "peculiar" to the untrained individual. We have noted this particularly in the case of visual-field defects, since they often produce a pattern of motor behavior, such as "hugging" the wall of a corridor while walking through it, that is actually adaptive in terms of preventing mishaps in the unsighted field but that may be misinterpreted in many ways. Without a grasp of the underlying neurology, it may be difficult to understand how a person can be blind for only half of the visual world. Thus, if the neurological or neuropsychological assessment discovers a hemifield or quadrantal visual defect, the consequences of this defect in terms of the behaviors that might be associated with it should be communicated to the staff. Similarly, patients with movement disorders, particularly Huntington's disease patients, may engage in patterns of motor behavior that may appear bizarre or at least purposeful, when in fact they are neither. It is not unusual for a Huntington's disease patient to strike someone near him with an adventitious movement, or lose his balance and fall over onto another person. These movements and accidents are entirely involuntary, and do not imply carelessness or belligerent intent on the part of the patient. The peculiar gait and posturing of these patients, although bizarre in appearance, are actually the involuntary manifestations of a severe dysfunction of the motor components of the central nervous system. Again, orientation of the staff involving the characteristics of these movements and their underlying neurology can be very helpful.

Summary

In this chapter, we have demonstrated that neuropsychological assessment can be used productively in the planning of rehabilitation programs as well as in its more traditional diagnostic role. We discussed the general principles upon which most neuropsychological test batteries are structured, and then presented one particular system, the Halstead-Reitan battery, in detail. We then provided an extended discussion of how one may go from the test data to the actual planning of programs. It was indicated that one's initial considerations might involve the extent of the impairment in general and the patient's premorbid level of function. Whereas extent of impairment can be readily estimated by global indices of dysfunction, such as the impairment index, estimation of the premorbid level can be problematic. However, it is obviously crucial to know what the capacities of the patient were prior to the acquisition of brain damage in order to plan a reasonable treatment program. It was suggested that certain nontest indices of premorbid function, such as education level, may be quite helpful in this regard. One of the most useful aspects of standard neuropsychological test batteries is that they provide a profile of the patient's deficits and preserved abilities. Quantitatively oriented batteries, such as the Halstead-Reitan, also provide information concerning relative degrees of preservation and impairment on various types of psychometric scales. These profiles are useful not only to identify areas of dysfunction needing restoration but also to identify resources the patient may have to cope with in the dysfunctional areas and in deciding whether to attack the observed deficit directly or to detour around it. For example, if the patient has a specific deficit that is mild or moderate in severity, and other relatively intact functions, a direct attempt might be made to remediate that deficit. However, if the specific deficit is very severe or the patient does not have compensatory resources, various detour strategies might be more advisable.

It was suggested that one major consideration in the planning of rehabilitation programs relates to the functional asymmetries of the cerebral hemispheres. While some patients have primarily "left-hemisphere" difficulties involving language specifically or sequential-analytic abilities in general, others have "right-hemisphere" deficits such that visual-spatial, pattern perception, and synthetic abilities are affected. The point here is that there can be no one training regime for "brain-damaged" patients, and that one point of natural cleavage in the planning of programs might involve the distinction between right-hemisphere and left-hemisphere cognitive deficits. However, that is only a beginning and it is necessary to become more specific when planning for the individual patient. This level of planning requires more detailed application of neuropsychological theory and often calls for the analysis of highly specific syndromes. Furthermore, the hemisphere asymmetry idea does not always provide what is needed to plan for the patient with diffuse brain damage. Thus, we provided some discussion of how one might proceed from a neuropsychological test protocol to treatment of relatively specific deficits. The areas chosen for this discussion included memory, visual-spatial abilities, language, conceptual reasoning, and perception. However, it must be emphasized that these areas are also general, and it is often necessary to identify the specific aspects of the area that are impaired. For example, it is rare that a patient loses all memory abilities. Far more frequently there is a preservation of remote memory with impairment of recent memory. Sometimes the memory defect is modality specific and might involve only verbal or nonverbal material.

It is important to emphasize that the level of specificity of a neuropsychological disorder is often unique, and since no well-described program for treatment is described in the literature, the therapist must improvise. However, such improvisation can often be clearly guided by the results of the neuropsychological assessment. This uniqueness has presented a problem for this area of treatment such that perhaps one of the major works in this field, Luria's restoration-of-function book (Luria, 1948/1963), is essentially a compilation of theoretical material with an extensive series of individual case examples. It is therefore not yet possible to write a "rehabilitation text" in which specific treatment programs are prescribed for a wide variety of neuropsychological profiles. However, a number of treatment modules are available in the area of memory training, and there are also the programs provided by the NYU group in visual-spatial disorders (block–design training) and visual scanning. Needless to say, there are numerous methods reported in the literature for treatment of the various aphasic syndromes (e.g., Albert, Goodglass, Helm, Rubens, & Alexander, 1981).

We might conclude with the consideration that neuropsychological assessment should not be viewed as the only tool used in rehabilitation planning. In the previous chapter, we discussed the importance of behavioral assessment, to which we might add daily observation of the patient on the ward or at home, review of the patient's history, and more casual contact with the patient through conversation and observation. Often valuable information can be gained from the patient's family, co-workers, and friends. However, neuropsychological assessment plays the unique role of providing detailed information concerning the level and pattern of the patient's deficit without which it is difficult, if not impossible, to reach a conceptual understanding of the nature of the disorder. If one wishes to engage in treatment based on this level of understanding, then neuropsychological assessment is obviously a necessity. However, it may not be necessary for general milieu therapy or for a classical behavior therapy approach in which the major consideration would be the environmental contingencies that maintain observed behavior.

6

THE WICHITA PROGRAM

Quantitative Neuropsychological Test Results and Case Illustrations

As described in Chapter 4, we were involved in a program for young brain-damaged adults that took place in a sheltered workshop setting in Wichita, Kansas. One of the major aspects of the program involved assessing the usefulness of neuropsychological tests, specifically the Halstead-Reitan battery, in rehabilitation planning. The research design of the program called for the longitudinal evaluation of an experimental or treated group and a control or untreated group. The experimental group was pretested with the Halstead-Reitan battery, then took part in a 6-month training program at the sheltered workshop, and was later retested. The control group received the preliminary testing, was retested six months later, but was not treated. The aspect of the research reviewed in this chapter has to do with whether there was a greater change in ability level, as measured by the neuropsychological tests, in the experimental group than in the control group.

Quantitative Studies: Subject Selection

At the onset of the project, a number of rather strict criteria were used in subject selection for the experimental and the control groups. Subjects had to be within the age range of 18 to 55 years. They had to have sustained some form of brain damage or acquired some form of brain disease after their sixteenth year. They had to be ambulatory at the time of entrance into the program. On the basis of school records and other available information, we had to ascertain that the potential subject was of at least average intelligence prior to acquisition of the brain disorder. In effect, we were attempting to rule out mentally retarded individuals who also sustained brain damage during adulthood. The most desirable subjects appeared to be those with traumatic head injuries, but individuals with static, nonprogressive brain disease were also acceptable. For example, an individual who sustained a stroke as a young man was accepted into the program. We did not want to accept individuals with such progressive brain disorders as multiple sclerosis. The research was aimed at investigating the effects of rehabilitation on degree of improvement in adjustment among brain-damaged individuals. These effects are impossible to evaluate when any improvements gained are compensated for by the progression of the lesion.

The intent of the program was to work with individuals with substantial, demonstrable impairment of function. Thus, individuals having what we considered to be mild impairment were not accepted. We also screened out a number of subjects who may have been substantially functionally impaired, but their inability to maintain competitive employment was clearly related to only minimal brain dysfunction accompanied by significant emotional and/or characterological difficulties. The presence of emotional or characterological difficulties in themselves did not rule out participation in the research. Many of the subjects we accepted had very significant psychiatric difficulties. However, they also had substantial impairment associated with their brain damage.

Because of an initial difficulty in recruiting subjects, our criteria had to be relaxed to some extent. One subject was admitted to the program having had a traumatic head injury in his sixteenth year. However, he also had a history of borderline mental retardation, was a member of a minority group, and had experienced severe cultural deprivation throughout his life. Thus, although his disabilities were no doubt attributable to more than his head injury, the head injury did produce identifiable brain dysfunction. We also accepted two subjects with what we viewed as mild to moderate brain dysfunction. Their major difficulty was a disability with regard to maintaining competitive employment. However, emotional and characterological problems were also strongly involved, and their job difficulties may have been based more on those factors than on their brain dysfunction. However, aside from these exceptions, the experimental and control groups did meet the criteria indicated above; that is, they were all severely disabled on the basis of adult-onset brain dysfunction.

The program involved thirty-two subjects at the outset, but ten of them left prematurely. Two clients left apparently owing to lack of interest, while the other eight were terminated because of significant health or emotional adjustment problems. The average age of the clients was 34.2 years, with ages ranging from 19 to 59 (one client, aged 59, was admitted and thus constituted an exception to the criterion of maximum age of 55). The majority of clients were age 30 or younger. Therefore, the sample can reasonably be described as a group of young adults. There were 28 males and 4 females in the original group; 14 clients were married, 13 were single, and 5 were divorced. The average amount of education prior to brain injury was 12 years. The time between acquisition of brain damage and admission to the program varied from one to six years, with an average of four years. This factor is quite significant, since it suggests that we were working with subjects with chronic lesions that had been in existence beyond the normally expected period of natural recovery. A systematic evaluation of social class was not accomplished. However, most of the clients probably would have been classified as lower middle class.

The final sample of experimental subjects who received the rehabilitation program and the neuropsychological assessment (before and after it) consisted of 22 individuals. There were 19 cases in the control group, who did not receive any kind of systematic rehabilitation work between the two neuropsychological assessments. The two groups were roughly comparable in regard to neurological diagnoses. Of the 22 experimental subjects, 15 had traumatic head injuries, 2 had encephalopathy of unknown etiology, 1 patient had a brain abscess, 2 had sustained strokes, and 2 subjects had episodes of cerebral anoxia. Of the 19 control subjects, there were 11 cases of traumatic head injury, 3 subjects had encephalopathy of unknown etiology, 3 had strokes, and there were 2 cases of cerebral anoxia.

Comments Regarding Acquisition of the Samples

Since we were forced to relax our criteria for admission, it is apparent that we experienced difficulties in recruiting a sufficient number of cases. Although the rate of adult-acquired brain damage in the general population is unknown to us, it would appear unlikely that the fundamental reason for our difficulty was a shortage of brain-damaged people in need of and desiring rehabilitative services. Nevertheless, despite extensive efforts, we did have problems in acquiring enough subjects.

It might be worthwhile to describe briefly what recruiting efforts were made. We had fairly widespread television and newspaper coverage for the program. We had continuing contact with the Department of Vocational Rehabilitation. Letters and sometimes personal contacts were made to service and health-related facilities, such as hospitals, community mental health centers, and family service agencies, which might provide services for brain-damaged adults. Physicians in the specialty areas of neurology, neurosurgery, and rehabilitation were contacted by letter, and also in many cases by telephone, to make them aware of the availability of the program to their patients who were appropriate candidates for it. Extensive contact was maintained with the Veterans Administration Hospitals in Wichita and in Topeka, Kansas. The VA did, in fact, send us some very suitable subjects. It was known at the outset that appropriate clients would not simply arrive without a concerted effort to actively recruit subjects and to make the general public and professional community aware of the extensive services being offered. An added inducement was that there was no fee for any of the services, including neuropsychological assessments and the rehabilitation program itself. These were paid for by grant funds.

Setting

The program presented here was developed in response to what we perceived as a need for more imaginative transitional rehabilitation programs for young adult, brain-damaged individuals. We were fortunate to gain the cooperation of the Kansas Elks Training Center in Wichita, a nonmedical facility supported by the Kansas Elks Clubs and devoted to prevocational training and job placement for handicapped young adults. The average population of the center was about 130. About half of the clients were mentally retarded, whereas the other half consisted primarily of young people, mostly with emotional problems. There were usually a number of clients with sensory impairment, and a substantial number of the clients had multiple handicaps. One section of the center served as an extended sheltered workshop primarily for multiply handicapped clients who were not considered to have the potential for being able to function in competitive employment. However, most of the clients received job evaluation and training with the goals of competitive employment and independent or semi-independent living. The general orientation of the program was toward development of work habits and work skills, rather than toward job training. The major training vehicle was an extensive work area in which clients did subcontract work of a varied nature for local industries. The administration of the center maintained an aggressive program for obtaining subcontracts from businesses in the Wichita area.

Within this general setting and structure, a special room was designated for our program for brain-damaged clients. The room was large and contained several stations for various work-related or prevocational training-related activities. Eventually, our clients also worked in other parts of the center, but this room always remained as headquarters for the program. We had a staff of one part-time and three full-time employees. At any particular time, there were 7 to 12 clients enrolled in the program. The staff served primarily as trainers but also assisted in data management and analysis.

Research Procedures

As indicated, the design of the study was a test–retest procedure comparing the experimental group with the control group. The dependent measures consisted of various scores from the Halstead-Reitan neuropsychological test battery. During the 6-month interval, the test materials were not available to either group and were in fact located in a different place from the training facility. The tester did not participate in any of the training. There did not appear to be any reason to assume that there was any systematic difference between the experimental and control groups. The control subjects were potentially eligible for admission to the program, and did not enter it for reasons other than their neurological condition. In fact, several control subjects did subsequently enter the program and were treated as both experimental and control subjects in the data analysis. Since the population of young adult brain-damaged individuals in Kansas is low, we were unable to utilize a matched-pair design. However, we could not uncover any evidence for systematic bias.

Before and after scores on a large number of measures obtained from the neuropsychology battery were tabulated separately for the experimental and control groups, and differences were evaluated statistically. The primary statistical method used was the t test for paired comparisons, since each subject's second testing was compared with the first testing. A result was said to be favorable in regard to demonstrating the efficacy of the program when a statistically significant change in the direction of improvement was noted in the experimental group, but not in the control group. In the case of neuropsychological tests, there is no substantive question regarding the direction of a change in regard to improvement. This point is mentioned because in the case of other kinds of psychological tests, particularly bipolar personality scales, the direction of improvement is sometimes debatable. However, with neuropsychological tests, there seems to be little question that increases in WAIS scores and decreases in errors and performance time for the various Halstead tests all reflect improvement.

For convenience of presentation, results are reported under three categories: index variables, WAIS variables, and Halstead Battery variables. The index variables are the global measures of extent of impairment, the IQ's, and the Average Impairment Rating. The results for these measures are presented in Table 4. This table contains many statistically significant findings, both for the experimental and the control groups. All the IQ scores showed significant improvement, but in two instances (Verbal and Full Scale) the level of significance was higher in the experimental group than it was in the control group. On the Average Impairment Rating measure, only the experimental group showed a statistically significant change.

These findings, first of all, provide some interesting data regarding the samples where noted, in which the IQ values for both groups are all in the 80s, reflecting the presence of dull normal-level intelligence. The average Impairment Ratings are all above 2.01, our cut-off for the moderately impaired range of functioning. Taken together, these scores strongly suggest that both groups contained a preponderance of substantially impaired people. Looking at this situation from the other side, since it is known that all the subjects in the study had documented brain lesions, it would appear that the tests chosen appear to be sensitive to the behavioral consequences of these lesions. These global indices thus appear to be sensitive to brain damage in a general way.

With regard to the changes, it was not surprising to find that both groups improved on the various IQ measures. When retesting is done with the same forms over a relatively brief time interval, some component of the improvement found must be attributed to the effects of prior experience with the tests. It is more interesting when significant improvement occurs in one group and not in another. In such a case, it is less likely that the change can be attributed solely to prior experience with the tests, when this experience is the same in both groups. This situation obtained in the case of the Average Impairment Rating. A highly significant change was noted in the experimental group, but not in the control group, which finding is made more interesting by the fact that the tests that contribute to the Average Impairment Rating have been shown to be more sensitive to brain dysfunction (Reitan, 1959b) than the Wechsler scale. If one accepts these premises, that differential change between groups is of more interest than change in

Table 4
Index Variables

	Ex-pre		Ex-post			C-pre		C-post		
Variable	*M*	*SD*	*M*	*SD*	*t*	*M*	*SD*	*M*	*SD*	*t*
VIQ	85.3	15.7	89.1	15.9	−3.42[b]	86.8	13.0	89.7	15.8	−1.75[a]
PIQ	81.5	14.6	85.1	14.6	−2.24[a]	81.8	20.4	85.8	18.8	−2.19[a]
FSIQ	82.6	13.8	86.6	14.4	−3.47[b]	83.7	14.2	87.3	16.0	−2.42[a]
A.I.R.	2.91	0.72	2.65	0.86	3.22[b]	2.53	0.65	2.42	0.70	1.29

[a] $p < .05$.
[b] $p < .01$.

both groups, and that the Average Impairment Rating is a more sensitive instrument than the IQ measures, then the obtained results look quite favorable in regard to the efficacy of the training program. In essence, subjects who went through the program improved on a global measure sensitive to brain dysfunction, whereas subjects who did not go through the program showed no such improvement.

The WAIS results by subtest are presented in Table 5. A rather mixed picture emerges here. In the case of the experimental group, there were five significant changes (Comprehension, Similarities, Digit Span, Picture Completion, and Block Design). The control group obtained two statistically significant changes (Comprehension and Block Design). All significant changes, in both groups, were in the direction of improved performance. Again, the results tend to favor the experimental group, in terms of number of significant differences. There appears to be no discernible pattern to the changing and nonchanging subtests. In regard to the control group changes, Comprehension is a Hold subtest, while Block Design is a Don't Hold subtest (Wechsler,

Table 5
WAIS Subtest Variables Expressed in Scaled Scores

Subtest variable	Ex-pre		Ex-post			C-pre		C-post		
	X	*SD*	*X*	*SD*	*t*	*X*	*SD*	*X*	*SD*	*t*
Information	7.6	2.8	8.0	2.9	−1.57	8.2	2.1	8.3	2.9	−0.23
Comprehension	8.0	3.8	9.0	3.5	−2.11[a]	7.4	2.1	8.8	2.9	−3.00[b]
Arithmetic	6.0	2.5	6.3	2.3	−0.80	7.1	3.1	7.7	3.6	−1.40
Similarities	8.0	3.8	9.0	3.3	−2.01[a]	8.1	3.7	8.7	3.8	−1.13
Digit Span	6.2	3.5	7.2	3.9	−1.97[a]	7.6	2.7	7.5	3.1	−0.23
Vocabulary	8.0	3.4	8.2	3.8	−1.04	7.9	3.1	8.3	2.9	−0.75
Digit Symbol	4.5	2.4	4.8	2.4	−0.88	5.5	2.9	5.9	2.9	−0.81
Picture Completion	8.3	1.8	9.1	2.6	−2.57[b]	8.2	3.0	8.4	1.9	−0.36
Block Completion	8.3	1.8	9.1	2.6	−2.57[b]	8.2	3.0	8.4	1.9	−0.36
Block Design	7.4	3.5	8.3	3.5	−2.32[a]	6.6	4.2	8.0	4.2	−3.72[b]
Picture Arrangement	7.0	3.2	7.5	3.2	−1.23	6.7	3.0	7.0	3.0	−1.12
Object Assembly	6.3	3.5	6.8	3.5	−1.19	7.2	4.4	7.9	5.0	−1.60

[a]$p < .05$.
[b]$p < .01$.

1944). The additional subtests that changed significantly in the experimental group include two Don't Hold subtests (Similarities and Digit Span) and one Hold subtest (Picture Completion). Thus, Wechsler's distinction between those tests that are more or less sensitive to changes in brain function does not appear to be related to those tests that did not show change in the present study. Regarding the "Babcock Hypothesis," it may be noted that the Vocabulary score did not change in either group, and thus the view is supported that Vocabulary is one of several stable measures of intellectual functioning.

The results for the component tests of the Halstead Battery are presented in Table 6. It will be noted that not all of the tests administered are reported on here. The reason for this is the lack of development of satisfactory quantitative scoring systems for some of the procedures. The results here clearly favor the experimental group. Significant changes were noted for four of the tests, with an additional one changing at a borderline significant level ($p < .10$). In the case of the control group, the only significant changes noted were for the Finger Tapping Test, but they were changes for the worse (i.e., reduced tapping speed on retesting).

In general, we can say that the Halstead Battery tests seem to be more sensitive to those factors that differentiate the groups than are the WAIS measures. However, alternative explanations for the changes noted can be offered and discussed. First, it could be said that the changes were found mostly on the Halstead Battery tests, because these tests are less reliable than the WAIS subtests. In this regard, Boll (1981) has reviewed the literature concerning reliability studies with the Halstead-Reitan battery and concluded that the battery as a whole is a reliable procedure when applied to a wide spectrum of individuals, including brain-damaged patients, normals (Matarazzo, Weins, Matarazzo, & Goldstein, 1974; Matarazzo, Matarazzo, Weins, Gallo, & Klonoff, 1976), and schizophrenics (Klonoff, Fibiger, & Hutton, 1970). However, even if these tests are relatively unreliable, this would not explain the differential findings between groups, nor the fact that all the changes in the experimental group were in the same direction—that of improvement. A second criticism might be that the Halstead Battery tests are easier the second time than is the WAIS. That is, the information derived from the first testing with the Halstead tests helps more on the second testing than is the case for the WAIS. First, there is no logical reason to assume that this distinction exists. In neither case were correct solutions told to the subject following failure, and there do not appear to be sufficient differences in the structure of the two procedures to explain the discrepancy. For example, Object As-

Table 6
Halstead Battery Variables

Variable	Type of measure	Ex-pre		Ex-post			C-pre		C-post		
		M	*SD*	*M*	*SD*	*t*	*M*	*SD*	*M*	*SD*	*t*
Category Test	Number of errors	83.1	32.5	65.3	35.7	4.34[b]	83.8	36.0	79.1	33.0	1.17
Tactual Performance Test	Total time	24.6	6.4	20.4	7.7	3.33[b]	23.8	6.1	24.2	7.5	−0.34
TPT—Memory	Number recalled	4.9	2.5	5.9	2.4	−2.11[a]	4.5	2.0	4.9	2.7	−0.78
TPT—Location	Number located properly	1.4	2.2	2.4	2.6	−1.79[c]	1.1	1.5	1.7	2.1	−1.05
Speech Perception	Number of errors	20.9	13.2	16.5	11.2	3.02	15.6	11.1	15.3	11.1	0.11
Rhythm Test	Number of errors	10.4	6.5	9.1	6.7	1.67	9.2	3.9	8.5	4.1	0.78
Trail Making B	Time in	248.7	134.4	235.3	163.4	0.53	188.0	82.8	202.9	111.2	−0.99
Tapping—Dom	Taps per 10 seconds	36.5	9.5	39.6	9.9	−1.53	43.8	11.6	41.1	9.6	2.07[a]
Tapping—Non-Dom	Taps per 10 seconds	34.2	9.5	33.7	10.0	0.39	41.6	6.3	37.0	9.0	2.35[a,d]
Aphasia	Number of errors	20.5	12.9	18.5	14.0	1.20	14.1	10.2	14.0	8.2	0.04

[a] $p < .05$.
[b] $p < .01$.
[c] $p < .10$.
[d] Posttest $\overline{X}$ score was worse than pre-test $\overline{X}$ score.

sembly and the Tactual Performance Test are both construction tasks, yet there was no change in Object Assembly performance, but a substantial change in Tactual Performance Test scores in the experimental group. Second, the relative easiness on repetition would again not explain the differences between groups.

A third possible criticism is that the experimental group subjects were specifically trained to do better on the tests, and the improvement noted may have no generalizability to behavior in other than the test setting. The question of generalization will have to be answered through application of test situation independent procedures, such as the follow-up described in a later chapter. Although it is true that the experimental subjects were trained to develop abilities that we believe the tests measure, the test material was not available to them during the program, nor were similar materials. Furthermore, the training was not done by the same person who administered the test batteries. The individuals engaged in the actual rehabilitation training of the subjects were not skilled administrators or interpreters of the Halstead-Reitan battery, and knew of it only through reading reports, and possibly some of the research literature. It is highly unlikely that they could have intentionally or unintentionally trained the patients simply to do better on the tests.

If one accepts our ruling out of the above alternatives, the explanation of the findings that is most probable is that they reflect improvement in functioning in the experimental group as compared with the control group. Since the two groups were comparable in regard to a number of apparently relevant variables, the improvement may reasonably be attributed to the effects of the training program.

An examination of the pattern of improvement suggests that it occurred mainly on the more complex tests. Significant improvement was not found for Finger Tapping or the Aphasia Test, both of which are tests of relatively simple functions. The one exception to this rule is the Trail Making Test, which we tend to view as being a complex test. It is possible that improvement was not found here because of the crucial role of motor speed, and as we see from the Finger Tapping results, speed was not significantly increased in either group. From these limited data, it might be suggested that the training program does more in the way of increasing general adaptive capacity than it does in regard to enhancing specific abilities. However, closer analysis of individual cases might reveal some instances of substantial specific ability improvement that is concealed by the group-based statistical analyses presented here.

Summary and Conclusions for the Quantitative Data

A method was presented of generating training program recommendations for individuals with brain lesions, on the basis of performance on a battery of neuropsychological tests. The study was not designed to assess the efficacy of these recommendations in individual cases, nor was the actual training program based solely on the neuropsychological tests, although they did play a large role. There was, however, an opportunity to evaluate in a general way improvement in various abilities as a function of taking or not taking a 6-month training program at the Kansas Elks Training Center. It was found that there was significant improvement in a number of cognitive and perceptual abilities in the trained but not in the untrained control group.

In view of these findings, the following conclusions may be drawn:

1. Individuals with static brain lesions can show substantial improvement in a number of abilities, but this improvement is less likely to occur in the absence of some systematic rehabilitation training program.
2. The battery of neuropsychological tests developed by Halstead and Reitan appears to be more sensitive to changes in functioning than does the WAIS.
3. The study has shown that the training program appeared to lead to improvement in general adaptive capacity. However, further research is needed to assess specific neuropsychological test-based recommendations in terms of specific outcomes.

Case Illustrations

Our experience with the Wichita program provided the suggestion that it may be conceptually useful to plan rehabilitation programs based on what is known about functional cerebral-hemisphere asymmetries. Of course, the NYU group has emphasized retraining of right-brain-damaged patients, while speech therapy may be viewed as retraining of functions usually associated with the left-hemisphere. In the case of the Wichita program, the initial neuropsychological assessment provided a recommendation as to whether the client should receive a left-hemisphere-oriented or right-hemisphere-oriented program. The attempt was then made to utilize the many educational facilities available in the Elks Center to implement either of these program types. As indicated previously, a major difficulty area is the patient with bilateral brain damage. Patients of this type do not generally have marked disorders of either the left- or right-hemisphere type but do have some components of these disorders. The nature of these components may be highly variable, such that sometimes speech is more affected than visual-spatial abilities, although sometimes the reverse occurs. Another complexity is that the neurological disorders that produce diffuse brain damage are frequently of a different nature from those associated with lateralized or localized disorders. Since the phenomenon of type-locus interaction exists, the behaviors associated with diffuse brain damage can be quite different from what is seen with more localized damage. The disorders associated with lateralized brain dysfunction are generally penetrating head injury, stroke, and certain types of brain tumors. Diffuse brain damage may be associated with many disorders including close-head injury, cerebral vascular disease, such various presenile and senile dementias as Alzheimer's disease, and a variety of other processes.

Despite these many complexities, it seemed useful for us to work with this triumvirate of cerebral-involvement types. The planning for right-hemisphere and left-hemisphere cases, as we will see, is reasonably straightforward, whereas the planning for the diffuse cases has to be much more exploratory. Various tasks must be tried with the aim of hitting on a program that is optimally effective. We will present brief descriptions of general characteristics of the three types as they appeared in the Wichita program, and detailed case reports of three patients, each representing one of these types. In presenting this material, we would caution the reader that although these cases are representative, there are many individual difference factors that make each case unique.

The Left-Hemisphere Client

The individual who sustains significant brain damage primarily involving the left hemisphere suffers a double deficit, particularly if he is right-handed. There is, first of all, the aphasia, but there is often loss of function of the preferred hand as well. That loss of function may be partial or complete, and it may involve either sensory or motor abilities, or a combination of the two. Less frequently, there is loss of vision in the right visual field. Thus, the patient with left-hemisphere brain damage often demonstrates a combination of symptoms including aphasia, right hemiplegia or hemiparesis, and a right homonymous hemianopia. Within this symptom picture there is a great deal of variability, depending upon the type of lesion, its age, and the specific region of the left hemisphere that is involved. Most notably, there are numerous disorders that go under the general heading of aphasia. In general, however, the more anterior the lesion, the more likely it is that the individual will have problems with speech fluency; whereas the more posterior the lesion, the greater the likelihood that there will be speech comprehension difficulties.

The rehabilitation of left brain-damaged patients traditionally has two components. One of them is speech therapy and the other is training of the left hand to take over some preferred hand functions, notably writing. These two components may be related, since training of the left hand may assist the right hemisphere in taking over language functions. Looking at assets, the left-hemisphere patient often has well-preserved nonverbal abilities. The capacities to interpret visual-spatial configurations, construct assemblies, and recognize faces are frequently well preserved. Thus, if there is sufficient preservation of function of the right upper extremity, these patients often do well at manual activities that involve putting together assemblies, as long as the level of verbal instruction is not particularly complex.

It may be pointed out that the documentation of the presence of aphasia in the left brain-damaged patient is not thought to be sufficient for purposes of planning a speech–therapy-oriented rehabilitation program. It is necessary to acquire detailed information regarding the nature of the aphasia. Some important points are whether or not the patient has fluent speech, whether he can repeat what is said to him, whether he can comprehend auditory and visual information, and whether some channel of communication is available. With regard to the latter point, some aphasic patients who cannot communicate well with speech can do so by writing. Some aphasics cannot understand spoken speech well, but can read. This kind of information is crucial to rehabilitation planning.

A Client (Gerald) with Left-Hemisphere Brain Damage. Gerald was selected because he is close in age and general level of impairment to the right-hemisphere case to be presented later. He was a 17-year-old male with 11 years of education who sustained a traumatic head injury from a penetrating gunshot wound a year prior to involvement in the research program. Figure 24 presents a summary of Gerald's test performance on the Halstead-Reitan series of tests which were administered just prior to his entering the 6-month rehabilitation program. It shows that on the 12 major tests of the Halstead series he obtained an Average Impairment rating of 2.50, indicating a cerebral dysfunction classification of moderate to severe impairment. On 91.7% of the 12 tests, his performance was in the impaired range. He was strongly

GERALD (1)
SUBJECT IS RIGHT HANDED.
SUBJECT IS RIGHT FOOTED.
SUBJECT IS RIGHT EYED.
SUBJECT DOES NOT HAVE CROSSED EYE-HAND DOMINANCE.

WIDE RANGE ACHIEVEMENT TEST
READING = 1.0 GRADE
SPELLING = 2.6 GRADE
ARITHMETIC = 4.4 GRADE
11 YEARS OF EDUCATION

NAME OF TEST	RATING
HALSTEAD CATEGORY	2.
FORM-BOARD, TIME	2.
FORM-BOARD, MEMORY . . .	1.
FORM-BOARD, LOCATION . .	3.
SPEECH PERCEPTION	4.
RHYTHM	2.
TAPPING SPEED	2.
TRAILS-B	4.
DIGIT SYMBOL	3.
APHASIA SCREENING	3.
SPATIAL RELATIONS	2.
PERCEPTUAL DISORDERS . .	2.
AVERAGE RATING =	2.50
PERCENT OF RATINGS IN THE IMPAIRED RANGE =	91.7%

RATING OF 0=SUPERIOR,
1=AVERAGE,
2=MILDLY IMPAIRED,
3=MODERATELY IMPAIRED,
4=SEVERELY IMPAIRED,
5=VERY SEVERELY IMPAIRED.

NAME OF WAIS SUBTEST	SCORE
INFORMATION	5.
COMPREHENSION	5.
ARITHMETIC	3.
SIMILARITIES	3.
DIGIT SPAN	2.
VOCABULARY	3.
DIGIT SYMBOL	6.
PICTURE COMPLETION	9.
BLOCK DESIGN	6.
PICTURE ARRANGEMENT	9.
OBJECT ASSEMBLY	6.
WAIS VERBAL I.Q. =	67.
WAIS PERFORMANCE I.Q. =	83.
WAIS TOTAL I.Q. =	72.

AVERAGE SCORE ON EACH SUBTEST IS 10 AND AVERAGE IQ IS 100. HIGHER VALUES ARE ABOVE AVERAGE AND LOWER ONES ARE BELOW AVERAGE.

Figure 24. Neuropsychology report—a client with left-hemisphere brain damage before treatment.

right-handed, right-footed, and right-eyed. On the Wide-Range Achievement Test, he earned a Reading grade of 1.0, a Spelling grade of 2.6, and an Arithmetic grade of 4.4. These results were obtained despite his 11 years of schooling. Figure 24 also shows a summary of his performance on the WAIS. He obtained a Verbal IQ of 67, a Performance IQ of 83, and Full Scale IQ of 72. His aphasic symptoms and perceptual disorders are reported at the bottom of Figure 24.

The neuropsychologist reported that Gerald was functioning in the borderline defective range of general intelligence and had mildly impaired abstraction and complex problem-solving skills. Other deficits were noted in the areas of auditory discrimination and motor speed. There was frank expressive and possibly a receptive aphasia manifested by word-finding difficulty and difficulties with spelling, reading, enunciation, formation of sentences, and calculation. There was also a problem with right-left discrimination. Related to the aphasia was a Verbal IQ that was 16 points lower than the Performance IQ.

With regard to differential functioning of the cerebral hemispheres, the aphasia provided good evidence for primary involvement of the left hemisphere. There was not a significant discrepancy between the right and left sides in motor speed, but there was a wide discrepancy in tactile functions. It was likely that the major brain damage was posterior to the frontal lobes and probably in the temporo-parietal area of the left hemisphere. His language and language-related functions were severely impaired, but he did fairly well on many right-hemisphere functions including his ability to deal with spatial relations and constructional-type tasks. None of his WAIS verbal subtests was higher than a scaled score of 5, but 2 of the 5 performance subtests reached a scaled score of 9. He had difficulty manipulating a pencil and in general his dyskinesia with his right hand probably accounted for part of his lowered score on Digit Symbol, Block Design, and Object Assembly. The subject's relatively good performance on the more complex tests, such as the Category Test and the Tactual Performance Test, was a positive prognostic indicator. For him to have done as well on these tests indicated that there must have been substantial brain tissue remaining that was not damaged. From the evaluation, it seemed that the best approach would be to work with the speech disorder directly. From the indications of relatively well-preserved higher-level cognitive abilities, Gerald seemed to have the resources necessary to benefit from traditional speech therapy and a program concentrating on language training in general. There was a feeling that, if there was good recovery of speech function, he would probably have a number of vocational alternatives open to him. If there was not substantial improvement of his speech and language abilities, he probably could do reasonably well in an employment situation involving perceptual-motor tasks. In order for him to do assembly-type work, he would have to learn to compensate for the mild dyskinesia of his right hand. On the basis of the evaluation, it was felt that we had some understanding of Gerald and his problems that could not be learned from a clinical interview and from knowledge of his history, or from neurological studies and hospital records. The aphasia was obvious and was apparent in a face-to-face clinical interview, but it was much more severe than would be obvious from ordinary conversation. In evaluating the prognosis for recovery from aphasia, much depends on the preservation of nonverbal cognitive functions. His scores on the WAIS were not too contributory to answering this question, although his scaled score of 9 on Picture Arrangement was a positive sign. His relatively good performance on the more complex tests suggested a good overall prognosis for the return of language functions and a good prognosis for rehabilitation in general.

Gerald's performance on the Halstead series of tests suggests strategies of rehabilitation for him. His rehabilitation was designed to emphasize retraining in speech, other language-related abilities, and academic skills. He had formal speech therapy sessions; he was given daily remedial work in academics, such as reading, spelling, and arithmetic, and in every area of the rehabilitation program, the effort was focused on the development of communicative skills, both receptive and expressive. His spatial-relations abilities seemed to be fairly well intact, and he did not receive any specific training in this area. He was involved in doing some subcontract work each day for about 3½ hours to give him experience in dealing with a work environment and also to help him to compensate for the dyskinesia, in an effort to improve the use of his right hand. Because of his attentional and auditory discrimination problems, he was given attentional retraining through learning the Morse Code. Because of the markedly reduced tactile sensitivity with his right hand, he was given training in tactile recognition of objects. On the Tactual Performance Task he was not able to insert any of the blocks into their slots in 4.0 minutes.

With this case and others, the Halstead evaluation seemed to give us very useful information not only in terms of diagnosis of cognitive deficits and remaining skills and abilities, but it gave us some basis for planning an appropriate rehabilitation program. We were also able to get some estimate of prognosis. After six months of the rehabilitation program that was briefly outlined above, Gerald was again evaluated with the Halstead series of tests, and the summary scores are presented in Figure 25. His improvement in functioning after six months of rehabilitation can be seen in comparing Figures 24 and 25. His verbal IQ on the WAIS increased from 67 to 74; his performance IQ from 83 to 90; and his Full Scale IQ increased from 72 to 80. However, the latter change only reflected an increase from a borderline defective to low-dull normal level of functioning.

The postrehabilitation neuropsychological testing reflected mild improvement. His abstraction and problem-solving abilities, which were not greatly impaired initially, were now slightly better. However, the aphasia and right-sided tactile disorders persisted. The aphasia was slightly improved but still clearly present. On the Aphasia Screening Test he received a rating of 3 on both occasions, although on the first administration he had 36 errors, and on the second testing he had 26 errors. Academic skills improved somewhat in reading, spelling, and arithmetic, but they were still far below the subject's educational level. The therapist who worked with this client felt that he made greater gains in academic skills than his scores would suggest. Similarly, discrimination ability improved but remained mildly

```
GERALD (2)
SUBJECT IS RIGHT HANDED.
SUBJECT IS RIGHT FOOTED.
SUBJECT IS RIGHT EYED.
SUBJECT DOES NOT HAVE CROSSED EYE-HAND DOMINANCE.

WIDE RANGE ACHIEVEMENT TEST
READING =  2.8 GRADE
SPELLING =  3.3 GRADE
ARITHMETIC =  5.7 GRADE
11 YEARS OF EDUCATION
```

NAME OF TEST	RATING
HALSTEAD CATEGORY	1.
FORM-BOARD, TIME	2.
FORM-BOARD, MEMORY . . .	1.
FORM-BOARD, LOCATION . .	1.
SPEECH PERCEPTION . . .	3.
RHYTHM	2.
TAPPING SPEED	3.
TRAILS-B	4.
DIGIT SYMBOL	4.
APHASIA SCREENING	3.
SPATIAL RELATIONS	1.
PERCEPTUAL DISORDERS . .	3.
AVERAGE RATING	2.33
PERCENT OF RATINGS IN THE IMPAIRED RANGE =	66.7

RATING OF 0=SUPERIOR,
1=AVERAGE,
2=MILDLY IMPAIRED,
3=MODERATELY IMPAIRED,
4=SEVERELY IMPAIRED,
5=VERY SEVERELY IMPAIRED.

NAME OF WAIS SUBTEST	SCORE
INFORMATION	5.
COMPREHENSION	5.
ARITHMETIC	4.
SIMILARITIES	7.
DIGIT SPAN	6.
VOCABULARY	1.
DIGIT SYMBOL	4.
PICTURE COMPLETION . . .	10.
BLOCK DESIGN	9.
PICTURE ARRANGEMENT . . .	12.
OBJECT ASSEMBLY	6.
WAIS VERBAL I.Q. =	74.
WAIS PERFORMANCE I.Q. =	90.
WAIS TOTAL I.Q. =	80.

AVERAGE SCORE ON EACH SUBTEST IS 10 AND AVERAGE IQ IS 100. HIGHER VALUES ARE ABOVE AVERAGE AND LOWER ONES ARE BELOW AVERAGE.

Figure 25. Neuropsychology report—a client with left-hemisphere brain damage after treatment.

impaired. What we appeared to have were a number of small increments in functioning in the context of continued significant impairment of left-hemisphere functions. On the first testing, his performance on 91% of the tests was in the impaired range and on the second testing there was a decrease to 66.7% of the scores in the impaired range. It can be seen from a comparison of the two tests' administrations that the Average Impairment Rating decreased from 2.50 to 2.33, which is in the direction of improvement. Within the context of overall improvement of functioning, there was also some evidence on the second administration of some decrease in functioning in some areas. There was found to be an increase in tactile suppression and decreases in motor speed and grip strength on his right side. Tapping speed on the finger tapping test was down from 47 to 41, and grip strength was down from 46 to 35 KG with his right hand. However, about six weeks prior to the second testing, the patient developed a small brain abscess that required additional surgery. It seems likely that these decreases in performance were a result of the abscess and the ensuing operation. It is possible, of course, that without this additional complication, the overall improvement may have been somewhat greater than was observed.

After Gerald completed the six months of rehabilitation, he returned to the public schools to work on his GED, and after a year of study was able to achieve his goal. Even though he was not gainfully employed, he was showing a rather mature and competent adjustment to his environment. Since leaving the rehabilitation program, he had gotten his driver's license back, and was driving quite responsibly and independently. He had a very active social life in which he was completely responsible for initiating social interaction. He was dating several girls and reported to the interviewer that he was not ready to become involved in a serious relationship. He played the guitar competently and organized a band of handicapped individuals who performed on a semiprofessional basis. His goals were somewhat unrealistic in the sense that he still entertained the desire to go on to college and become a physician or a dentist. Even prior to the traumatic head injury, he probably did not have the intellectual potential to be able to achieve a goal such as medicine or dentistry. His aphasia and diminished language skills made such a goal even more unrealistic. However, he was still young and he had great determination, and little could be gained by trying to discourage him from such an ambition. When seen at the follow-up, he certainly had the ability to function in competitive employment as long as he did something of a manual nature that did not require complex verbal expression or the use of high-level academic skills. The original estimate of prognosis based upon the first evaluation was substantiated at the second evaluation and the follow-up. Gerald's case demonstrates a fairly typical pattern for left-hemisphere cases, particularly in regard to their relatively favorable prognoses.

The Right-Hemisphere Client

The right-hemisphere cases are neuropsychologically opposite to the left-hemisphere ones: they are not aphasic; they usually have fairly intact language, verbal, and academic skills (except, on occasion, for certain aspects of arithmetic), but unlike their counterparts, they are maximally impaired in any type of motor performance requiring an appreciation of the spatial characteristics involved in the task. On psychological tests they tend to do very poorly on performance subtests of the WAIS; they have difficulty copying simple geometric designs.

In any type of work involving assembly or visual-motor processes, they tend to be severely limited. Despite the fact that they retain use of their preferred hand in most cases, they are nevertheless deficient in many types of performance tasks. Right-hemisphere cases may have a significant loss of spatial orientation to the environment that is more fundamental than simple motor loss as such. Many have difficulty finding their way around in the world, and they have a major problem orienting themselves to problem-solving situations when the tasks are of a manual nature. Sometimes they have difficulty in perceiving non-verbal objects, notably faces.

We now offer a brief presentation of a right-hemisphere case, who, except for a severe ideomotor dyspraxia, was a good representative of such cases.

A Client (Dale) with Right-Hemisphere Brain Damage. When he was 20 years old and a sophomore in college, Dale sustained what was diagnosed as a severe cerebral vascular accident with a left hemiparesis. Two years after the stroke, he was evaluated as a potential candidate for the program; he was fully ambulatory but he walked slowly and with an unsteady gait. He had almost no use of his left arm, and his language skills seemed to be intact. Figure 26 presents summary scores of his performance on the test battery based on the first evaluation. The reader should compare this summary with the summary scores of the previous case. Dale showed fairly well-preserved language functions on a clinical level, and his scores on the tests reflect this. He functioned in the normal range on both the aphasia-screening test and the speech-perception test, and his scores on the verbal section of the WAIS were much better than his scores on the performance section (Verbal IQ 94, Performance IQ 66). On Information and Vocabulary, two subtests of the WAIS that are highly dependent on language functioning, he obtained scores significantly above average and close to his premorbid level. On the Wide-Range Achievement Test he was found to be reading at the 12.8 grade level, and spelling at the 11.2 grade level. The only area in which he was impaired was Arithmetic, where he obtained a grade level of 4.4. Also, on the Arithmetic subtest of the WAIS he was able to earn only a scaled score of 3. Well-retained reading and spelling, but with impaired arithmetic abilities, are not infrequent relationships in right-hemisphere cases. The basis of the impairment here seems to be on the level of impaired spatial-relations ability and does not appear to represent any impairment of number concept as such. With the right-hemisphere cases, the difficulty with arithmetic may have to do with the problem of spatial ordering of numbers (Levin, 1979). Figure 26 shows that Dale's subtest

DALE (1)
SUBJECT IS RIGHT HANDED.
SUBJECT IS LEFT FOOTED.
SUBJECT IS LEFT EYED.
SUBJECT DOES HAVE CROSSED EYE-HAND DOMINANCE.

WIDE RANGE ACHIEVEMENT TEST
READING = 12.8 GRADE
SPELLING = 11.2 GRADE
ARITHMETIC = 4.4 GRADE
13 YEARS OF EDUCATION

NAME OF TEST	RATING
HALSTEAD CATEGORY	3.
FORM-BOARD, TIME	4.
FORM-BOARD, MEMORY	0.
FORM-BOARD, LOCATION	3.
SPEECH PERCEPTION	1.
RHYTHM	2.
TAPPING SPEED	4.
TRAILS-B	4.
DIGIT SYMBOL	4.
APHASIA SCREENING	1.
SPATIAL RELATIONS	2.
PERCEPTUAL DISORDERS	3.
AVERAGE RATING	2.58
PERCENT OF RATINGS IN THE IMPAIRED RANGE =	75.%

RATING OF 0=SUPERIOR,
1=AVERAGE,
2=MILDLY IMPAIRED,
3=MODERATELY IMPAIRED,
4=SEVERELY IMPAIRED,
5=VERY SEVERELY IMPAIRED.

NAME OF WAIS SUBTEST	SCORE
INFORMATION	13.
COMPREHENSION	4.
ARITHMETIC	3.
SIMILARITIES	9.
DIGIT SPAN	12.
VOCABULARY	12.
DIGIT SYMBOL	3.
PICTURE COMPLETION	7.
BLOCK DESIGN	5.
PICTURE ARRANGEMENT	6.
OBJECT ASSEMBLY	4.
WAIS VERBAL I.Q. =	94.
WAIS PERFORMANCE I.Q. =	66.
WAIS TOTAL I.Q. =	81.

AVERAGE SCORE ON EACH SUBTEST IS 10 AND AVERAGE IQ IS 100. HIGHER VALUES ARE ABOVE AVERAGE AND LOWER ONES ARE BELOW AVERAGE.

Figure 26. Neuropsychology report—a client with right-hemisphere brain damage before treatment.

scores on the performance section of the WAIS were uniformly poor and much lower than his subtest scores on the verbal section. It should also be kept in mind that he was strongly right-handed and used his preferred hand in working the manual problems on the WAIS. In Dale's performance on the Halstead-Reitan Battery, his Average Impairment Rating was 2.58, which reflected moderate impairment. The neuropsychologist noted that Dale was functioning with substantially impaired abstraction and problem-solving abilities in the dull-normal range of general intelligence. Other deficits noted were in the areas of attentional capacity, motor speed, and various perceptual abilities. Dale had a hearing loss in his right ear, and marked loss of sensory and motor functioning on his left side.

With regard to differential functioning of the cerebral hemispheres, the presence of a hemiparesis and hemianesthesia on his left side suggested primary involvement of his right-cerebral hemisphere. In that there was no substantial disturbance of language functions, we suspected that his left hemisphere was reasonably intact although, in conditions as severe as this, it is likely that some diffuse effects have generalized to the left hemisphere. Severity of damage to the right hemisphere seemed to be quite pervasive in that it involved sensorimotor functions as well as higher-level behavioral functions in the area of spatial relations ability. The latter inference was based on the large discrepancy between Verbal and Performance IQs, with the Performance IQ being 22 points lower.

The findings from the neuropsychological evaluations suggested a retraining program involving work on attentional capacity, abstract thinking, complex problem-solving abilities, and particularly intensive work on spatial relations ability and perceptual-motor functioning in general. In the initial evaluation, there was also a suggestion of ideomotor dyspraxia, which had to be further evaluated.

Throughout his rehabilitation, Dale was involved for at least 3 hours a day in doing tasks of a vocational nature. The jobs were typical of those found in most sheltered workshops, including a host of mailroom activities, such as folding, stapling, stuffing envelopes, and labeling. He worked on a variety of subcontract projects for local industries which included such activities as assembling, labeling, packaging, construction of wooden boxes, and ribbon-and-bow making. All these tasks required him to perform in three-dimensional space. These work activities tapped many areas in which Dale was impaired, and we felt that exposure to them provided retraining of these functions. Since these activities require correct orientation to the task, the focusing of attention, sequential nonverbal reasoning, analyzing spatial demands, and performing the correct visual-motor operations, a number of right-hemisphere functions are exercised.

The work environment also called for such high-level functions as decision making, judgment, and planning ability. These work assignments gave Dale an opportunity to practice these skills in a real-life situation. Some of these work assignments were in different workshop areas and rooms of the Training Center, and interroom and intraroom movements gave him an opportunity to reorient himself to his environment. In the beginning, Dale was very spatially disoriented and could not find his way around the Training Center, but as we worked with him and provided opportunities for him to learn, his orientation did improve to the point where he began to be able to find his way around. When needs arose to deliver small items to staff members in another room or to obtain items from other areas, Dale was often selected to carry out this task to give him learning opportunities to practice orientation to his environment. This type of *in vivo* retraining of adaptive functions was backed up with more specialized and specifically designed retraining programs.

A "Memory Board" was used quite extensively with Dale in retraining his spatial-relations skills. It consisted of a large floor plan of a factory (also one of a house) divided into typical rooms and areas of a factory, as well as a box containing assorted objects, such as paper clips, buttons, bolts, keys, and so forth. A typical training procedure would be to have him locate one or several items and "deliver" them, for example, to the secretary's office on the floor plan. To complete the task successfully, he would first of all have to select the correct items. Once the items had been correctly found, he would have to locate the designated place on the floor plan and place the object in the right "room." Systematic observation of his performance on this task showed that he improved with practice. In general, errors would decrease as a function of practice. The task had merit in that its complexity could be varied quite easily. Dale might be asked to select one object and to deliver it to the appropriate place or person. At a higher level he might be asked to "deliver" several objects to different rooms in sequence. The task also seemed to require not only spatial abilities but also attention and visual tracking.

Along with spatial-relations retraining, we had Dale practice a variety of tasks that required visual tracking, orientation, and various visual-motor operations. We had him practice copying geometric figures, and gave him assignments involving writing or printing as part of his spatial-relations retraining. We had him do a number of activities that involved visual searching of a stimulus configuration to locate the correct stimulus. With regard to visual tracking and discrimination, we would use such stimulus material as a printed page full of numbers, and he would be directed to go through them as quickly as possible and circle, for example, all "7s" that appeared in the group (i.e., a cancellation task). Here, as on other tasks, we noticed that speed and accuracy tended to increase over time. We had him practice on a pursuit rotor to track a spot that was revolving with a hand-held stylus. Practice on the Jensen Memory Board also involved spatial-relations training in addition to the memory and conceptual processes involved. Dale was also involved in the recreational program which provided retraining of spatial-relations skills. We encouraged him to play table tennis for visual tracking and visual-motor coordination.

Many of the training devices described above under the rubric of spatial-relations training also provided training of attentional capacity. Performance of the tasks required attention to both the task and to the person making the request. The learning of the Morse Code was used to retrain Dale's auditory attention. We set up a sending and receiving situation for Dale in which he and another client would alternate sending and receiving messages. This kind of device also seems to be a useful task for the language-impaired client, since it involves using the individual letters of the alphabet on one level of complexity, words on another level, and whole messages on still another. The task also involved writing down the messages that were sent, and this was felt to be good practice for both the left- and right-hemisphere cases. The right-hemisphere cases receive practice in putting the letters and words into appropriate spatial continuity. Visual attention was retrained in a variety of tasks, many of which have already been noted. In addition, there was one task that involved a visual display of a stimulus pattern which required the pressing of a key when the particular pattern would occur in a matrix of several visual displays.

In retraining problem-solving skills, Dale was exposed to many situations in the training environment, in work, recreation, and other areas that required him to master some solution to a novel task. He received training on forming nonverbal concepts with the Progressive Matrices test, which we used as a method of training critical and conceptual skills. He also received training in arithmetic skills. Since his language functions were fairly well intact, he did not receive speech therapy as such. However, his language intonation was worked on to some extent. Perhaps because of damage to his right hemisphere, Dale's speech tended to be flat, monotonic, and without rhythm or changes in inflection, thus requiring work in this area as well. He also had much difficulty with words that had to do with spatial orientation, such as "on," "under," "above," "between," and "next to."

Dale had an ideomotor dyspraxia in addition to the other impairments noted above, and early in his training program this deficit was further evaluated, and retraining techniques initiated. The ideomotor dyspraxia was found to be quite pervasive and included the inability to use tools and other objects meaningfully. For example, he responded to a hammer as if it were a stick with a piece of heavy metal on the end and not a tool for driving a nail into a piece of wood. The staff used imitation as the primary technique for training the functional use of objects and tools. One technique was to have several tools and objects on a table in front of the examiner. The staff member would pantomime the motor operations for one of the objects or tools on the table and Dale would be asked to select the object to which the activity in question belonged Another type of retraining activity on this task involved the staff member pointing to two objects and asking Dale how they might be used in relation to each other. For example, a screwdriver and a screw would be pointed to, and he would be asked to pantomime the activity associated with these objects. Through these types of experiences Dale seemed to assimilate some purposeful skilled activities into his motor repertoire, and they seemed to carry over to the other situations in the training environment.

Figure 27 presents the neuropsychological summary scores from Dale's postrehabilitation testing. These scores, in comparison with the earlier ones, suggested that there had been some improvement in performance. There was a decrease in the Average Impairment Ratings from 2.58 to 2.25, which is in the direction of improved functioning. His ability to deal with spatial relations had improved. This change was seen mainly on the performance subtests of the WAIS. His Performance IQ had risen from 66 to 80, and there was a particularly striking change on the Object Assembly subtest which went from a scaled score of 4 to one of 9. He was now capable of better abstract reasoning and complex problem solving than he was previously (note the im-

DALE (2)
SUBJECT IS RIGHT HANDED.
SUBJECT IS LEFT FOOTED.
SUBJECT IS LEFT EYED
SUBJECT DOES HAVE CROSSED EYE-HAND DOMINANCE.

WIDE RANGE ACHIEVEMENT TEST
READING = 17.7 GRADE
SPELLING = 13.6 GRADE
ARITHMETIC = 5.7 GRADE
YEARS OF EDUCATION

NAME OF TEST	RATING
HALSTEAD CATEGORY	2.
FORM-BOARD, TIME	3.
FORM-BOARD, MEMORY . . .	1.
FORM-BOARD, LOCATION . .	4.
SPEECH PERCEPTION . . .	0.
RHYTHM	1.
TAPPING SPEED	2.
TRAILS-B	4.
DIGIT SYMBOL	4.
APHASIA SCREENING	1.
SPATIAL RELATIONS	2.
PERCEPTUAL DISORDERS . .	3.
AVERAGE RATING	2.25
PERCENT OF RATINGS IN THE IMPAIRED RANGE =	66.7%

RATING OF 0=SUPERIOR,
1=AVERAGE,
2=MILDLY IMPAIRED,
3=MODERATELY IMPAIRED,
4=SEVERELY IMPAIRED,
5=VERY SEVERELY IMPAIRED.

NAME OF WAIS SUBTEST	SCORE
INFORMATION	13.
COMPREHENSION	6.
ARITHMETIC	7.
SIMILARITIES	9.
DIGIT SPAN	14.
VOCABULARY	12.
DIGIT SYMBOL	4.
PICTURE COMPLETION . . .	8.
BLOCK DESIGN	6.
PICTURE ARRANGEMENT . . .	8.
OBJECT ASSEMBLY	9.

WAIS VERBAL I.Q. = .
WAIS PERFORMANCE I.Q. = .
WAIS TOTAL I.Q. = .

AVERAGE SCORE ON EACH SUBTEST IS 10 AND AVERAGE IQ IS 100. HIGHER VALUES ARE ABOVE AVERAGE AND LOWER ONES ARE BELOW AVERAGE.

Figure 27. Neuropsychology report—a client with right-hemisphere brain damage after treatment.

provements on the Halstead Category and the Tactual Performance Test Ratings). The status of his left side did not change greatly. His grip was a bit stronger, but he still could not use his left hand to do any manipulative task. Functions that remained stable were verbal skills, which were fairly normal to begin with, and psychomotor speed, which remained quite slow. His ability to calculate mentally had improved, and arithmetic achievement had improved by more than one-year level (from a 4.4 to a 5.7 grade level). On the second testing, he had not gotten worse in any area tested; he remained stable in certain areas, and there had been improvement in others. At the second evaluation, he did things in a more organized manner. He could deal more adequately with spatial-relations problems, and he was more capable of learning from experience. The latter point was exemplified by his performance on the Tactual Performance Test. On the first testing, his performance did not improve at all over the three trials of this test. On the second testing, he showed initial impairment, taking the full 10 minutes without completing the task. However, on the third trial he completed the task in less than 4 minutes (i.e., he correctly inserted all 10 blocks).

Dale was kept in the experimental rehabilitation program for another 6 months to see if any more improvement could be brought about. At the end of this 6-month period, he was again administered the Halstead series of tests and his performance remained virtually unchanged from the second testing. On the second and third evaluations, his Average Impairment Rating was identical (2.25), and in each case 66.7% of the performance scores were in the impaired range.

In contrast to the left-hemisphere case discussed earlier, at follow-up, Dale was found to be quite marginally adjusted. He was certainly not capable of functioning in competitive employment, but following termination from the program, he was placed in a sheltered workshop, and he continued to be fairly productive in that setting. He lived at home with his father. He had a very limited social life, and most of his socialization contacts came directly or indirectly from his placement at the sheltered workshop.

Comparisons

On the global measures of impairment on the Halstead tests these two cases were fairly similar initially. The left-hemisphere case had an Average Impairment Rating of 2.50; 91.7% of his scores were in the impaired range, and he had WAIS Verbal, Performance, and Full Scale IQs of 67, 83, and 72, respectively. The right-hemisphere case had an

Average Impairment Rating of 2.58; 75% of his scores were in the impaired range, and he had WAIS Verbal, Performance, and Full Scale IQ's of 94, 66, and 81, respectively. They both showed improvement on these global measures on their postrehabilitation assessment with the neuropsychological tests. Comparison of these two sets of postrehabilitation global impairment scores would suggest that both subjects had either a comparable degree of residual disability, or if anything the *right*-hemisphere case had slightly less impairment (his IQ scores were better and he had a slightly lower average rating). However, in actual reality, the *right*-hemisphere case was significantly more disabled than the left-hemisphere case. The left-hemisphere case was felt to be much more socially and vocationally competent than his counterpart, yet their neuropsychological test scores did not reflect this obvious differential competence levels. The problem is how to account for this discrepancy between test performance and nontest measures of adaptive functioning. There may be factors other than neuropsychological variables that account for it; factors such as personality and emotional integration. Indeed, in these two cases the left-hemisphere client, at least on a clinical level, had better emotional integration than his counterpart and his higher functional level may have arisen from his better emotional organization. However, his better emotional organization may reflect the operation of neuropsychological variables, namely, left-hemisphere brain dysfunction may give rise to less emotional disorganization than comparable damage in the right-cerebral hemisphere. The presence of the right-hemisphere "indifference" syndrome (Gainotti, 1972) may be related to this matter. There is another possibility to account for this state of affairs, one that is subjectively compelling based on our observations of left- and right-hemisphere-damaged patients, and that is that damage to right-hemisphere functions is much more generally disabling for the individual than comparable damage to left-hemisphere functions. The Halstead global measures assume equality of impaired scores on left- and right-hemisphere functions. For example, a rating of 3 on Speech Perception (a left-hemisphere function) is taken as comparable to a rating of 3 on Spatial Relations (a right-hemisphere function) and they contribute equally in the computation of the Average Rating. They are taken to be arithmetically equal, but they may not be behaviorally equal in terms of the adjustment difficulty that each would give rise to. A rating of 3 on Spatial Relations may deserve a higher "weight" than the same rating on Speech Perception, in regard to adaptation to the environment, because these two functions may not be equally disabling to the individual. We are inclined to seriously consider this interpretation of the data, and, if it is so, rehabilitation specialists may have been underplaying the importance that these nonverbal abilities play in a person's adaptation to the environment. In any event, our right-hemisphere case had more adjustment problems than our left-hemisphere case, and this situation was not atypical.

A Client (Connie) with Diffuse Brain Damage. Our next case illustration concerns Connie, a 22-year-old single woman whose junior year in college was interrupted by a severe case of encephalitis. Connie was not able to continue her studies following her release from the hospital owing to the residual brain damage she sustained. She was referred to our program from a state mental hospital, at which she had been a patient for the two previous years. During the 2-year period, the hospital had placed her on a number of trial visits, but she was never able to make any kind of a nonhospital adjustment. Figure 28 presents her neuropsychological summary scores. The test picture is one of diffuse impairment of all functions. She received an Average Rating of 2.75 (which reflects moderate impairment), and on the 12 major tests of the Halstead series, 83% of her scores were in the impaired range. She was functioning in the borderline defective range of general intelligence and had severely impaired abstraction and complex problem-solving abilities. Deficits were noted throughout the series of neuropsychological tests and included the areas of auditory discrimination, attentional capacity, motor speed, and spatial relations ability. Despite the subject's 12 years of education, she was reading and spelling at the mid-seventh grade level, and doing arithmetic at the mid-fifth grade level. Her execution of the Greek Cross was diagnostic of right-hemisphere damage and indicative of impaired spatial relations abilities.

Based on this assessment, a rehabilitation program was designed for Connie. Efforts were directed toward general intellectual enrichment and improvement of academic skills. She needed retraining in spatial relations skills, visual-motor retraining, and her marked attentional impairment needed an intensive rehabilitation focus.

Whenever there was a case like Connie's with diffusely impaired functions, anything that such a person is exposed to in a rehabilitation program is potentially of benefit. In other individuals, where there is circumscribed loss of functions, there is a possibility that some of the rehabilitation experiences may have little or no benefit to the person. For example, speech and language training for a classical right-hemisphere case would usually have little value for the person. On the other hand, the diffusely impaired individual may profit from any type of rehabilitation training, no matter what the cognitive demands of the

CONNIE (1)
SUBJECT IS RIGHT HANDED.
SUBJECT IS RIGHT FOOTED.
SUBJECT IS RIGHT EYED.
SUBJECT DOES NOT HAVE CROSSED EYE-HAND DOMINANCE.

WIDE RANGE ACHIEVEMENT TEST
READING = 7.5 GRADE
SPELLING = 7.8 GRADE
ARITHMETIC = 5.3 GRADE

NAME OF TEST	RATING
HALSTEAD CATEGORY	4.
FORM-BOARD, TIME	4.
FORM-BOARD, MEMORY . . .	2.
FORM-BOARD, LOCATION . .	4.
SPEECH PERCEPTION	2.
RHYTHM	4.
TAPPING SPEED	2.
TRAILS-B	4.
DIGIT SYMBOL	1.
APHASIA SCREENING	1.
SPATIAL RELATIONS	3.
PERCEPTUAL DISORDERS . .	2.
AVERAGE RATING =	2.75
PERCENT OF RATINGS IN THE IMPAIRED RANGE =	83.3%

RATING OF 0=SUPERIOR,
1=AVERAGE,
2=MILDLY IMPAIRED,
3=MODERATELY IMPAIRED,
4=SEVERELY IMPAIRED,
5=VERY SEVERELY IMPAIRED.

NAME OF WAIS SUBTEST	SCORE
INFORMATION	7.
COMPREHENSION	7.
ARITHMETIC	7.
SIMILARITIES	10
DIGIT SPAN	7.
VOCABULARY	5.
DIGIT SYMBOL	8.
PICTURE COMPLETION . . .	8.
BLOCK DESIGN	3.
PICTURE ARRANGEMENT . . .	7.
OBJECT ASSEMBLY	6.
WAIS VERBAL I.Q. =	84.
WAIS PERFORMANCE I.Q. =	76.
WAIS TOTAL I.Q. =	79.

AVERAGE SCORE ON EACH SUBTEST IS 10 AND AVERAGE IQ IS 100. HIGHER VALUES ARE ABOVE AVERAGE AND LOWER ONES ARE BELOW AVERAGE.

Figure 28. Neuropsychology report—a client with diffuse brain damage before treatment.

tasks are. Since the deficits are in many cognitive and perceptual areas, any task involving language or nonlanguage functions could be a potentially fruitful area for retraining. In Connie's case she was equally exposed to tasks designed to enhance language and language-related abilities, spatial-manipulative skills, and higher-level cognitive functions and attentional processes. Throughout her retraining program, she was involved in productive work assignments for a minimum of 3 hours a day. These work assignments were changed frequently to allow maximum exposure to new adaptive situations. However, she spent somewhat more time in the leather area than in other areas because she had unusual talents and interest in these activities. It was also felt that these interests and skills could lead to later competitive employment in leather goods manufacturing. This hope materialized, and following termination from the program, she landed a job, with minimal help, in the manufacture of leather goods. After eighteen months of employment, she obtained a substantial salary increase. In the interim, she had moved from a foster home situation to almost completely independent living (she shared an apartment with a roommate). Another subsequent follow-up indicated a very successful adjustment, and she seemed able to exercise very good judgment in her handling of financial and other matters.

In addition to the work and recreation–socialization programs, this young woman received a great deal of individualized retraining in language and academic skills, and individualized retraining on spatial-relations tasks and in abstract thinking and complex problem-solving. She had daily work in reading and arithmetic skills, reading comprehension, vocabulary and knowledge enrichment. These sessions involved interpersonal retraining as well as retraining of specific cognitive skills. Initially she worked with the conventional reading workbook materials that are used routinely by teachers in the primary grades. Typically she would read a section of material (e.g., a short article), she would answer questions about the material she had read, and then would be further questioned about the material. In addition to the retraining of reading and comprehension skills, these sessions gave the therapist an opportunity to test her on what she had read in such a way as to call forth the need to make logical inferences and to deal with abstractions. She received training on the pursuit rotor for developing visual-motor skills. She received spatial-relations training on the "Memory Board," the visual-acuity test, and the Jensen Board. She was worked with very extensively on memory and conceptual reasoning through work on the Jensen Board and particularly the Progressive Matrices test. In all these areas and tasks, complexity was varied. When she was able to master lower level demands on these tasks, the complexity was increased. There was a gradual improvement noted in her performance in all these areas, and with it there was improvement in her emotional and social integration. At the outset, in view of her severe and chronic maladaptive difficulties, it was not anticipated that a successful vocational adjustment could be made or indeed any adjustment outside of a structured situation. Toward the end of the program, the staff became somewhat more cautiously optimistic about possible competitive employment for her.

This client was retested with the Halstead series of tests six months after the first evaluation and following her rehabilitation program. Figure 29 presents a summary of these scores, which can be compared with the earlier scores and ratings from the first evaluation. Her Average Impairment Rating had decreased from 2.75 to 2.17, which is in the direction of improvement. The percentage of ratings in the impaired range decreased from 83.3% on the first evaluation to 75% on the second. There was a general increase in overall intellectual effec-

CONNIE (2)
SUBJECT IS RIGHT HANDED.
SUBJECT IS RIGHT FOOTED.
SUBJECT IS RIGHT EYED.
SUBJECT DOES NOT HAVE CROSSED EYE-HAND DOMINANCE.

WIDE RANGE ACHIEVEMENT TEST
READING = 7.9 GRADE
SPELLING = 7.8 GRADE
ARITHMETIC = 3.9 GRADE
12½ YEARS OF EDUCATION

NAME OF TEST	RATING	NAME OF SUBTEST	SCORE
HALSTEAD CATEGORY	3.	INFORMATION	7.
FORM-BOARD, TIME	2.	COMPREHENSION	9.
FORM-BOARD, MEMORY . . .	2.	ARITHMETIC	6.
FORM-BOARD, LOCATION . .	3.	SIMILARITIES	12.
SPEECH PERCEPTION	1.	DIGIT SPAN	7.
RHYTHM	3.	VOCABULARY	8.
TAPPING SPEED	2.	DIGIT SYMBOL	10.
TRAILS-B	4.	PICTURE COMPLETION . . .	8.
DIGIT SYMBOL	1.	BLOCK DESIGN	5.
APHASIA SCREENING	2.	PICTURE ARRANGEMENT . . .	8.
SPATIAL RELATIONS	1.	OBJECT ASSEMBLY	10.
PERCEPTUAL DISORDERS . .	2.		
AVERAGE RATING =	2.17	WAIS VERBAL I.Q. =	90.
PERCENT OF RATINGS IN		WAIS PERFORMANCE I.Q. =	87.
THE IMPAIRED RANGE =	75.%	WAIS TOTAL I.Q. =	88.

RATING OF 0=SUPERIOR,
1=AVERAGE,
2=MILDLY IMPAIRED,
3=MODERATELY IMPAIRED,
4=SEVERELY IMPAIRED,
5=VERY SEVERELY IMPAIRED.

AVERAGE SCORE ON EACH SUBTEST IS 10 AND AVERAGE IQ IS 100. HIGHER VALUES ARE ABOVE AVERAGE AND LOWER ONES ARE BELOW AVERAGE.

Figure 29. Neuropsychology report—a client with diffuse brain damage after treatment.

tiveness. On the first testing she was functioning in the borderline defective range, and on the second testing she was functioning at a dull-normal level. There had been some elevation of the IQ levels and there had been improved performance on some of the neuropsychological tests. While her capacity for abstract reasoning continued to be substantially impaired, she did much better on the second testing than on the first. Improvement was more dramatic on tasks involving nonverbal, manipulative problem solving. She did not show the tactile suppressions found previously. Mild improvements were also noted in the areas of auditory discrimination and attentional capacity. Her execution of the Greek Cross suggested improvement of her spatial relations abilities, and this finding generally coincided with her improved functioning on the performance subtests of the WAIS.

This client provides a nice illustration of the exploratory nature of rehabilitation of patients with diffuse brain damage. In her case, a number of modalities were used, until we came upon leather work—a task she could do and that was vocationally relevant. While the neuropsychological tests suggested a wide range of more or less substantial deficits, she did improve, and seemed to benefit significantly from being in the program.

Concluding Comments

These case histories suggest that there is some basis for constructing rationally planned rehabilitation programs based, at least in part, on the concept of functional hemisphere asymmetries. Our clients did improve clinically and psychometrically, although we cannot prove that the improvement would have been greater or lesser had some other approach been attempted. Each of these clients was a significantly impaired individual with major structural brain damage: a severe head trauma, a stroke, and a severe infectious illness. These individuals certainly did not return to "normal" premorbid function, but at least two of the three cases made adequate social and vocational adjustments.

Perhaps two major points were learned from review of these cases. First, there is what seems to be a special adjustment problem in the case of the patient with severe right-hemisphere brain damage. In informal discussions with other clinicians, we have learned that this phenomenon is not uncommon, but it is not well understood. However, the role of affective factors seems to be very important. The "indifference" syndrome of the right-hemisphere patient, in addition to the difficulty he may have in perceiving affective response in the form of facial expression, may be highly critical factors. The second point has to do with treatment of the patient with diffuse brain damage. We often find it difficult to know what to do with these patients, since there is no specific deficit or symptom that can be targeted. In the case of the patient with localized brain damage, the symptom almost automatically defines the rehabilitation approach. Things are clearly not that straightforward with the diffusely brain-damaged patient. Our advice is to first look at the general level of performance, as may be determined by such indices as the Average Impairment Rating, and try out a number of modalities at that level. Using a gradient of adaptive challenges approach, complexity can be increased once a suitable modality has been identified. Considerations as to what the client likes to do, and what is educationally or vocationally relevant, seem to be the major ones. The important point is not to use a modality simply because it is available, without regard to whether or not it is appropriate. The shifting around of modalities that was employed with our last case seemed to be useful, because we finally hit on the leatherwork as something the client liked to do and that ultimately got her a job. Flexibility, which seems to be the watchword in the case of the individual with a diffuse lesion, may not be as desirable in the case of clients with focal lesions, who sometimes require more structure in their programming to ensure that they adhere to retraining tasks they may not particularly like to do because they involve their deficits. Thus, the aphasic patient may not particularly enjoy his speech therapy, but he should be encouraged to pursue it. However, when such a specific, maladaptive symptom is not present, a somewhat looser approach seems more potentially fruitful, as we have illustrated with our last case.

7

THE WICHITA PROGRAM:

Follow-up Studies

The clients who completed the rehabilitation program were followed up from 6 months to 1 year after their final assessment with the Halstead-Reitan Battery. A few of them were lost to the follow-up procedures, and so we were able to secure information on only 16 program participants and 6 of the subjects who did not receive any formal rehabilitation during the 6-month interim period between the two testing sessions. The general purpose of the follow-up was to determine if there was any differential improvement in the adjustment level of the subjects who did and did not go through the program. As indicated in Chapter 6, program subjects showed significantly more improvement in their test scores than the control subjects. The follow-up represented an attempt to learn if the superior test performance was correlated with superior adjustment to the environment. However, any findings derived from these studies must be taken cautiously in view of the very limited number of control subjects.

The follow-up studies of these 16 program and 6 control subjects included face-to-face interviews with the subject and a spouse, relative, or someone who had firsthand knowledge of the subject's day-to-day life activities. The follow-up interviewing was generally done in the subject's home by an individual with extensive experience at a training center for handicapped young adults. A 96-item adjustment interview schedule was designed to gather information about the subject's adjustment in several areas of functioning. The informant was questioned about the subject's behavior on each of these 96 items, and the interviewer wrote a behavioral description of the degree to which the subject had or did not have this particular behavior within his or her repertoire. In addition to this survey, the interviewer administered the Vineland Social Maturity scale to the informant as a means of deriving some objective measure of social adjustment. The informant was administered selected items from the Mooney Problem Checklist: he or she circled those problems on the check list that seemed characteristic of the subject. The subject was also administered these same items from the Mooney Problem Checklist and was asked to indicate those problems on the list that seemed to be characteristic of his or her adjustment. The interviewer wrote a narrative summary of the information gained, including her impressions of the subject, the informant, and the subject's life situation.

Two raters who were uninformed of the identity of the subjects were supplied the behavioral and descriptive data from the interview schedule, the narrative summary, the results of the Vineland Social Maturity Scale, and the two Mooney Problem Checklists (the one by the informant and the one by the subject). On the basis of this information they rated each subject's level of adjustment on a 7-point scale in the following 7 adjustment areas:

1. Self-care
2. Socialization
3. Cultural Awareness
4. Activity Level
5. Recreation
6. Higher-level independent behaviors
7. Economic Dependence-Independence

The two raters each had several years of experience working in a prevocational training program. Ratings were made independently and blindly with regard to whether or not the subject went through the program. Prior to making the actual ratings, the judges received training in the use of the 7-point rating scale and each of the 7 adjustment areas was explained to them. These two raters achieved a high degree of agreement in their ratings of the 16 experimental and 5 of the control subjects. The Pearson Product Moment correlation of .84 was significant at the .01 level and indicated adequate interrater reliability.

The next analysis concerned the difference between ratings of the experimental and the control clients on the 7 adjustment areas. Ratings from interview data are presented in Table 7. When there was a discrepancy between ratings of the two raters, the average of the two ratings was used. The sum of the ratings for any individual over the 7 adjustment areas could theoretically vary from 7 to 49. The higher the numerical rating, the better the adjustment. The sum of the ratings for the 7 areas of adjustment revealed a mean for the experimental group of 38.5, and a mean for the control group of 27.6. A student's t of 3.10

Table 7

Ratings Using a 7-Point Scale on Seven Adjustment Areas for Experimental and Control Clients

Adjustment areas	Experimental clients																Control clients				
	1	2	3	4	5	6	7	8	9	10	11	12	13	14	15	16	1	2	3	4	5
Self care	7	7	7	6	7	4	7	7	7	7	7	7	6	6	7	6	4	7	5	7	7
Socialization	7	5	7	6	4	3	5	7	4	6	7	5	7	5	7	7	4	4	5	6	2
Cultural awareness	3	7	6	4	4	1	4	6	3	5	6	5	7	3	7	6	3	3	3	5	3
Activity (gross)	7	5	7	6	5	4	6	7	7	6	7	6	7	5	7	5	2	5	4	3	5
Recreation	6	6	6	7	3	4	7	6	6	6	7	4	7	5	7	4	2	5	2	3	5
Higher level independence	5	6	6	5	4	1	6	7	6	6	7	6	7	5	5	5	2	6	7	3	3
Economic dependence–independence	5	5	4	3	2	2	3	6	7	2	4	2	6	4	7	1	1	2	1	5	2

($df = 19$) was obtained, which was significant at less than the .01 probability level. This indicated a highly significant difference in favor of the experimental clients on this overall measure of adaptive competence.

The finding of better adaptive competency for the experimentals did not hold up when the two groups were compared on the Vineland Social Maturity Scale. Scores on the Vineland Social Maturity Scale are presented in Table 8. Using raw scores from the Vineland, the experimental group had a mean of 90.29, and the control group had a mean of 86.33. This difference was not found to be statistically significant ($t = 1.05$, $df = 2$, $p > .05$). It was expected that results from the Vineland would correlate with the other measure of general adjustment. However, the particular instructions used in scoring the Vineland may have neutralized any real difference between the two groups. The scorer was told not to give credit for any item unless she was absolutely sure that the particular behavior was well within the repertoire of the individual. Perhaps this sanction led to an arbitrary restriction of the range which did not allow any real differences to emerge, if indeed there were real differences.

There was no difference between experimental and control clients in regard to their admitting to personal adjustment problems on the Mooney Problem Checklist. Scores on the Mooney Problem Checklist are presented in Table 9. The mean number of self-admitted problems for the experimental group was 25.35, and 29.33 for the control group. This difference resulted in a student's t of .2262, which is not significant ($df = 21$, $p > .05$).

Although there was no difference between the experimental and control clients in their awareness and admission of problems, the informants of the clients tended to see the situation otherwise. The informants of control cases tended to see them as having more adjustment problems than their counterparts. The experimental clients had a mean number of problems of 19.62, as seen by their informants; the control clients, as seen by their informants, had a mean number of problems of 39.33. This difference resulted in a student's t of 2.869 ($df = 17$); this value is significant at the .01 level. Although the experimental and control clients did not differ significantly in their perception or self-report of problems, the *informants* assessed the situation quite differently in favor of the experimental clients. The informants for the control clients saw them as experiencing more difficulties than the informants of the experimental clients. Also of interest is a comparison of the *client* experience (and report) of difficulty and their informants' assessment of the same situation. There is a trend for the

Table 8
Scores for Experimental and Control Clients on the Vineland Social Maturity Scale

	Experimental clients																	Control clients				
Vineland Scale	1	2	3	4	5	6	7	8	9	10	11	12	13	14	15	16	17	1	2	3	4	5
Basal score	66	65	66	77	89	70	77	79	94	84	87	80	79	66	77	79	72	66	77	74	76	72
Total score	95	73	92	89	96	75	93	95	98	101	102	96	90	75	92	91	87	87	88	95	83	78
Age equivalent	18.0	8.8	16.5	15.0	18.3	9.3	17.0	18.0	19.0	20.0	21.0	18.3	15.5	9.3	16.5	16.0	13.8	13.8	14.4	18.0	11.7	10.3
Social quotient	72	35	66	71	96	40	73	72	86	80	84	73	62	37	66	65	55	55	58	72	47	41

Table 9
Scores for Experimental Clients, Control Clients, and Informants on the Mooney Problem Checklist

	Experimental clients																	Control clients					
	1	2	3	4	5	6	7	8	9	10	11	12	13	14	15	16	17	1	2	3	4	5	6
Client	43	14	15	60	14	25	10	25	54	31	19	3	42	33	9	2	32	4	25	23	29	63	32
Informant	15	22		12	12	44	31	37	18			11	7	19		10	17	32	23	56	18	67	40

program clients to admit to experiencing more difficulty (25.35 problems) than their informants' assessment of their adjustment (19.62 problems). The opposite situation held for the control clients. Here there is a trend for them to admit to fewer difficulties (29.3) than their informants' assessment of them (39.33 problems). This discrepancy suggests that the control clients had a relatively deficient awareness of their difficulties. There is much more agreement between self- and informant assessment of the experimental clients. Here there is even a tendency for the informant to ascribe *fewer* problems to the experimental client than the clients actually endorsed themselves.

Although evidence supporting the greater adaptive superiority of the experimental subjects is hardly dramatic, there is a clear trend pointing to greater adaptive competency for the experimental subjects as opposed to the control subjects at the time of follow-up. The direction and magnitude of the trend seem to parallel the differences observed in the two groups in their performance on the neuropsychological tests. This correspondence between test and behavioral variables lends validity to the tests themselves. Also, since there were some nontest indicators of superior improvement by the subjects who went through the program as compared to those who did not, it is reasonable to assume that the effects of the rehabilitation went beyond mere improvement of the clients' ability to perform on psychological tests.

We can add a certain amount of general information concerning the adjustment of the members of the experimental and control groups. None of the six control subjects was gainfully employed during the time of the follow-up, either on a part-time or on a full-time basis. One control subject was living full-time in an institutional setting; one was working productively in a workshop setting; one was living with her husband doing only marginally well in taking care of housekeeping duties; two younger control subjects continued to be economically dependent upon their parents. Of the 16 experimental subjects, three were found to be working on a full-time basis in job situations that provided complete economic independence. One of the three was working in a feed lot; a second was working in a filling station doing attendant work and minor mechanical repairs; and a third was working in the manufacture of leather goods and making enough money to provide for her needs. Among the others, another experimental subject was doing part-time farm labor on a fairly frequent and competent basis. A younger experimental subject, who had benefited as much or more than anyone else in the program, was not employed, but he was pursuing schooling to enable him to earn his GED (his schooling had been interrupted by the traumatic head injury which he had sustained). The information from the follow-up interview suggested that this young man was employable, although he did not want to pursue employment at the time because of his schooling. Another experimental subject was receiving on-the-job training that could eventually lead to full-time employment. Two of the experimental subjects were working productively in a sheltered workshop setting and their placements were considered to be of a permanent nature. Another two of the experimental subjects were living with their spouses and were receiving disability payments. Four of the experimental subjects were found to be living with their parents at the time of the follow-up studies and they were completely economically dependent. Three of these four were felt to have some competitive employment potential following further rehabilitation; also, the parents needed some help in resisting their tendency to allow their sons to remain completely dependent. One experimental subject, a middle-aged woman with a left hemiparesis, was living alone and maintaining herself with the help of interested neighbors. The last experimental subject we have follow-up information on, a young woman, was living independently and taking care of her three children. Although she was economically dependent, she was handling her maternal responsibilities quite well. She had taken some further vocational training and at the time of the follow-up she was entering a prevocational program that was designed to lead to job placement.

Perhaps the most serious shortcoming of the rehabilitation program was the failure to involve family members of the clients in a significant way in assisting in the rehabilitation programs. It became quite clear, particularly at the follow-up, that in many cases family and friends of the subjects needed changes in their attitudes and interaction patterns to allow continued growth to occur. The characteristic pattern that emerged in some subjects' families was the tendency to encourage dependency and to provide a protective attitude when the subjects' status never did or no longer required such a degree of protectiveness. In some family situations, the emotional and psychosocial make-up of the family may unfortunately "require" the continual fostering of a dependent status for the subject.

Summary

It was possible to follow up 16 of the subjects who had gone through the Wichita rehabilitation program and six of the subjects who were used as controls. An interviewer's ratings, accomplished on a blind basis, indicated that the subjects who went through the program were better adjusted in various aspects of adaptive function than were the controls. However, no difference was found on the Vineland Social Maturity Scale or the Mooney Problem Checklist when it was filled out by the subjects. When the Mooney Checklist was filled out by informants, there were significantly more problems listed for the controls than for the subjects who went through the program. Whereas the informants for the controls saw them as experiencing more problems than the subjects themselves admitted to, the program subjects tended to report more problems than did their informants. Thus, there seemed to be some congruence between the neuropsychological test and follow-up findings, suggesting more improvement for the subjects who went through the rehabilitation program than those who did not. However, the limited number of control subjects obviously makes this conclusion a very tentative one.

8

BRAIN DAMAGE AND EMOTIONAL DIFFICULTIES

Clinical Considerations

Most, if not all, illness has emotional consequences for the patient and those around him or her. In the case of the brain-damaged patient, one can expect to find many of the emotional concomitants of illness in general, but there appear to be some special consequences as well. In this chapter, we will explore some of these consequences, primarily with case examples. Conceptually, we may think in terms of three forms of interaction between brain pathology and emotional reaction. There is first the impact of the premorbid personality on the individual's reaction to brain damage; second, there are the direct consequences of the lesion itself; and third, there is the manner in which the individual reacts to the illness and the deficits it produces. With regard to the first point, clinical evidence suggests that many people, after sustaining brain damage, develop exaggerated forms of predamage personality characteristics. In general, the nature of the individual's personality and character often plays a significant role in regard to determining the nature of the postdamage outcome. However, the premorbid personality is not the only consideration. As we will see, certain types of brain lesions and lesions in certain areas of the brain have particular implications for the individual's affective life. It also seems that over and above premorbid and neuropathological considerations, there are major individual differences regarding how people react to illness in general and brain damage in particular.

The area we are now entering is obviously a complex one. What is it that determines the behavioral consequences of brain damage? It is clearly not the lesion alone, since individuals with similar lesions respond differently, particularly in the area of the emotions. Some remain distressed and dysphoric over long periods of time, while others accept their disorder with equanimity and go on to do the best they can in order to adjust to their difficulties. Behaviors observed that are initially thought to be consequences of the brain damage turn out to be characteristics the individual exhibited before the brain damage was sustained. In some cases, the brain damage only impairs the individual's capacity to modulate the expression of that characteristic. There are often exceedingly complex symptom pictures in which there are various combinations of bona fide organic deficits and functional elaborations of those deficits. For example, an individual who has had a head injury may develop not only neurological symptoms associated with it but hysterical symptoms as well. There are individuals who suffer from blatant, severe psychopathology prior to their brain damage. In these cases, it becomes exceptionally difficult to differentiate between what is attributable to the brain damage and what is attributable to the preexisting psychopathology. Particular neurological conditions seem to be associated with particular forms of psychopathology. For example, there appears to be a strong association between Huntington's chorea and schizophrenic symptomatology (Whittier, 1977).

One could go on at length considering the various combinations and intertwinements of neurological and psychiatric disorder. However, we will select instead certain topics, describe the pertinent problems, and provide illustrative case material. The first problem to be reviewed concerns the relationship between the laterality of the lesion and adjustment to brain damage. As we found in our own study and as others found, the right-hemisphere patient often does less well in rehabilitation programs than does the left-hemisphere patient. Part of the problem may be that we know less about rehabilitating right-hemisphere functions than we know about language-related left-hemisphere functions. Another possible reason for the discrepancy is that right-hemisphere lesions tend to be larger or more advanced than left-hemisphere lesions at the time when they come to medical attention, probably because the symptoms are more subtle in the case of right-hemisphere brain damage. However, it would appear that the problem goes beyond these considerations. There is a fairly extensive literature concerning the nature of affective response to right-hemisphere brain damage. There appear to be two major findings. One of them is that patients with right-hemisphere brain damage often exhibit a *euphoria*—the term traditionally used to describe it is "douce indifference", an apparent lack of concern for the nature of one's condition (Alajouanine & L'hermitte, 1957; Critchley, 1953; Gainotti, 1972).

In our clinical experience, the absence of an appropriate amount of depression is common, although the presence of a frank euphoria is rare. The other commonly observed syndrome associated with right-hemisphere brain damage is *denial* (Weinstein & Kahn, 1955, 1959; Weinstein, Cole, Mitchell, & Lyerly, 1964). Sometimes there is explicit verbal denial of a disability, such as paralysis of the left arm or leg. Technically, the syndrome is known as *anosognosia*, a lack of awareness of an illness.

These clinical and research findings suggest that the patient with right-hemisphere brain damage, for some reason, does not develop appropriate affective responsivity. This view was dramatically supported in a study by Morrow, Vrtunski, Kim, and Boller (1979) in which it was shown that patients with right-hemisphere brain damage showed literally no affective response, as measured by galvanic skin response to emotional stimuli, whereas even aphasic left-hemisphere subjects showed some degree of response. This important finding strongly suggests that damage to the right hemisphere significantly diminishes the capacity to respond to life events with appropriate affect. Clearly, it becomes difficult to rehabilitate an individual who may deny that he has a disability, or who seems to have little concern for the fact that he is disabled. It is interesting to note that the NYU group described in Chapter 6 works largely with right-hemisphere stroke patients but has a high success rate. Denial and indifference syndromes may be less ubiquitous than we think or perhaps the NYU group has developed techniques that counteract these tendencies.

The emotional impact of the cognitive difficulties associated with right-hemisphere brain damage must be considered. One aspect of this disability has to do with the lack of stability of the perceptual world. The right-hemisphere patient readily becomes bewildered by the complexities of navigating in space and thereby dealing effectively with the ordinary activities of daily life. There is sometimes a deficit in regard to recognition of familiar faces (*prosopagnosia*) in right-hemisphere patients, and this difficulty may lead to an inability to interpret the facial expressions of others. Thus, the patient cannot respond appropriately to the nonverbal affective responses of others. Although the left-hemisphere patient may experience a loss of capacity to communicate verbally, the right-hemisphere patient loses at least some of the ability to communicate nonverbally, thereby depriving him of affective contact with others.

The severe psychiatric difficulties of the right-hemisphere patient are well illustrated in the case of a 55-year-old man who sustained a stroke, which left him with a left hemiplegia. The idea that all right-hemisphere patients develop euphoria or denial is clearly contraindicated by this case. This patient sustained his stroke shortly after a minor auto accident. After hospitalization for the acute phase of the illness, he was transferred to a rehabilitation center, but rehabilitation efforts were unsuccessful. The patient developed a serious depression and made several suicidal threats. He was then referred to a psychiatric facility and placed on several psychotropic medications. Nevertheless, his condition worsened; he became less able to ambulate and became more depressed. It was reported that he developed paranoid ideation and began using obscene language. Ultimately the patient was sent to a psychiatric hospital, where he received the battery of neuropsychological tests shown in Figure 30. It will be noted that the tests were administered about two years after the patient had his stroke.

In his report, the clinical neuropsychologist noted that the patient did appear to have right-hemisphere brain damage, and that his gen-

```
WARNING - 6 TESTS WERE NOT ADMINISTERED.
LATERAL DOMINANCE EXAMINATION.
SUBJECT IS RIGHT HANDED.
SUBJECT IS LEFT EYED.
SUBJECT HAS CROSSED EYE-HAND DOMINANCE.

PERCEPTUAL DISORDERS EXAMINATION   RIGHT    LEFT    WIDE RANGE
TACTILE SUPPRESSIONS                17%       0%    ACHIEVEMENT TEST
AUDITORY SUPPRESSIONS               25%      50%    READING = 13.5 GRADE
VISUAL SUPPRESSIONS                  0%     100%    SPELLING = 07.6 GRADE
FINGER AGNOSIA ERRORS               20%     100%    ARITHMETIC =  .  GRADE
FINGER TIP WRITING ERRORS           45%     100%    11 YEARS OF EDUCATION
LEFT HOMONYMOUS HEMIANOPIA
DICHOTIC LISTENING ERRORS            4%      88%

   NAME OF TEST            RATING          NAME OF WAIS SUBTEST        SCORE

   HALSTEAD CATEGORY . . . 5.              INFORMATION . . . . . . . . . 13.
   FORM-BOARD, TIME  . . . 5.              COMPREHENSION . . . . . . . . 14.
   FORM-BOARD, MEMORY. . . 5.              ARITHMETIC  . . . . . . . . .  8.
   FORM-BOARD, LOCATION  . 5.              SIMILARITIES  . . . . . . . . 10.
   SPEECH PERCEPTION . . . 4.              DIGIT SPAN  . . . . . . . . .  7.
   RHYTHM  . . . . . . . . 3.              PICTURE COMPLETION  . . . . . 13.
   TAPPING SPEEC . . . . . 5.              BLOCK DESIGN  . . . . . . . .  0.
   TRAILS - B  . . . . . . 5.              PICTURE ARRANGEMENT . . . . .  5.
   DIGIT SYMBOL  . . . . . 5.              OBJECT ASSEMBLY . . . . . . .  0.
   APHASIA SCREENING . . . 3.
   SPATIAL RELATIONS . . . 5.              WAIS VERBAL I.Q. =               109.
   PERCEPTUAL DISORDERS  . 4.              WAIS PERFORMANCE I.Q. =           61.
   AVERAGE RATING =        4.50            WAIS TOTAL I.Q. =                 88.
   PERCENT OF RATINGS IN                   AVERAGE SCORE ON EACH SUBTEST IS 10
   THE IMPAIRED RANGE =  100.00%           AN AVERAGE IQ IS 100.  HIGHER
   RATING OF  =SUPERIOR,                   VALUES ARE ABOVE AVERAGE AND LOWER
   1=AVERAGE,                              ONES ARE BELOW AVERAGE.
   2=MILDLY IMPAIRED,
   3=MODERATELY IMPAIRED,
   4=SEVERELY IMPAIRED,
   5=VERY SEVERELY IMPAIRED.
```

Figure 30. Neuropsychology report of a depressed right-hemisphere stroke patient.

eral level of performance was in the severely impaired range. It appeared, clinically, that even though the stroke occurred some time in the past, it still looked as though the patient was suffering from acute brain damage. It was noted that the patient had a left homonymous hemianopia, and that he might benefit from visual-field retraining. It was suggested that he not be spoken to, or shown things, in the area subtended by his left visual field.

Subsequent psychiatric evaluations of this patient contained comments concerning poor memory, judgment, and insight. The patient continued to have a depressed mood. However, following a course of counseling sessions, the depression was somewhat relieved, and the patient's motivation improved. Ultimately, plans were made to discharge the patient in the care of his wife, who was quite devoted to him. Thus, although this man apparently could leave the hospital to return to a very supportive environment, he nevertheless had a very difficult time in dealing with the consequences of his stroke. His difficulty occurred despite extensive rehabilitation efforts and his supportive wife.

It is, of course, not clear that this patient would have had a less difficult time had his stroke been in his left hemisphere. Indeed, there was some evidence that his depression was developing prior to the stroke. Nevertheless, in our experience the case is not at all atypical of patients we have studied who sustained right-hemisphere strokes. It is perhaps most important to point out that this patient, despite the fact that his lesion involved the right hemisphere, had a serious, potentially life-threatening depression following the stroke. One cannot assume that euphoria–denial syndromes will always permanently accompany right-hemisphere brain damage; the reverse may occur. Psychotherapy and possibly medication (the patient was taking Sinequan, an antidepressant) may be helpful to patients of this type.

In the case of the patient with right-hemisphere brain damage, one possibility is that the euphoria–denial syndromes are acute phenomena, and last only as long as the pathological process is active. In regard to anosognosia, it seems that this condition lasts for only a brief period following the stroke or other acute lesion. Following this phase, both right- and left-hemisphere cases may develop reactive depressions associated with the life-threatening event they have experienced and the obvious loss of function. To illustrate this point, we will present the case of a left-hemisphere stroke patient.

The patient was a 62-year-old man who suffered a stroke that left him with a right hemiparesis. There was no mention in the medical records of an aphasia resulting from the stroke. About ten months after this episode, the patient was admitted to a psychiatric facility not for the stroke but for a severe depression. He made an apparently serious threat to shoot his wife and himself. The neuropsychological tests were administered shortly after the psychiatric admission, but eleven months after the stroke. The results are presented in Figure 31. It will be noted that the patient was of average intelligence, but had difficulties with tasks involving psychomotor speed, abstract reasoning, and visual-spatial abilities. On the aphasia examination, the only difficulties were in the areas of enunciation and right-left orientation. Clearly, he was not aphasic in the usual sense in which the term is used. Scores on tapping and grip strength suggested that the right hemiparesis was quite mild. These findings would indicate that the patient had a mild stroke, that he had made a good recovery, or both. Nevertheless, he developed such a severe depression following the stroke that he required hospitalization for it. He complained of weight loss, dizzy spells, and loss of sexual interest and capability.

There is obviously much resemblance between the left-hemi-

LATERAL DOMINANCE EXAMINATION
SUBJECT IS RIGHT HANDED.
SUBJECT IS RIGHT EYED.
SUBJECT DOES NOT HAVE CROSSED EYE-HAND DOMINANCE.

PERCEPTUAL DISORDERS EXAMINATION	RIGHT	LEFT
TACTILE SUPPRESSIONS	8%	8%
AUDITORY SUPPRESSIONS	25%	25%
VISUAL SUPPRESSIONS	0%	0%
FINGER AGNOSIA ERRORS	0%	0%
FINGER TIP WRITING ERRORS	20%	25%

WIDE RANGE
ACHIEVEMENT TEST
READING = 10.2 GRADE
SPELLING = 07.2 GRADE
ARITHMETIC = 05.3 GRADE
8 YEARS OF EDUCATION

NAME OF TEST	RATING
HALSTEAD CATEGORY . . .	4.
FORM-BOARD, TIME . . .	2.
FORM-BOARD, MEMORY . .	1.
FORM-BOARD, LOCATION .	4.
SPEECH PERCEPTION . . .	3.
RHYTHM	2.
TAPPING SPEED	4.
TRAILS-B	5.
DIGIT SYMBOL	4.
APHASIA SCREENING . . .	3.
SPATIAL RELATIONS . . .	2.
PERCEPTUAL DISORDERS .	2.
AVERAGE RATING =	3.00
PERCENT OF RATINGS IN THE IMPAIRED RANGE =	91.67%

RATING OF 0=SUPERIOR,
1=AVERAGE,
2=MILDLY IMPAIRED,
3=MODERATELY IMPAIRED,
4=SEVERELY IMPAIRED,
5=VERY SEVERELY IMPAIRED.

NAME OF WAIS SUBTEST	SCORE
INFORMATION	10.
COMPREHENSION	9.
ARITHMETIC	9.
SIMILARITIES	8.
DIGIT SPAN	6.
VOCABULARY	11.
DIGIT SYMBOL	3.
PICTURE COMPLETION	9.
BLOCK DESIGN	5.
PICTURE ARRANGEMENT	7.
OBJECT ASSEMBLY	4.
WAIS VERBAL I.Q. =	97.
WAIS PERFORMANCE I.Q. =	88.
WAIS TOTAL I.Q. =	93.

AVERAGE SCORE ON EACH SUBTEST IS 10 AND AVERAGE IQ IS 100. HIGHER VALUES ARE ABOVE AVERAGE AND LOWER ONES ARE BELOW AVERAGE

Figure 31. Neuropsychology report of a depressed left-hemisphere stroke patient.

sphere and the right-hemisphere case. Both had strokes following which there ensued serious depressions, with suicide contemplated. It is very difficult to generalize on the basis of a small number of case examples, but our experience has been that the pattern represented by these cases is not at all atypical. Thus, regardless of whether the patient has right- or left-hemisphere brain damage, at least in the case of stroke, one should be alert to the possibility of the emergence of a serious depression. If the patient has a right-hemisphere lesion, euphoria and denial may be observed during the acute phase of the illness, but these symptoms will most likely disappear and may be replaced by depression. The left-hemisphere patient may be depressed even during the acute phase of the illness, particularly if aphasia is present.

The General Problem of Depression

Depression is a natural and expected consequence of serious illness. Brain damage is really always serious in nature, and one would expect to find some degree of reactive depression associated with it. As we have seen in the two cases mentioned above, this reaction is sometimes more than what would be expected, and the patient may become psychotically, suicidally depressed. Depression is naturally common among patients with such progressive terminal illnesses as Huntington's disease. It is well known that there is a high incidence of suicide among Huntington's disease patients (Dewhurst, Oliver, & McKnight, 1970; Oltman & Friedman, 1961; Reed & Chandler, 1958; Wilson & Garron, 1979). Even Huntington himself (Huntington, 1872) defined a feature of the disorder as "a tendency to that insanity that leads to suicide" (p.320). In attempting to understand the depression of the brain-damaged patient, it is not always possible to distinguish between "normal" reactive depression and some direct consequence of the brain lesion itself. Some forms of brain damage, notably a condition called *pseudobulbar palsy*, are associated with an abnormal affective state in which there is sudden spontaneous laughing and crying. However, this condition is generally viewed as an exceptional one in which the laughing and crying simply emerge spontaneously and bear no relationship to the content of the patient's ongoing thoughts.

In our experience, we have seen some brain-damaged patients who appear to develop agitated depressions. They pace, wring their hands, and often become verbally or physically belligerent. They tend to have sleep disturbances and generally appear to be uncomfortable. This state is not atypical among patients with Korsakoff's syndrome, and may be seen in other patients as well. The etiology of this kind of state is unclear, but something like it is often seen in alcoholics and in some individuals with affective disorders not associated with structural brain damage. Himmelhoch, Mulla, Neil, Detre, and Kupfer (1976) have referred to this condition as a "mixed mood state" because it reflects a combination of anxiety and depression. On the basis of reports from individuals who have episodes of this state, it is extremely uncomfortable.

In the understanding of depression in individual brain-damaged patients, it is important to distinguish between whether an affective disorder was or was not present prior to acquisition of the brain lesion. In the first case, the reaction to the lesion is grafted onto a previously existing depressive character structure, if not a frank affective disorder. After all, individuals with such conditions as manic-depressive illness or unipolar affective disorders can have strokes, head injuries, and other brain disorders as well. In the case of the not previously depressed patient, the depression is often no more than a normal grief reaction to some loss of mental or physical function. As in the general psychiatric process of diagnosis, it is also important to determine the type of affective disorder present. Rather than depression there may be mania, which may also be episodic or stable. The depression may be of the agitated or retarded type. In essence, for any kind of rational treatment and rehabilitation approach, the nature of the affective disorder should be ascertained.

The reasons for this kind of careful and detailed evaluation are related to treatment considerations. Different forms of affective disorder are treated differently. Even within the realm of pharmacological treatment, patients with different subtypes of affective disorder show differential responses to the various medications (Goodwin, Cowdry, & Webster, 1978). There are several classes of medication used for treatment of affective disorders generally described as tricyclic antidepressants, tetracyclic antidepressants, monoamine oxidase (MAO) inhibitors, energizers, and antimanic medications. The choice of the particular drug class seems associated with the type of affective disorder such that, for example, patients with bipolar affective disorders often respond well to lithium, whereas patients with unipolar disorders generally do better with the tricyclic or tetracyclic medications. The application of behavioral therapies, such as psychotherapy or cognitive behavior therapy, to depression, or the combined approach involving behavioral therapies with medication, must also take specific diagnosis into consideration. For example, Lewinsohn (1979) has indicated that his behaviorally oriented approach to treatment of depression is not effective for individuals with bipolar (manic-depressive) disorders.

From the point of view of rehabilitation, depression is an exceedingly important issue. Sometimes, observed cognitive, perceptual, and motor deficits may not be consequences of the brain lesion, but of the depression associated with it. There is much interest currently in the problem of so-called pseudodementia (Wells, 1979), a condition that looks like the intellectual changes typically associated with dementia. When the patient is successfully treated for depression, the "dementia" often goes away. Thus, there seems to be little point in utilizing expensive neurodiagnostic procedures or engaging in an extensive cognitive retraining program with a patient when the cognitive deficits are depression related and better treated with psychotherapy or medication. Of course, some patients are both demented and depressed. In these cases, it is usually necessary to deal with both matters. It should not be assumed that the depression cannot be treated because of the dementia, nor should it be assumed that the depression will automatically disappear as the patient learns to cope with his intellectual deficits. A comprehensive rehabilitation program should include treatment of both cognitive and affective disorders.

The Problem of Psychosis Associated with Brain Damage

The issue of psychosis in the brain-damaged patient is a complicated one. The patient may have been psychotic before acquisition of the brain lesion, and we are then faced with a combination of the two problems. On the other hand, the effects of the lesion may be so traumatic that they induce a reactive psychosis; that is, a severe acute disorganization of the personality. The terms *organic psychosis* and *psychotic organic brain syndrome* are also sometimes used in psychiatric circles. Whereas the determination of whether an organic brain syndrome is psychotic or nonpsychotic appears to be based on the severity of the disorder, there are typical psychotic-type symptoms associated with certain brain disorders. For example, in the case of general paresis there are often visual or auditory hallucinations as the predominant symptoms. Most often, however, the emergence of psychotic symptoms represents some interaction between the brain damage and the premorbid character. For example, an individual with an obsessive-compulsive personality structure may develop paranoid symptoms following brain damage.

It frequently happens that the early symptoms of certain brain disorders are diagnosed as functional psychoses. Perhaps the most widely studied example of this tendency is the case of Huntington's chorea. This illness often presents with a combination of behavioral symptoms some of which are typical of patients with dementia, but some of which show a great resemblance to the behavior of schizophrenics. Whittier (1977), in his scholarly analysis of the psychopathology of Huntington's chorea, cites a study of Garron (1973) in which he indicates that such symptoms as inappropriate sexual behavior, delusions of grandiosity, delusions of persecution, and hallucinations are seen in patients with Huntington's chorea. Often these symptoms precede the onset of the choreic movements themselves. Of course, the issue of the extent to which these symptoms are attributable to premorbid tendencies that have become disinhibited by the progression of the disease process has not been fully resolved. In addition to Huntington's chorea, the early manifestations of multiple sclerosis often lead to the diagnosis of schizophrenia.

Often, the intertwinements between psychosis and structural brain damage are of particular clinical interest. For example, we treated a patient who was probably premorbidly a schizoid individual given to episodes of severe depression. During one of these episodes, he attempted suicide with carbon monoxide. He was resuscitated, but there was a residual of massive, permanent brain damage. In dealing with this patient, it was clear that what we had to cope with was an interaction between the kind of person he was before the episode, and the severely dementing effects produced by carbon monoxide poisoning. The usual situation in patients who have survived carbon monoxide intoxication involves an interaction between the premorbid depression, which is characteristically of psychotic proportions, and the effects of the toxic agent. The confabulatory, sometimes seemingly delusional verbalizations of the patient with Korsakoff's syndrome may well represent an interaction between those personality characteristics that stimulated the excessive use of alcohol and the direct effects the alcohol and thiamine deficiency have on the brain. The propensity for confabulation may, in some cases, precede the onset of the Korsakoff's syndrome.

It seems clear that the presence of a psychosis can present serious obstacles to successful rehabilitation. The patient's behavior may be so bizarre and disorganized that any retraining effort may prove to be impossible. Delusional beliefs or disturbing hallucinations may also make for obvious difficulties in rehabilitation. Thus, the treatment of the psychosis becomes a pertinent issue. If retraining can be carried on while the psychosis is present, it certainly should be. The process of retraining itself may aid the patient in gaining better contact with the environment and in reestablishing his identity. However, sometimes the psychotic symptoms are sufficiently severe to rule out any possibility of getting the patient to cooperate for a systematic retraining program. In these cases, it is necessary to treat the psychosis first. Most clinicians would admit, perhaps begrudgingly in some cases, that the best treatments currently available for controlling psychotic behavior are the various psychotropic medications: in particular, the phenothiazines and related compounds. To the more behaviorally or psychodynamically oriented reader, it may be pointed out that the use of medication should not be negated because of our biases. There is little point in not relieving, through the use of medication, the painful symptoms the patient is experiencing because of biases against somatic therapies. On the other hand, to the more pharmacologically oriented clinicians, it must be emphasized that medication should be used to prepare the patient for rehabilitation and not to substitute for it. Thus, it is often useful to maintain the patient on medication while the rehabilitation program is ongoing.

There is an extensive literature regarding the application of behavior therapy to patients with functional psychoses (Hersen & Bellack, 1978). The research in this area attests to the effectiveness of such methods as token economy, modeling, assertiveness training, and the like with regard to reducing symptomatology in schizophrenic patients. However, the extent to which these methods are applicable to the organic psychoses has not been thoroughly evaluated. Whereas the Wichita program provided data indicating that operant methods were effective with the brain-damaged clients treated, there was only one client who was both brain damaged and frankly psychotic. As it turned out, she did not do well in the program. It would appear *a priori* that for such programs to be effective, one would have to deal with both the psychotic behavior and the behavior that can reasonably be attributed to the brain lesion itself. As indicated, some if not all of the psychotic behavior may be treatable with medication, but sometimes medication in combination with behavioral treatment is needed. The operant treatment of psychotic symptoms is something of an art, and at times has to take into consideration the nature of the underlying psychopathology as well as the overt behavior.

We would now like to present two cases, not necessarily as rehabilitation successes, but simply as illustrations of situations in which the patient has a structural brain lesion and a functional psychosis. The first patient was a fifty-year-old man with a long psychiatric history. Indeed, his first psychiatric symptoms were noted in the military, some 25 years prior to our evaluations of him. He was hospitalized for most of the intervening years. His behavior at the time of evaluation was characterized by withdrawal, poor memory, poor judgment, and remnants of paranoid thinking. He required much assistance in order to maintain adequate grooming; on the ward, he responded only when spoken to; on occasion, he experienced auditory hallucinations. In essence, the patient presented the picture of what is typically seen among chronic schizophrenic patients with long histories of institutionalization. However, there was a history of one grand mal seizure suffered at some time in the past while the patient was in a state hospital.

The seizure history and the poor memory led to a neurological consultation for the patient. A CT scan was taken and proved abnormal. The abnormality was suspected to be a tumor in the posterolateral aspect of the right frontal lobe: it was felt to be either a meningioma or a small glioblastoma. The electroencephalogram was normal, but the neurological examination yielded several positive findings; the pupils were sluggish, and the patient had a mild astereognosis. In view of these relatively soft neurological signs but the very abnormal CT scan the patient was referred for neurosurgical consultation. An arteriogram was done, which was normal. This finding indicated that the patient probably did not have a tumor, so no neurosurgery was performed. He remains hospitalized. The results of the neuropsychological tests are presented in Figure 32.

The report indicated that the patient was functioning at a very impaired level. The test data actually suggested that the left hemisphere was functioning less well than the right. The patient was motorically slower, more dyskinetic and weaker with his right hand than he was with his left. These findings are somewhat puzzling in view of the fact that the lesion was apparently in the right-hemisphere. We do not have an explanation of this discrepancy except to say that if the patient did have a tumor we may have been seeing the effects of pressure against the noninvolved hemisphere. What is perhaps more pertinent is the overall picture. One would not normally expect a severe and pervasive dementia of the type exhibited by this patient on the basis of a small, focal lesion. It is highly likely that what we are seeing is a combination of the effects of the lesion and the schizophrenic process. In our experience, this kind of picture is not at all atypical among long-term chronic schizophrenic patients, many of whom look severely demented on neuropsychological tests even in the absence of identifiable structural brain lesions.

The second patient was a thirty-year-old man with a military ser-

```
WARNING - 6 TESTS WERE NOT ADMINISTERED.
LATERAL DOMINANCE EXAMINATION.
SUBJECT IS RIGHT HANDED.
SUBJECT IS RIGHT EYED.
SUBJECT DOES NOT HAVE CROSSED EYE-HAND DOMINANCE.

PERCEPTUAL DISORDERS EXAMINATION    RIGHT   LEFT    WIDE RANGE
TACTILE SUPPRESSIONS                   0%    17%    ACHIEVEMENT TEST
AUDITORY SUPPRESSIONS                  0%     0%    READING = 04.6  GRADE
VISUAL SUPPRESSIONS                    8%     0%    SPELLING = 04.3  GRADE
FINGER AGNOSIA ERRORS                 70%    80%    ARITHMETIC = 01.9 GRADE
FINGER TIP WRITING ERRORS             40%    50%    12 YEARS OF EDUCATION
DICHOTIC LISTENING                    50%    56%

    NAME OF TEST                 RATING     NAME OF WAIS SUBTEST        SCORE

    HALSTEAD CATEGORY . . . . .    5.       INFORMATION . . . . . . . .    8.
    FORM-BOARD, TIME  . . . . .    5.       COMPREHENSION . . . . . . .    4.
    FORM-BOARD, MEMORY  . . . .    5.       ARITHMETIC  . . . . . . . .    3.
    FORM-BOARD, LOCATION  . . .    5.       SIMILARITIES  . . . . . . .    8.
    SPEECH PERCEPTION . . . . .    5.       DIGIT SPAN  . . . . . . . .    4.
    RHYTHM  . . . . . . . . . .    3.       VOCABULARY  . . . . . . . .    5.
    TAPPING SPEED . . . . . . .    4.       DIGIT SYMBOL  . . . . . . .    0.
    TRAILS-B  . . . . . . . . .    5.       PICTURE COMPLETION  . . . .    2.
    DIGIT SYMBOL  . . . . . . .    5.       BLOCK DESIGN  . . . . . . .    3.
    APHASIA SCREENING . . . . .    4.       PICTURE ARRANGEMENT . . . .    0.
    SPATIAL RELATIONS . . . . .    3.       OBJECT ASSEMBLY . . . . . .    1.
    PERCEPTUAL DISORDERS  . . .    4.
    AVERAGE RATING =               4.42     WAIS VERBAL I.Q. =                74.
    PERCENT OF RATINGS IN                   WAIS PERFORMANCE I.Q. =           54.
    THE IMPAIRED RANGE =         100.00%    WAIS TOTAL I.Q. =                 64.
    RATING OF 0=SUPERIOR,
    1=AVERAGE,
    2=MILDLY IMPAIRED,
    3=MODERATELY IMPAIRED,
    4=SEVERELY IMPAIRED,
    5=VERY SEVERELY IMPAIRED.
```

Figure 32. Neuropsychology report of a patient suspected of having a brain tumor.

vice-connected disability for schizophrenia. His psychiatric symptoms persisted after discharge from the service, and periodic rehospitalizations were necessitated by episodes of what were generally described as bizarre behavior. On admission, he was withdrawn, usually spoke incoherently, and smiled and laughed inappropriately. The basic reason for the multiple admissions was that after discharge the patient could not cope effectively with independent living and would decompensate. During one hospitalization, a CT scan was performed and was abnormal. The abnormalities were in the left temporal lobe and in the posterior temporal-anterior occipital region, and were thought to be arteriovenous malformations. Neurosurgical consultation indicated that the lesions were inoperable. The neuropsychological test results are present in Figure 33. In his interpretation, the neuropsychologist suggested the presence of left-hemisphere brain damage, but again, the Average Impairment Rating of 3.00 suggests a greater degree of dementia than would be expected on the basis of two focal vascular lesions.

While these two cases can in no way be characterized as "rehabilitation successes," they do point to the difficulties one can expect to encounter with patients who are both brain damaged and schizophrenic. In both cases, the severity of intellectual impairment was far greater than what might have been expected on the basis of the structural brain lesion alone. It is not relevant here to deal with the venerable issue of whether or not schizophrenia is in fact an organic brain disease. However, if one sympathizes with the biological views of schizophrenia, it can be said that these patients have two brain disorders to cope with: one of known etiology and the other of unknown etiology. The interaction between a structural lesion and a "schizophrenic brain" is surely a complex matter about which very little is known.

LATERAL DOMINANCE EXAMINATION
SUBJECT IS RIGHT HANDED.
SUBJECT IS RIGHT EYED.
SUBJECT DOES NOT HAVE CROSSED EYE-HAND DOMINANCE.
SUBJECT IS RIGHT FOOTED.

PERCEPTUAL DISORDERS EXAMINATION	RIGHT	LEFT
TACTILE SUPPRESSIONS	0%	0%
AUDITORY SUPPRESSIONS	0%	0%
VISUAL SUPPRESSIONS	67%	0%
FINGER AGNOSIA ERRORS	0%	0%
FINGER TIP WRITING ERRORS	35%	35%
DICHOTIC LISTENING ERRORS	48%	16%

WIDE RANGE
ACHIEVEMENT TEST
READING = 13.5 GRADE
SPELLING = 14.4 GRADE
ARITHMETIC = 06.5 GRADE
13 YEARS OF EDUCATION

NAME OF TEST	RATING
HALSTEAD CATEGORY	3.
FORM-BOARD, TIME	5.
FORM-BOARD, MEMORY . . .	3.
FORM-BOARD, LOCATION . .	4.
SPEECH PERCEPTION	2.
RHYTHM	2.
TAPPING SPEED	4.
TRAILS-B	3.
DIGIT SYMBOL	3.
APHASIA SCREENING	2.
SPATIAL RELATIONS	3.
PERCEPTUAL DISORDERS . .	2.
AVERAGE RATING =	3.
PERCENT OF RATINGS IN THE IMPAIRED RANGE =	3.00

RATING OF 0=SUPERIOR,
1=AVERAGE,
2=MILDLY IMPAIRED,
3=MODERATELY IMPAIRED,
4=SEVERELY IMPAIRED,
5=VERY SEVERELY IMPAIRED.

NAME OF WAIS SUBTEST	SCORE
INFORMATION	8.
COMPREHENSION	7.
ARITHMETIC	10.
SIMILARITIES	9.
DIGIT SPAN	9.
VOCABULARY	11.
DIGIT SYMBOL	5.
PICTURE COMPLETION	8.
BLOCK DESIGN	5.
PICTURE ARRANGEMENT	10.
OBJECT ASSEMBLY	7.
WAIS VERBAL I.Q.	93.
WAIS PERFORMANCE I.Q. =	81.
WAIS TOTAL I.Q. =	87.

Figure 33. Neuropsychology report of a patient with an arteriovenous malformation.

As we have suggested, the best treatment for brain-damaged patients who are also schizophrenic usually involves a combination of pharmacological and behavioral methods. In our experience, the pharmacological methods are not as effective when the psychosis is not a true schizophrenia, but is directly associated with the structural neurological disorder. For example, the delusions sometimes seen in patients with organic brain syndromes of various types do not seem to be effectively treatable with phenothiazines and related medications. Caution should be exercised in using various pharmacological agents with brain-damaged patients, since, in some cases, the nature of the neurological disorder might contraindicate the use of certain drugs. As a general rule, the brain-damaged patient requires less medication than the nonbrain-damaged patient in order to achieve some given effect. We have seen some severely brain-damaged patients developing marked extrapyramidal side effects with very small doses of antipsychotic medication. The behavioral methods used may vary greatly, depending on the patient's symptoms and the goals of the treatment. The fact that the patient has a brain lesion should not rule out the use of such methods as token economy, single-subject design programs, assertiveness training, or any of the other currently available behavior-therapeutic methods.

Perhaps the neuropsychologically oriented behavior management and retraining methods reported in this book can be productively applied to chronic schizophrenic patients who do not have documented structural brain damage. Would it help the schizophrenic patient if he or she were retrained in memory abilities, conceptual abilities, motor functions, and the other aspects of behavior focused on in the programs described here? We do not have the answer to that question, but it may be noted that there is a great deal of current interest in the notion that schizophrenia may have its basis in some kind of dysfunction of brain laterality. Some believe that the dysfunction involves the left-hemisphere (e.g., Gur, 1978); others believe that it is a kind of "disconnection syndrome" (Beaumont & Dimond, 1973). Based on her conceptualization of the problem, Gur (1979) recommends a course of treatment for the schizophrenic in which one first deals with the left-hemisphere dysfunction and then goes on to attempt to teach the patient to place reliance on the right hemisphere. The "left-hemisphere" stage of the treatment has to do with correction of logical errors in interpreting reality, whereas the right-hemisphere phase has to do with focusing on the more intuitive and emotional aspects of behavior. To the best of our knowledge no one has systematically employed this strategy with schizophrenic patients, but the laterality approach does provide a conceptual framework for treatments that might be employed in the future. According to Myslobodsky and Weiner (1978) the hemisphere–asymmetry concept even has implications for pharmaceutical treatment since certain drugs may have an unequal influence on each hemisphere. In summary, although no one has yet (as far as we know) directly applied a hemisphere–asymmetry concept, either behaviorally or chemically, to the treatment of schizophrenia, the concept of there being some disturbance of brain laterality in this condition may eventually lead to such treatment.

There remains the matter of schizophrenia-like or "schizophreniform" conditions seen in some of the neurological disorders. We have already mentioned the case of Huntington's disease, in which psychotic-like symptoms are sometimes seen preceding the onset of motor symptoms. Before seeking some esoteric reason for this occurrence, we should be reminded that a psychosis is often a form of solution to a problem, and represents the individual's best adjustment to an overwhelmingly stressful situation. In the case of Huntington's disease, most individuals with the genetic loading for this disorder are aware that they are at a great risk of acquiring it. Thus, they are frequently under great stress from the point in life at which they are able to comprehend what may happen to them. Under these circumstances, the appearance of psychotic symptoms is not surprising. Furthermore, Boll, Heaton, and Reitan (1974) showed on the basis of MMPI data that patients with Huntington's chorea had essentially the same degree and type of emotional disturbance as did patients who were equally disabled due to other types of brain damage.

Horenstein (1970) provides numerous examples of how various paranoid and delusional symptoms are explainable on the basis of focal brain damage. In the area of vision, the patient may perceive stimuli arising from homologous portions of the visual field as "different" from each other. Visual function may be grossly defective, but the patient will deny that there is anything wrong with his vision. Many misinterpretations, distortions in reproduction, and misperceptions may occur, although the patient may persist in denying that anything is wrong. The patient with constructional apraxia, being uncertain of his spatial orientation and unaware of his situation, may attribute his errors and illness to vaguely defined external agents. This complex disorder easily leads to the development of delusions and hallucinations. These kinds of phenomena, which may technically be classified as psychotic behaviors, are apparently directly attributable to the primary consequences of the lesion, and need have nothing to do with the premorbid personality. Although Horenstein mentions that some investigations have indicated a relationship between anosognosia and a predisposing personality type, he indicates that the evidence is less than convincing.

In summary, it would appear that many of the psychotic reactions seen in association with brain damage in people who were not schizophrenic before the damage was sustained are of two major types. The first type seems to be a stress or "catastrophic reaction" developed in response to the often accurately perceived dire consequences of the illness. The second type has to do with the consequences of alterations in perceptual function or visual-spatial relations that are directly attributable to the structural brain lesion itself. These latter symptoms are sometimes self-limiting and may disappear with recovery. Thus, for example, when the patient's constructional apraxia resolves, the paranoid-appearing delusions developed in association with it may disappear as well. A special problem emerges here for the obvious reasons that schizophrenic symptoms will almost of necessity interfere with rehabilitation efforts, and our capability of resolving such symptoms is limited because of our lack of knowledge about the causes and treatment of schizophrenia.

Disorders of Activation Associated with Brain Damage

We have used the term *activation* to describe the series of psychiatric disorders under consideration now, rather than the term *impulse disorder*. The reason for this choice is that we are dealing with a bipolar phenomenon. Brain damage may produce impulsiveness and hyperactivity, but it may also produce loss of socially appropriate impulsiveness, and formerly normally active individuals may become severely inert and apathetic following certain types of brain damage. As in other areas, a change in activity level following brain damage may represent an interaction with the premorbid personality, or it may be a direct consequence of the lesion. We will begin our discussion with the more familiar problem of impulse disorder, or hyperactivity, and proceed to discuss several disorders associated with hypoactivity.

The Hyperactivity Disorders

The hyperactivity or impulse disorders may conveniently be divided into two major types. One type is episodic in nature, whereas the other is a more or less steady-state condition. Of course, all impulsivity by definition occurs in episodes, but while some individuals seem to be in a perpetual state of engaging in such episodes, others engage in impulsive activity rarely, and often unpredictably. The reader will no doubt make the association between the latter condition and epilepsy. The condition that most often produces these episodic, impulsive outbursts is what is commonly called *psychomotor* or *temporal-lobe epilepsy*. "Out of a clear blue sky" the patient may suddenly engage in some purposeful-appearing act of behavior—sometimes, but rarely, violent behavior. After the episode ends, the patient may have no recollection of it. In cases of true temporal-lobe epilepsy, the purposive-appearing behavior is actually a seizure, and there is an electroencephalographic abnormality accompanying it. There is no particular reason to go into the intricacies of diagnosing temporal-lobe epilepsy here. The treatment for the disorder is somatic in nature. Usually, anticonvulsant drugs, such as Dilantin, can successfully control the incidence of episodes. When the condition is severely disabling and the patient is refractory to drug therapy, certain surgical procedures are sometimes effective.

From the point of view of rehabilitation, the patient with steady-state impulsive behavior presents more of a challenge. There appears to be a strong association between impulsiveness and a disorder of attention. The classic example is the hyperkinetic, learning-disabled child whose fundamental problem is often an exceedingly brief attention span. Because of this failure of attention, these children are highly distractible, and make overt responses to almost any stimulus in the field. They cannot filter out distracting stimulation to focus their attention to the task at hand. K. Goldstein and Scheerer (1941) referred to this phenomenon as "stimulus bond" or "forced responsiveness." The individual is compelled to respond to any stimulus that confronts him. Sometimes this condition is termed *hypermetamorphosis*—an excessive tendency to pay attention to and respond to every visual stimulus. The new Diagnostic and Statistical Manual of the American Psychiatric Association (DSM-III) refers to this condition in children as an "Attention Deficit Disorder with Hyperactivity." An attention disorder may also exist in the absence of hyperactivity, but it is rare.

There are several neurological conditions that are known to give rise to impulse disorders. Sometimes open-head injuries result in this condition, particularly when the frontal lobes are involved and there is an epileptogenic focus. The reader may be familiar with this problem in postencephalitic children. As Jervis (1959) graphically described the behavior, "It is marked by episodes of overactivity, restlessness, impulsiveness, assaultiveness, and wanton destruction" (p. 1305). There is a classic condition first described in animals, known as the Klüver-Bucy syndrome, which results from bilateral removal of the temporal lobes. The symptoms include, among other things, rage reactions, increased sexual activity (often of an inappropriate nature), and hypermetamorphosis. Human analogues of the Klüver-Bucy syndrome are rare, and we ourselves have not had any experience with postencephalitic children. However, in the case of the head-injured patient, it is our opinion that the presence of a steady-state impulse disorder represents an interaction between the lesion and the premorbid personality. We treated one head-injured patient who was perpetually assaultive, verbally and physically. When we looked at his history, it was apparent that he was premorbidly a violence-prone individual with a relatively long record of episodes of violent, impulsive behavior. After his head injury in a car accident, he showed little moral constraint in his behavior, and being physically strong, seriously injured several people on the ward in which he was hospitalized.

The treatment of patients of this type is difficult. The use of medication is generally not effective, as these patients, in our experience, do not respond well to phenothiazines or other antipsychotic agents. With cases who are extremely violent, such as the head-injured patient just described, the only solution seems to be to maintain light sedation. We treated our patient with Benadryl with some effect in that the episodes of assaultiveness were reduced. Typically, these patients must be treated in some form of institutional setting, since they cannot function in normal society without getting into significant legal or moral difficulties. In essence, whereas our emphasis on cognitive and perceptual retraining is certainly important with these patients, the first treatment goal must, of necessity, be resocialization. The patient must become at least tolerable in normal society. Thus, the rehabilitation setting has to become a microcosm for the social world in which the patient can be encouraged to learn to respond in socially acceptable ways. In many respects, the treatment can take the form of the manner in which one brings up a child. Since these patients often do not have the cognitive mechanisms necessary to impose controls internally, it is often necessary to make sometimes heroic efforts to place the patient under environmental control. Often a major rehabilitation aim is that of inducing some form of self-control. In the interest of achieving this goal, we have found it sometimes necessary to use reinforcers, such as cigarettes if the patient is a smoker, that we would generally prefer not to use.

Another important aspect of rehabilitating these patients has to do with the problem of attention. As indicated, these patients are typically exceedingly distractible, or stimulus bound, and they may not focus attention on anything for more than a fleeting moment. If the patient can be taught to respond in some systematic manner to environmental contingencies, the next step would appear to be attention training. Generally, the paradigm for such training involves a vigilance task which requires increasing time lengths of sustained observation, and inhibition of response until the appropriate moment or until the appropriate stimulus configuration appears. Feedback should be provided in the form of social or physical reinforcement. For more impaired patients, social reinforcement may not be effective initially, and it may be necessary to use food, cigarettes, or some other form of material reinforcer. The nature of the task can vary greatly. The patient can be asked to monitor the movement of a pendulum and press a key when it passes specific points in its arc. He may be asked to respond only when a specific configuration of lights appears on a panel. We tried using a pursuit rotor with an aphasic patient, not to treat the aphasia, but to teach the patient that he can do better if he slows down and acts less impulsively while attending to the task.

In doing this type of training with the impulsive patient, certain trainer characteristics are clearly desirable. The trainer must "stand out" in the patient's perceptual field. His or her impact must be made known. This is no place for a passive or "nondirective" relationship with the patient. The patient must be actively engaged in order to get his attention, and he must be constantly checked and brought back to the task if he becomes distracted. Indeed, one of the training methods might involve attempting to lengthen the period of time during which the patient is left to his own devices and can function without trainer intervention. Initially, however, the trainer has to "come on strong" and make what is expected clear to the patient. The role of structure at the beginning of the training is crucial.

There is a practical aspect to the treatment of this kind of patient that might be mentioned here: these patients do take up a great deal of staff time. Having one perpetually assaultive patient on a ward can occupy much of the staff's time simply to control that patient. Sometimes, this disproportionate allocation of time serves as a disadvantage to the other patients on the ward. Although we have no solution to this problem, we can offer the common-sense remark that impulsive patients should not be treated on wards that are crowded or not adequately staffed. To do so is not only unfair to the other patients, but encourages the use of aversive treatment, such as seclusion and restraint, for the violent patient.

With an impulsive patient who is not dangerously assaultive, one can focus on resolving discordant social interactions and learning socially appropriate behavior. The patient who impulsively bolts down his food can be instructed to eat more slowly. The patient who uses obscene language inappropriately can be taught in what situations it is and is not correct to use such language. The patient who gets into fights can be taught, perhaps through role playing or straight instruction, how to resolve matters amicably. Often, these patients have a lack of appropriate concern for such matters as dress, grooming, and personal hygiene. Instruction in these matters revolving around an activities of daily living orientation might be useful. In general, not all patients with impulse disorders are uncontrollably violent and aggressive individuals, since, more often than not, they have impulse problems and attentional deficits that are amenable to retraining. These patients may respond well to a social skill directed training program provided by an adequately trained and sized staff, especially when behavior–therapy concepts, particularly with regard to matters of stimulus control and contingency management, can be implemented. The general principle is that these patients have difficulty in self-control of their behavior, and that this difficulty may be disruptive socially. One basic aim of retraining should be to bring the behavior either under environmental control through drugs and contingency management, or under personal control through such techniques as instruction and role playing.

Epilepsy and Impulsivity

There seems to be a popular assumption concerning a relationship between epilepsy and impulsiveness. The suddenness and explosiveness of a convulsive seizure appears to have led to the belief that epileptics are explosive people. At a professional level, a literature has developed around the so-called epileptic or epileptoid personality (Pichot, Lempérière, & Perse, 1955). In discussing this matter, it is first important to point out that epilepsy is often a symptom rather than a discrete disease entity. Individuals can have seizures following head trauma and stroke. Some brain–tumor patients have seizures, as do some patients with metabolic disorders. However, there is a clinical group consisting of individuals who just have seizures and no other symptoms. The etiology of the seizures is unknown, and so the condition is often referred to as *idiopathic epilepsy*. Sometimes it is popularly known as genuine epilepsy, in order to distinguish it from those conditions in which the epilepsy is a symptom of some known disorder.

Our observations suggest that many impulsive epileptics tend to be individuals who were impulsive before they acquired the epilepsy. Typically, these patients are young men who developed their seizure disorders in association with head trauma. When one investigates the circumstances surrounding the acquisition of the head trauma, they often involve the exercise of poor judgment in a situation in which greater caution should have been exercised; such as those accidents resulting from careless driving of motorcycles, or driving while intoxicated. The acquisition of the brain damage appears to potentiate the impulsiveness in these cases; these patients often present significant rehabilitation problems. Although they look like ideal candidates for retraining, they may leave the hospital or retraining facility abruptly. One phenomenon that occurs in epilepsy which may not be a manifestation of the premorbid personality is the so-called build-up. The patient becomes increasingly tense and irritable for perhaps several days preceding a seizure. Observant clinicians can almost predict when the patient is going to have a seizure on the basis of the build-up. If the patient has the seizure, it seems to discharge the irritability, and there is a return to a state of normal temperament.

A component of the personality difficulties epileptics may have can clearly be a reaction to the attitudes of others toward the disorder. Epileptics are often legally restricted from certain activities, such as driving a car or operating certain kinds of equipment. In institutional settings, epileptics may not be permitted to use the swimming pool or work in certain shops. One can understand why the epileptic might become depressed and angry under these circumstances. Thus, whereas the notion of an epileptic personality has not proven to be a productive concept, many epileptics do have psychiatric problems associated with their premorbid personalities or with their reaction to the restrictions and sometimes negative attitudes associated with their disorder. The rehabilitation of brain-damaged patients with epilepsy, therefore, may be somewhat limited by the restraints, imposed by the disorder, on activities engaged in, or in vocational choice. Furthermore, it is often necessary to deal in some way with psychiatric problems epileptics tend to develop on the basis of these considerations. Whereas most epilepsy is treatable as far as the seizure disorder goes, the person working with the epileptic should be aware that epilepsy is frequently not as benign a condition as one might think.

A specific type of impulse disorder seen in a small number of brain-damaged patients is hypersexuality. Sometimes the making of inappropriate sexual advances is among the first manifestations of a neurological degenerative disorder. There seems to be some tendency for patients with frontal-lobe lesions to develop these conditions, but we have also seen it in patients with subcortical lesions. Hypersexuality is part of the Klüver-Bucy syndrome. In our opinion, hypersexuality does not seem to be strongly associated with the premorbid personality. The symptom is often seen in middle-aged individuals who led highly respectable lives prior to acquisition of the illness. One could say that the brain damage releases some well-suppressed tendency that was always present; however, such an inference appears to be speculative. As in other impulse disorders, inappropriate sexual behavior may be more or less episodic or steady state. The temporal-lobe epileptic may commit a sexual offense during a seizure, or the postencephalitic adolescent may do so on impulse. On the other hand, there are some patients whose behavior is sexually inappropriate a good deal of the time. While this condition is relatively uncommon, when it does occur it typically presents major treatment and management problems. There is a danger of the occurrence of fights, especially when the sexual object is unselective with regard to gender.

Although there is no specific treatment for this symptom, the experienced physician knows that the drug Mellaril has some capacity to reduce libido. Behavioral methods, such as use of time-out following the occurrence of an advance, are sometimes useful. However, our experience indicates that these methods do not tend to be effective for more than very brief time periods. It should be emphasized that hypersexuality is sometimes a major adaptive problem for the patient and those around him. Sometimes, it is the only behavior that prevents the patient from leaving an institutional setting, and sometimes it places the patient in physical danger. It is therefore generally advisable to make extensive efforts, through chemical and behavioral means, to try to deal with the problem. Unfortunately, we do not have definitive treatments, and the therapist has to employ his or her ingenuity and expertise to devise a workable treatment program.

The Hypoactivity Disorders

The experienced clinician working in a psychiatric setting is well aware that the problem with most patients is that they do too little rather than too much. The apathy and withdrawal of the chronic schizophrenic patient is well known to most mental health workers. In the case of the clients in the Wichita program, most of them shared this symptom. They tended to be living at home, and were not working or in school. In essence, hypoactivity was a much greater problem than hyperactivity. Hypoactivity or apathy is, of course, a complex behavior that has many sources. At its extreme, there is a condition Luria (1973) calls the apathico-akinetic-abulic syndrome, in which "Patients usually lie completely passively, express no wishes or desires and make no requests; not even a state of hunger can rouse them to take the necessary action" (p. 198). There is also a condition known as impulse apraxia, in which the patient has a great difficulty in initiating any action. These syndromes tend to be associated with massive damage to the frontal lobes. Extreme conditions of this type are much more frequently seen in milder forms, and represent varying degrees of apathy and inactivity. They should be separated from the apraxias in which the problem is not one of lack of motivation but of being unable to execute skilled acts in some manner.

One common characteristic of the brain-damaged patient is an inability or unwillingness to initiate behavior. The patient may not speak for lengthy periods of time unless spoken to, or he may not engage in activities unless encouraged to do so. This characteristic occurs so frequently that it is almost taken for granted, so that for a rehabilitation program to be successful, it is necessary to maintain a high level of structured, stimulated activity. Even satisfactory assessments of brain-damaged patients require active, vocal examiners.

This problem is brought up in the context of emotional or psychiatric disorders because there is a widespread belief that many of the signs of apathy and motor retardation in brain-damaged patients are really manifestations more of depression than of structural brain lesions. The extent to which depression is a significant explanatory factor is a matter of controversy, but, nevertheless, it is generally wise to rule it out before assuming that the problem is neurological in nature and can therefore be treated with the kinds of cognitive and perceptual retraining methods described in this book. If the opportunity avails itself, probably the best method for ruling out depression is a trial with an antidepressant drug. If the depression lifts and the patient's cognitive functioning does improve, retraining will probably be more successful if the depression is relieved.

The matter of depression is of particular importance in the case of the brain syndromes associated with old age. Whereas a certain portion of the elderly population develops senile dementia, other members of that population develop a *pseudodementia*—a reduction in cognitive function primarily associated with the commonly occurring depression of old age. In recent years, geriatric psychiatry has emerged as an independent speciality that is largely concerned with the borderline between the organic dementias and the functional psychiatric disorders, particularly the affective disorders (Gurland, Dean, Cross,

& Golden, 1980). With regard to rehabilitation of the elderly, it is usually worthwhile to treat the depression, if one is found, as well as the cognitive disorders that may be present. Hypoactivity or apathy is such a commonly occurring symptom associated with brain damage that it is difficult to make any specific remarks about it. Some general suggestions for management of the problem may nevertheless be offered. It is a widely held belief that sustained hypoactivity engenders deterioration in and of itself—aside from the deterioration that may be attributable to the disease process that engendered the hypoactivity. The atrophy that occurs in muscles as a function of disuse is commonly observed in bedfast patients. Perhaps the analogy between muscle function and mental function may not be too farfetched. Disuse of intellectual abilities may well contribute to a degeneration of those abilities. In view of this consideration, many treatment programs now emphasize a high-activity level. When the program takes place in an institutional setting, the best strategy is probably to program the patient's activities like a normal day. The morning can be spent on personal hygiene and housekeeping, the middle of the day on various assignments, and the evening on recreation and leisure. In the case of brain-damaged patients, it is highly important to schedule the day in a structured manner. Although this is often easier said than done, several techniques have proven to be useful. In some cases, the ward can be locked during the day, not to keep the patients on it but to keep them out of it so that they do not have easy access to the sleeping and lounging facilities of the ward. Sleep is often a major problem. In the case of the patient who shows excessive fatigability, or who for some other reason goes to his bed whenever he is unsupervised, we have found the best approach is not to deprive the patient entirely of daytime sleep. Rather, naps may be scheduled, and the patient can sleep for reasonable periods of time during the day on a scheduled basis. Obviously, the use of various behavior–therapy techniques can be used to stimulate activity in individual patients, but these techniques seem to work best in a milieu that stresses activity.

Questions that often arise concerning activation of patients have to do not only with the matter of designing an active milieu but also with the problem of maintaining the patient at an optimal level of activation. In institutional settings, the promotion of a high-activity level on a ward can often be accomplished with a token economy system (Kazdin, 1977). In such a system, activity itself can be reinforced through providing tokens for not napping during the day and participating in ward programs and activities. The implementation of a daily schedule of activities is often useful. In our own experience, work assignments for compensation, field trips, and utilization of hospital recreational facilities have been helpful in regard to maintaining patients on a regular activity schedule. With regard to personal maintenance of a high activity level, the use of psychotropic medication sometimes produces a problem. Indeed, clinicians utilizing these drugs have been concerned about their sedating properties, and attempts were made to develop nonsedating or alerting neuroleptics. Davis (1980) points out that antipsychotic drugs are equally effective in counteracting psychotic symptoms regardless of whether they are sedating or nonsedating. He also provides a table listing the sedative action of the various antipsychotics. For example, trifluoperazine (Stelazine) has little sedative action while chlorpromazine (Thorazine) has a moderate amount. The clinician who wishes to control psychotic behaviors while maintaining the patient's alertness might consider using one of the less sedating neuroleptics.

It should be pointed out that some of the views expressed here run counter to the older treatment philosophies and to lay opinion concerning the functioning of institutions for the treatment of brain-damaged and psychiatric patients. Not too many years ago, the basic philosophy of many institutions could be well captured by the term "rest home." In the case of the patient with a chronic disorder, it would not be far from the truth to say that, in many cases, the institution was viewed as a benign setting in which the patient could live out his remaining years in peace. As new philosophies in psychiatry and rehabilitation medicine emerged, the trend changed drastically, and the new orientation involved active treatment and the goal of return to the community. In cases in which return to the community was not possible, institutionalization in the least restrictive setting became the goal. As part of this process, psychiatric institutions have become increasingly less restrictive, both in terms of physical restraints placed on patients' activities and of retaining patients on an involuntary basis.

Blunt affect, psychomotor retardation, and loss of initiative are perhaps the most universally observed symptoms of brain-damaged patients. As some clinicians have described it, there is a lack of élan or "sparkle" among these individuals. Perhaps the entire behavior–therapy-reinforcement-oriented thrust of the Wichita program revolved around this matter; otherwise, straight instruction might have been as effective as contingency management. Although straight instruction was not directly evaluated, the ordinary efforts that might normally have been made to rehabilitate our clients before they entered our program were surely unsuccessful. Whereas impairment of motivational mechanisms may be a direct consequence of at least certain types of brain damage, the indirect consequences may also play a major role. The brain-damaged patient may have severe difficulties with even the more routine activities of ordinary life. Behaviors that we perform automatically and take for granted can only be performed by brain-damaged patients at the expense of a great deal of effort. Dressing, speaking, recalling past events, performing activities that require simple motor skills, and many related behaviors can sometimes be accomplished by brain-damaged patients only by circuitous means and with great expenditures of energy. It is little wonder that this substantial increase in the amount of effort needed to function and survive often leads to major psychiatric difficulties. Perhaps more often than not, these difficulties eventuate in the giving up of normal activity and the acquisition of a sometimes apathetic state. The hypoactivity of the brain-damaged patient therefore often seems to represent a state of resignation associated with the frustrations stemming from the limitations in functional capacity produced by the brain lesion. Whereas there may sometimes be a direct destruction of the neural mechanisms that mediate affect, as in the case of lesions of the amygdala or other deep structures in the brain, it is usually not necessary to postulate such a defect. The hypoactivity may be completely explainable on the basis of the patient's reaction to his defect, and may, in fact, diminish substantially with adequate treatment and rehabilitation.

The Posttraumatic Syndrome

There is a set of symptoms commonly observed following head injury that, as a composite, is known as the *posttraumatic syndrome*. There is some question as to how much of this syndrome is accounted for by functional or psychiatric considerations, and how much of it is explainable on the basis of the structural lesion itself. Some feel that it is all psychiatric, while others believe it to be a combination of the two etiologies. Usually, the symptoms included in this syndrome are:

1. Headache
2. True vertigo
3. Dizziness
4. Tinnitus
5. Nervousness
6. Irritability
7. Diminished concentration
8. Impaired memory

9. Excessive fatigue
10. Difficulty in sleeping
11. Sexual difficulties
12. Diminished self-assuredness

Of course, not all of these symptoms appear in all cases, but frequently enough of them do appear together to say that the individual has the syndrome. Indeed, it is a common sequel of head injury and was commonly observed among soldiers who sustained brain wounds during the various wars. Its occurrence does not appear to be relatable to lesions in any particular portion of the brain but rather to both open- and closed-head injuries in general.

Many of the symptoms of the posttraumatic syndrome would appear to be stress related. Even organic-sounding symptoms, such as loss of memory, are often not neurologically or neuropsychologically documentable. For example, the loss of memory may in actuality be a complaint of loss of memory, which is a different matter. In this case, the patient feels that his memory is not as good as it used to be, although objective testing indicates that his memory is normal. This symptom may, in fact, be associated with symptoms of anxiety and diminished self-assuredness. The latter symptom is often one of the more distressing ones. The head–injury victim feels, sometimes unjustifiably, that he can no longer do what he used to do: he cannot work, do well in school, perform sexually, drive a car, participate in sports, or do many of the other things he felt certain about preinjury. There seems to be a pervasive feeling of being uncertain about coping with a variety of situations.

Whether the posttraumatic syndrome is neurological, psychiatric, or a combination of both, the only treatment is essentially psychiatric or behavioral in nature. From a psychiatric standpoint, the use of antianxiety medication might be attempted. In general, the treatment of choice would be counseling with emphasis on reassurance and encouragement to try things out. The concept of a gradient of adaptive challenges may be usefully employed. Here, the patient is gradually provided with tasks of increasing complexity. Thus, although he may eventually be brought to what his limits are, it is hoped that this transition can be handled in a way that avoids the kinds of catastrophic reaction that can occur when environmental demands become overly taxing.

For those who work with war veterans, the posttraumatic syndrome is probably one of the most commonly occurring and difficult problems for individuals who sustained brain injury during combat.

The trauma of war seems to interact with the brain damage in sometimes producing some major adjustment problems. This condition seems to be especially severe in individuals who have been prisoners of war. In recent years, it has been noted that veterans of the Vietnamese War frequently experience a "delayed stress reaction" in which, following a relatively normal adjustment to civilian life, there is a sudden appearance of symptoms that may include depression, high anxiety, a sleep disturbance involving traumatic dreams based on experiences in combat, and abuse of drugs or alcohol. Although the incidence of delayed-stress reactions in veterans who sustained brain wounds has not yet been documented, it would seem important to do so.

Hysteria

Hysteria is a psychiatric condition that is of particular interest to the neurologist, since the classical hysterical symptoms mimic neurological symptoms. Blindness, deafness, paralysis, loss of sensation in a limb, and related difficulties may all be either hysterical or neurological symptoms. This chapter is not the appropriate place to engage in a discussion of the diagnosis or the etiology of hysteria. However, there are cases in which there is an intermingling of organic and hysterical symptoms. Hysteria and organicity are not mutually exclusive alternatives; one can have both conditions. For example, we can mention the case of a patient we treated who sustained massive traumatic brain damage to the frontal lobes. The patient was left with a severe speech deficit, a spastic paralysis of one of his arms, and severe intellectual impairment. When he walked, he would do so with a staggering gait and would often almost lose his balance. The interesting aspect of this patient was that although he did suffer massive brain damage, and although he did have bona fide intellectual, speech, and motor symptoms, there was no neurological basis for the gait disturbance. Physically, the patient should have been able to walk reasonably well. Thus, we had a patient with brain damage and a hysterical symptom. These combinations are not uncommon, and must often be dealt with in the rehabilitation of brain-damaged patients. If the basis for hysteria is anxiety, these patients often have abundant reasons to be anxious. Often they have suffered a severe, life-threatening trauma such as a head injury or stroke. The idea that some of this anxiety may be converted into somatic symptoms should be understandable. One must be open to all possibilities, and not automatically attribute to the lesion itself all of the observed symptoms. These patients in particular require careful and detailed neurological examinations to delineate what can reasonably be attributed to the lesion from what cannot be so attributed.

Another common hysteric symptom is denial. The *belle indifférence* of the hysterical patient is well documented (Hécaen & Albert, 1978), as is the tendency of individuals with hysterical or hysteroid personalities to see the world in the best possible light, even when it is inappropriate to do so. The issue of hysterical denial in the brain-damaged patient is a complex one, and the etiology of denial symptoms may not be entirely psychogenic. Such investigators as Weinstein and Kahn (1955) and Kolb (1959) have made detailed studies of denial in brain-damaged patients, particularly in stroke patients. Elsewhere, we have discussed the syndrome of anosognosia, or lack of awareness of illness. While it may be initially difficult to treat a patient for a defect that he denies he has, Diller (1970) suggests that this difficulty can be overcome through paying careful attention to the patient and developing a good rapport with him. As we have also indicated, some of the denial syndromes are self-limiting in that they diminish or disappear as the acute phase of the illness resolves.

It would thus appear that in the case of the brain-damaged patient, hysteria may appear in two forms. It may involve the superimposition of a nonorganic symptom onto one or more organic symptoms, or it may exist in the form of denial of an organic symptom. The treatment of hysterical symptoms in brain-damaged patients, in principle, should be no different from what is done with nonbrain-damaged hysterics. Behavior–therapy techniques sometimes involve selective inattention to complaints about the hysterical symptoms, reward or praise for normal performance, and related techniques. We have seen at least temporary improvements following direct instruction. One of our patients would develop an impaired gait when he knew he was being observed by the staff but would walk normally on other occasions. Simply telling the patient to walk right would often improve his gait temporarily. Although we have not explored these methods, there is no reason not to use such techniques as hypnosis and suggestion to relieve hysterical symptoms.

Behavior-therapeutic approaches to the treatment of conversion symptoms stress differential attention and development of social skills. Differential attention often involves reinforcement of nonsymptom-related adaptive behavior with nonreinforcement of symptom-related behavior. For example, the patient's complaints about aches and pains may be ignored while adaptive social behavior is praised. There are several cases reported in the literature (Alford, Blanchard, & Buckley,

1972; L.H. Epstein & Hersen, 1974; Kallman, Hersen, & O'Toole, 1975) in which this philosophy was implemented in the form of individual behavior–therapy programs. For example, a patient with hysterical paralysis was praised for standing and walking over a series of sessions (Kallman *et al.*, 1975). The point of the social skills training is primarily that of providing the patient with gratifications in life other than those received from his symptoms. It would appear that in many cases, while differential-attention procedures can provide immediate symptom relief, in order for there to be a sustained effect, the patient must alter his life-style in a way that allows him to receive attention and support without having to rely on somatic symptoms. An alternative procedure to those mentioned above is the habit-reversal technique (Azrin & Nunn, 1973; Azrin, Nunn, & Frantz-Renshaw,1980) in which patients are first of all trained to be aware of the symptom (e.g., a tic), and then the symptom is taken out of the train of normal movements through establishment of a competing response. Thus, for example, if the symptom is backward head jerking, the competing response would be pulling the chin in and down.

The problem of hysteria seems to blend naturally into issues associated with malingering and the acquisition of factitious disorders. In the neurological realm, pseudoseizures and questionable complaints of loss of memory for specific episodes are not at all uncommon. The clinician is often called upon to make a determination regarding whether such phenomena are genuine or fabricated by the patient voluntarily. Again, patients with well-documented brain damage may malinger or develop factitious symptoms. A patient may fake a seizure when he is about to be discharged from the hospital against his will. Sometimes alcoholic patients with well-documented brain atrophy will claim they had a memory lapse for an episode in which they committed some antisocial act. We have seen numerous head-injured patients who claimed they could do things before their accident that they could not do following the accident. For example, we treated one patient who claimed he could read before his head injury but could not do so afterward. Although there are acquired alexias that develop as a result of traumatic brain damage, an examination of this patient's history revealed that he could never read well, and that his dyslexia was developmental in nature. In this area, the expertise of the neuropsychologist is crucial in regard to attempting to reconstruct the premorbid status, and in separating what the patient could and could not do before the acquisition of the brain damage. The important point is not to accept at face value the claim of the patient or the patient's family regarding premorbid capabilities. The point is obviously a significant one for rehabilitation. Retraining an individual with an acquired alexia is a different matter from attempting to teach an individual to read for the first time.

Whereas the claim of premorbid intactness of function sometimes has litiginous overtones, there is also a natural tendency to attribute all the patient's difficulties to the brain damage, and thereby to idealize his premorbid status. In the case of the parents of brain-injured children, this process can engender some relief from guilt concerning responsibility for the child's deficits. When the deficits can be attributed to some external event, such as an automobile accident, then the parents may feel assured that they are not responsible for their child's acquisition of those deficits.

Ideally, the specialist in rehabilitation of brain-damaged clients likes to work with individuals who have specific and well-documented neuropsychological deficits that can reasonably be attributed to some form of structural damage to the brain. It therefore becomes disconcerting when deficits appear and disappear depending on who is watching the patient or on what particular situation the patient is in at the time. In other words, when symptoms start looking as though they are hysterical in nature, the treatment team may experience embarrassment or even the unpleasant feeling that they were fooled by the patient. We once had contact with a patient with a long-standing diagnosis of multiple sclerosis. On one occasion, however, he was examined by a particularly astute neurologist who also reviewed the patient's history thoroughly. On the basis of this examination and review, it became apparent that the patient could not possibly have multiple sclerosis, and that there was in fact no organic basis for any of his symptoms. One might think that this news might have been greeted with great joy by the patient and staff because the patient did not have this dread disease, but that was not their reaction. The staff felt, in a way, "fooled" by the patient and became greatly concerned about how they could go about informing him that he did not have the disease, without producing a depression. This patient had organized his entire life, and that of his family, around having multiple sclerosis, and it was felt that the reorganization required could be stressful. Events of this type inevitably get the rehabilitation specialist involved in problems of a "functional-psychiatric" nature and tend to relieve one of any puristic notions one might initially have about working only with neuropsychological problems of brain-damaged patients.

Some General Considerations

Thus far, we have been discussing the interfaces among those psychiatric disorders that frequently interact with various organic neurological conditions. However, there is another aspect of the problem having to do with the relationship between treatment of brain-damaged patients and psychiatry in general. It should be pointed out that our use of the term psychiatry refers to the problem of disorders of a functional or emotional nature that are usually treated by various mental health specialists, including psychiatrists and clinical psychologists. The matter can be introduced by raising the question of whether or not brain damage in humans is primarily a psychiatric disorder that has a known neurological basis. After all, the diagnosis of "organic brain syndrome" is a psychiatric and not a neurological diagnosis. On the other hand, such conditions as aphasia are not generally considered to be within the purview of the mental health professions. Conditions of this type are generally evaluated by neurologists and speech pathologists, and sometimes by neuropsychologists. We know of no authority who would argue that aphasia and the related disorders are functional or emotional problems, granted that they may be associated with such problems. It is apparent that the problem raised here is theoretically and practically complex.

Our view of the matter is that a "mental health" approach to the brain-damaged patient is crucial in conceptualizing and influencing the manner in which the patient adjusts to his or her environment. From a conceptual standpoint, this adjustment is most adequately understood in terms of some theory of personality. The pioneer in dealing with personality aspects of the brain-damaged individual was K. Goldstein (1939, 1959a, 1959b), who had an intense interest in the problem of adjustment, or in how the patient comes to terms with the environment. In his descriptions K. Goldstein placed heavy emphasis on a form of self-actualization theory that was made more popular later on by Rogers (1951) and Maslow (1968). In essence, the brain-damaged patient is viewed as striving to make the best possible adjustment to the environment with his remaining capabilities. K. Goldstein (1959b) made a clear distinction between symptoms directly related to destruction of brain tissue and symptoms representing the reaction of the patient to the defect. The reader should review K. Goldstein's (1939, 1959b) and K. Goldstein and Scheerer's (1941) detailed clinical descriptions of the behavior of brain-damaged patients to obtain what probably still stands as the best clinical phenomenological descriptions of patients with various types of brain disorder.

A consideration of the symptoms representing the patient's reaction to the defect causes us to enter a realm of behavior that is somewhat different from the behaviors assessed with neuropsychological tests. We may observe such symptoms while the patient is taking tests, but they are not directly measured by the tests themselves. In this regard, probably the most outstanding behavioral feature is the catastrophic reaction. When the patient has to perform a task with which he cannot cope, he may develop a state that looks very much like an anxiety attack. A previously calm and friendly patient may abruptly become agitated and irritable. His face may flush, or he may become tearful. It is not necessary to postulate any previously existing psychopathology in the case of the catastrophic reaction. As K. Goldstein (1959b) put it, "the catastrophic reaction is not a conscious reaction to the failure but, rather, belongs intrinsically to the objective situation of the organism in failure" (p. 783).

Aside from the catastrophic reaction, the most prominent reactive symptoms relate to the protective mechanisms used by the patient to avoid having a catastrophic reaction. These mechanisms become more fully developed as time passes following the acquisition of the brain lesion. The patient may simply withdraw from social contact to avoid having to engage in conversations; he may develop the often noted orderliness of the brain-damaged patient, in which objects are always put in the same place, and other aspects of his life may become highly structured and ritualized. A denial syndrome may develop in which the patient deals with his difficulty by denying that it exists. Interestingly, Goldstein recognized the process of denial, although he did not like the term itself. The term *denial* implies a conscious, voluntary effort; a task that brain-damaged patients are often not able to perform. Rather, he viewed it simply as a new organization of behavior that occurs as a result of the organism's tendency to realize its preserved capacity. Although we may not be totally satisfied with this explanation, it suffices to bring out the point that the usual psychodynamic explanations of denial in neurologically intact individuals may not obtain in the case of brain-damaged patients in whom denial may be viewed as a coping effort that aids the patients in doing the best they can with their remaining resources.

K. Goldstein (1942) also pointed out that there are two basic ways of adjusting to defects: yielding and compensating. In yielding one gives in to the defect, while in compensating one builds some mechanism by means of which compensation for the defect is provided. Yielding is seen as somewhat "safer" in that it is a more natural reaction and does not demand much in the way of voluntary activity. For example, patients with cerebellar lesions may adjust by simply yielding to the lesion and allow their heads to remain in a tilted position. The compensating patient may attempt to develop mechanisms that allow him to restore a normal posture. With regard to rehabilitation, it would appear that the aim is sometimes to change a yielding mechanism to a compensating mechanism, and sometimes to get a faulty compensating mechanism to work more productively.

These considerations based on Goldstein's philosophy are offered because they are remarkable and unique examples of how one can understand the brain-damaged world as different from the one experienced before the brain damage was acquired. Whether or not one wishes to consider the symptoms representing the reaction to the defect as "psychiatric" in nature or not, they are a major source of individual differences in how people adjust to brain damage, and the postulation of such symptoms thereby "humanizes" our view of the brain-damaged patient. He or she becomes more than the sum of the cognitive, perceptual, and motor defects. Although neuropsychologists have not always paid close attention to these secondary symptoms, they are obviously crucial to any rehabilitation effort. The patient who denies or is not aware of a major disability, or the patient who withdraws to such an extent that he is unapproachable, presents major rehabilitation obstacles. These problems have to be dealt with as much as the cognitive and perceptual defects.

Our own observations have led us to stress the matter of how the patient is viewed by others. The major consideration appears to be whether the patient's deficits are overestimated or underestimated. Overestimation of the patient's disability often leads to a sometimes pathological dependency situation in which the family does everything for the patient, even those things that the patient is perfectly capable of doing for himself. In these situations, psychodynamic interpretations are probably appropriate in that the family often has a need to make the patient dependent, whereas the patient may have a need to remain dependent. In these situations, one of the more difficult clinical tasks is that of getting the family to allow the patient to be more independent, as well as that of getting the patient to give up his dependency. On the other hand, the underestimation of severity of the disability provides too much opportunity for catastrophic reactions. We can provide many vignettes of the type in which the achievement-oriented parent continues to make efforts to have his brain-injured son return to law school. In an attempt to please his parents or peers, the patient attempts to achieve so as to measure up to their expectations and finds himself in numerous situations with which he can no longer cope. In rehabilitation, it is important not to make the errors of overestimating or underestimating the severity of the disability. Setting unrealistically high rehabilitation goals for the patient can ultimately be devastating, whereas not expecting enough from the patient may ensure a level of adjustment substantially below the patient's potential.

In our experiences with many kinds of brain-damaged patients, we have found that any interaction with these patients on a sustained basis necessarily involves some form of dealing with their emotional lives. In this chapter, we have tried to deal with some of the major issues in a clinical, anecdotal manner. However, many of the problems presented should really be researched on a formal basis because they represent major clinical issues. More often than we might imagine, the psychiatric factors are the ones that are most crucial in determining rehabilitation outcome. There is a kind of implicit debate between psychiatrically oriented and neuroscience-oriented individuals regarding the nature of the crucial factors involved in the brain disorders. Whereas neuroscientists in general and neuropsychologists in particular have stressed the matter of various skills and abilities, the psychiatrically oriented group has stressed the issues of depression and the patient's reaction to his illness. What seems to be important here is to refrain from assuming an extreme view. There appear to be some individuals, particularly those with psychodynamic orientations, who do not appear to have a clear conceptual grasp of how damage to particular structures of the brain, in and of itself, can give rise to certain specific behavioral outcomes. They tend to interpret these behaviors, mistakenly we think, within a psychodynamic framework. Dementia becomes depression and anger; seizures become the patient's attempt to satisfy dependency needs by appearing ill in a dramatic way. On the other hand, some neuroscientists essentially completely neglect the problem of how emotional and social considerations can substantially influence performance on neuropsychological tests and on adaptive functioning in general.

Most of the time, emphasis is placed on how "psychiatric" or emotional considerations detract from the brain-damaged individual's level of functioning. However, there are many cases in which the influence of the patient's emotional life is positive in nature. Individuals with high morale, strong motivation, and the capacity to bounce back from adversity often do better than what would be expected on the basis of what is known about their brain disorders. In our experience, neuropsychological tests sometimes tend to overplay the individual's disabilities, and one is surprised to find people who do poorly on these tests, but who are continuing to function in the community in a reasonably acceptable manner. Chapman and Wolff (1956), in a classic study, found that brain–tumor patients did better in their communities than might have been expected. They reported that only massive brain damage impaired comprehension of commonly occurring social situations, suggesting that many brain-damaged individuals can be quite socially appropriate despite significant structural damage to the brain. Wells (1979) reported that whereas retention of social skills tends to be a clinical feature of true dementia, loss of such skills is often a sign of pseudodementia.

We would feel remiss if we did not remind the reader that an emphasis on psychiatric considerations should not be accomplished in a manner that neglects the possibility of an organic basis for such symptoms. Merikangas (1977) has reported several instances in which this situation turned out to be the case. Many of us may have had personal experiences with individuals who were treated psychiatrically but who in fact had physical illnesses that gave rise to psychiatric-looking symptoms. Symptoms that look like manifestations of anxiety may be based on a thyroid disorder. What appears as conversion hysteria may be multiple sclerosis. Sexual acting out in a middle-aged person may be the first sign of a presenile dementia. As Merikangas pointed out in this study, a mood disorder can be the presenting complaint based on a seizure disorder.

Summary

In this chapter, an attempt was made to present a clinical discussion of the role of emotional and personality characteristics in the assessment and rehabilitation of brain-damaged patients. In general, it was pointed out that the area is a complex one in which the status of the patient is affected not only by the lesion parameters but by premorbid personality, reaction to the brain disorder, and reactions of others toward it. Some brain-damaged patients have significant psychiatric disorders prior to the acquisition of the brain lesions, and, in some cases, the lesion acquisition is directly or indirectly associated with that disorder. Problems associated with depression, impulse control, apathy, anxiety-related symptomatology, and psychosis are frequently seen in brain-damaged patients. Again, these matters become exceedingly complex because of the various interactions among the lesion parameters, concurrently existing psychopathology, and reaction of the patient and those around him to the neurological disorder.

It is not our view that brain damage in humans is a psychiatric disorder in a narrow sense. In fact, the psychiatric-diagnostic nomenclature involving such terms as "organic brain syndrome" does not appear to have made a substantial contribution to conceptualization or treatment of the brain disorders. However, psychiatry has pointed out that brain-damaged people as human beings have emotional difficulties too, and that these difficulties may be amenable to psychiatric treatment. Unfortunately, when brain-damaged patients become institutionalized in psychiatric settings, it is often only these secondary emotional disorders that receive treatment—the primary cognitive, perceptual, and motor difficulties are viewed as untreatable. The major thrust of this book is that of demonstrating that many of these primary deficits are treatable through training methods developed within a neuropsychological rather than a psychiatric framework. It is hoped that the preliminary work presented here will ultimately eventuate in more active treatment programs in which brain-damaged patients are not treated simply for their agitation or depression, but for what are usually their most disabling and manifest problems: the direct consequences of the brain damage itself. However, brain-damaged patients often do have disabling psychiatric problems that require treatment along with their neuropsychological problems. In order to provide such treatment rationally, it is necessary to offer some conception of the kinds of psychiatric and emotional problems such patients often develop. At present, there is little research in this area, and we have had to rely primarily on clinical experience and case illustrations.

In conceptualizing the psychiatric aspects of brain damage (in the sense in which we have been using the term), we have leaned heavily on the work of K. Goldstein and Scheerer. As scientist-clinicians they have focused on an area that has been neglected by many contemporary neuropsychologists: the way in which the brain-damaged patient copes with his environment. In dealing with this matter intelligently, it seems necessary to adopt some form of personality theory. Goldstein and Scheerer devised what would be described in contemporary personality theory as a humanistic approach in which the individual strives for self-actualization through making the best adjustment possible to acquired defects. Goldstein and Scheerer were well aware that brain-damaged patients often have psychiatric problems, and in fact Goldstein conducted an extensive psychotherapeutically oriented practice for such patients. The descriptions of the adjustment patterns of brain-damaged patients that came from Goldstein and Scheerer's assessment and treatment activities remain, in our view, the most definitive material we have in this area.

With regard to the matter of rehabilitation planning and implementation, our experiences with both chronic patients and patients with recently acquired lesions suggests that one cannot work on the assumption that the patient has no other significant problems except those associated with the brain damage. When one reads the literature concerning rehabilitation of stroke, amnesic, or head–trauma cases, one might get the impression that what is being rehabilitated is always the major problem affecting normal functioning. We get the image of the normal person afflicted with his or her brain disorder. This position is often a reasonable one, but there are other considerations that frequently have to be taken into account. Many stroke patients, for example, often develop severe, long-lasting depressions. The amnesic patient is often an alcoholic who has a long history of physical and social deterioration. Aside from during wartime, head-injured patients often become head injured because of poor judgment associated with premorbid impulse disorders or problems of a characterological nature. In this case, we are not simply dealing with a head-injured patient, but a head-injured patient with an immature personality or a frank character disorder. These matters should not only be recognized but be acted on. Therefore, it is our view that any comprehensive rehabilitation program should have psychiatric support available in the form of consultation, and expertise on the part of the rehabilitation staff as well. Fortunately, many neuropsychologists are also trained as clinical psychologists, and many neurologists are also psychiatrists. These combinations of capabilities would appear to be ideal for individuals engaged in rehabilitation of brain-damaged patients.

9

THE CHRONIC HOSPITALIZED PATIENT

As we have seen, the patients in the Wichita program tended to be relatively young and generally had good rehabilitation potential from vocational and educational points of view. In this chapter, we will discuss the chronic patient in a hospital setting. These patients tend to be older than the Wichita clients and have typically been unemployed for long periods. The patients are also found in convalescent centers, nursing homes, VA domiciliaries, and similar institutions. We will report the experiences of Gerald Goldstein (one of the authors) with a ward program for patients with chronic brain damage. There are serious considerations regarding the treatment of this type of patient. They are recalcitrant to most of the standard forms of psychiatric treatment; they often do not respond well to psychotropic medications, nor do they appear to gain much from verbal therapies; they typically have become dependent on the institution in which they reside, and are reluctant to leave; they are frequently alienated from their home communities, and are often not accepted back home by spouses and relatives. Thus, they have a combination of neuropsychological and sociological difficulties that make anything but custodial care difficult. We would therefore like to explore some ideas about treatment of these patients in this chapter.

The Nature of Chronic Brain Damage

Chronic hospitalized patients usually carry the diagnosis of "organic brain syndrome," which may be associated with a variety of etiologies. This diagnosis usually connotes irreversible brain damage that gives rise to a number of behavioral symptoms of the type we have been discussing throughout this book. In effect, an organic brain syndrome can be viewed as the end result of a variety of agents that impair brain function. These agents may sometimes involve such processes as alcoholism or cerebral vascular disease. Sometimes head trauma is the responsible agent. Not infrequently the etiology is unknown, or at least not fully understood, as in such conditions as Huntington's disease and Alzheimer's disease. Sometimes the condition is relatively static, but sometimes it is progressive and will eventually cause death. The patients we are addressing ourselves to here are those individuals whose brain impairment has reached such severity that long-term if not permanent institutionalization seems indicated. Our observations have indicated that there are a number of classes of such patients, which we will proceed to describe briefly.

The Amnesic Alcoholic

Alcoholic patients who develop a complex of symptoms known as Korsakoff's syndrome frequently become institutionalized on a long-term basis. These patients are characterized by their severe impairment of recent memory. Typically, they cannot recall what they did a few moments previously, seem incapable of new learning, and may be disoriented with regard to time and place. Their illness is thought to be attributable to excessive use of alcohol and a thiamine-poor diet. Although the Korsakoff patient has been studied extensively (Butters & Cermak, 1980), little has been done in the way of rehabilitation, since this syndrome is generally viewed as a chronic, irremediable condition. Frequently, chronic Korsakoff patients are no longer active alcoholics, and may not drink heavily even if they have the opportunity to do so. In any event, they rarely have the mental resources needed to obtain alcohol independently, and friends and relatives usually restrain them from doing so when they are out of the hospital. Thus, the rehabilitation problem with the Korsakoff patient is more often that of restoration of memory than it is abstinence from alcohol. "Can we restore memory in the Korsakoff patient?" is an excellent but currently unanswered research question. Attempts have been made to restore memory in brain-injured individuals through such techniques as visual imagery (Lewinsohn, Danaher, & Kikel, 1977), with some success. However, there is very little research literature involving memory training with alcoholic Korsakoff's patients. Cermak (1975), working with such patients, was able to show that imagery-oriented training was better than rote and cued learning at improving learning and attention. Binder and Schreiber (1980) demonstrated the efficacy of memory training with recovering alcoholics, but they did not study alcoholic Korsakoff patients specifically. Although there is an extensive literature on memory training in normal and neurologically impaired populations, there has been very little done with alcoholic Korsakoff patients.

The Nonamnesic Alcoholic

One might be surprised at the number of individuals who become long-term residents of neuropsychiatric hospitals because of the severe degree of mental deterioration produced by chronic alcoholism. Although some of these people develop Korsakoff-type illnesses, most do not. Rather, they acquire a slowly progressive dementia that makes them increasingly incapable of adapting to the demands of a complex environment. The neuropsychological aspects of the condition have been well studied (Tarter, 1976), but, again, the problem of direct remediation of alcohol-related deficits has not really been attacked. Both the Korsakoff and the non-Korsakoff alcoholic have a multitude of problems, including alcohol-related organic and mental difficulties, the social alienation often created by excessive drinking, and the problems that may have been engendered by long-term institutionalization. It is our opinion that these individuals really require treatment for their dementia rather than for their alcoholism. We are familiar with the argument that, if you do not treat the alcoholism, the alcoholic will simply return to drinking following discharge, and any retraining will have been in vain. The problem is that the verbal, insight-oriented treatment often given to alcoholics may also be in vain. The best approach to such patients might involve controlling drinking through environmental manipulations while attempting to restore such functions as memory and problem-solving ability through active training.

There are, of course, numerous treatment programs for chronic alcoholics, many of which involve behavior–therapy-oriented approaches (P.M. Miller, 1978) or approaches that combine behavior therapy with other modalities. However, to the best of our knowledge, these programs do not utilize direct cognitive-perceptual remediation methods. It may be pointed out that there are, of course, cognitive deficits associated with acute alcoholism (Parsons & Farr, 1981) that diminish with detoxication and abstinence. We are not really addressing ourselves to these cases here, but rather to the very long-term chronic alcoholic who has developed a permanent alcoholism-related dementia.

The Atherosclerotic Patient

There are essentially two types of atherosclerotic chronic patients, although the distinction tends to blur with the passage of time. The distinction is between the patient who has had a stroke and the one who has not had a stroke. From a neurological standpoint, the post-stroke patient may have specific deficits associated with the side of the lesion, while the nonstroke patient will more likely have generalized intellectual impairment. However, what seems to happen over time in the stroke patient is that the symptoms directly attributable to the stroke, such as hemiparesis, diminish, whereas generalized impairment increases in correspondence with progressive deterioration of the vascular system as a whole. Many, if not most, stroke patients return to community living, and most people with cerebral vascular disease can function outside of a hospital or other institutional setting until relatively late in life. However, some remain in institutions for long periods of time, beginning at relatively young ages. We have seen such patients in their fifties.

Based on admittedly limited observations, we would say that it is often not the intellectual deficits that keep the atherosclerotic patient in the hospital but the associated psychiatric and sociological difficulties. Perhaps another reason for continued hospitalization is the persistence of active medical difficulies. Treatment of atherosclerotic patients who are medically stable can generally be accomplished effectively on an outpatient basis. The atherosclerotic patients we have seen who remained chronically hospitalized generally had some combination of active illness, such as uncontrollable hypertension with a severe psychiatric disorder.

In rehabilitating such patients, it is particularly important to use a multidimensional approach. It is first important to appreciate the fact that there is hope. Hypertension is usually controllable with recently developed drugs; the vasodilator drugs mentioned in Chapter 2 are sometimes useful for restoring some degree of function. There is a technique called *hyperbaric oxygenation* (Jacobs, Winter, Alvis, & Small, 1969) that seems to be useful for some patients with cerebral vascular disease. Diet and exercise seem to be effective in controlling the progression of vascular disease. In effect, there are steps that can be taken to treat cerebral vascular disease from a medical-organic standpoint. If the medical problems are stabilized, these patients may be good candidates for various types of training, as the NYU group has well demonstrated.

The emotional difficulties of vascular-disease patients tend to revolve around fear and depression, since some of them have had such severe life-threatening episodes as heart attack or stroke. These people are often fearful of having another episode, or depressed about the status of their physical condition. The patient who does not respond well to his medication or other somatic treatment may tend to become discouraged about his potential for functioning productively again, or even surviving. When these conditions become particularly severe, these people find their way into psychiatric hospitals, and the combination of their intellectual impairment and emotional difficulties may keep them hospitalized for long periods of time. Treatment efforts, particularly those involving use of medication, are compromised because the vascular disease may contraindicate the use of many of the psychotropic drugs. Furthermore, some of the drugs used to treat the vascular condition, particularly some of the antihypertensive drugs, may in themselves produce depression as a side effect. In attempting to rehabilitate these patients, one has to balance medications and activities in such a way that the patient remains physically stable while, at the same time, one reduces his depression and keeps him active and alert, which is often not easy.

The Patient with a Terminal Illness

Can one speak of rehabilitation for patients with terminal illnesses? With regard to diseases of the brain, there are many conditions for which there is no cure. These conditions include Huntington's disease, multiple sclerosis, and the presenile and senile dementias, such as Alzheimer's and Pick's diseases. What can be done for such patients? Although individuals with these illnesses can function in the community for varying lengths of time following the appearance of the first symptoms, it is usual that they eventually are admitted to various institutions. The question then is how should they be treated in these institutions. First and foremost, many of these patients will survive for many years following the onset of symptoms. These years should be as productive and happy for the patient as possible. Second, although the disease processes are not directly treatable in these cases, some of the symptoms can be relieved, usually with medication. Thus, for example, in the case of Huntington's disease, haloperidol or one of the phenothiazines can be used to reduce the involuntary movements. Third, these patients often develop emotional and behavioral difficulties that are treatable. We have observed that the reaction of staff members to these patients often is heavily influenced by the knowledge that they have a terminal illness. This knowledge understandably produces sympathy, but as Wright (1960) points out, sympathy has its positive and negative features. It is unfortunate that the generally desirable human characteristics of sympathy and compassion can sometimes be expressed in a counterproductive way. For example, the formation of an overindulgent relationship with a terminally ill patient may backfire, to the detriment of the patient. If the relationship has to be ended, for some reason, the patient may have a very adverse reaction. Working with terminally ill patients is difficult and requires development of the ability to be concerned and sympathetic, without being overindulgent.

The Head-Injured Patient

Long-term hospitalization of head-injured patients is rare but generally occurs when the head injury is so severe that the patient is left with extremely impaired cognitive and perceptual abilities. Often there is a physical disability as well, such as partial paralysis. We have seen several brain-wounded patients who spent long periods of time in institutions following recovery from the acute phase of the episode. Since many of the clients in the Wichita program had head injuries, we need not elaborate on this problem here. The only point we would make is that we have observed that head-injured patients often have behavior problems, frequently involving impulse control. This problem is a complex and interesting one since we do not really know whether the head injury caused the lack of impulse control or the lack of impulse control caused the head injury. We can describe two somewhat distinct patterns producing long-term hospitalization in head-injured patients. In one case, the head injury has produced so much dementia and physical disability that the patient would have great difficulty in adapting to anything but an institutional environment. In the second case, there may not be too much intellectual or physical impairment, but the head trauma seems to release some underlying behavioral disorder that becomes so prominent that it requires long-term psychiatrically oriented hospital treatment.

Chronic Brain Damage and Institutionalization

Do these patients have to live in institutions? If they do, do these institutions have to be hospitals? Why can they not be nursing homes, domiciliaries, or sheltered residences of the type now being developed for elderly people? Why can their families not take care of them? Why keep people in a hospital when there is no known treatment for their disease? These questions and others related to them appear to provide a basis for a revolution now taking place in patterns of care for the chronically ill patient. It is not our purpose here to review the extensive literature that has sprung up concerning the iatrogenic effects of institutionalization, and the ethical and legal dilemmas long-term institutionalization presents. Rather, the point we want to make is that the institution needs to be a consideration in regard to the rehabilitation approach taken to the patient. In attempting to rehabilitate the hospitalized brain-damaged patient, it is important to consider the fact that the patient is in a hospital, and it is also important to consider how long he has been hospitalized. Treatment cannot take place in a vacuum removed from its social context. For example, what would be the impact of perceptual-motor retraining on a patient who had not been off the grounds of an institution for the past fifteen years as contrasted with its impact on a recently head-injured patient who lives in the community and goes to a clinic for his training? Obviously, there would be a difference. We have learned that, in the case of the former patient, the fact of institutionalization must be dealt with in some way. Ideally, it should be dealt with in a manner that allows for gradual transition from an institutional to a noninstitutional life. If the patient must always live in a sheltered environment, we should attempt to seek the least restrictive alternative. We would suggest that one aim of all rehabilitation programs for hospitalized patients should be that of getting the patient out of the hospital. This aim should be adopted regardless of how long the patient has been in the hospital or how difficult placement in the community may be. Even in the case of severely brain-damaged patients, indefinitely continued hospitalization is not likely to be optimal treatment.

How does one get patients out of the hospital? One way is to make the hospital less attractive and community life more attractive. With regard to the first part of this procedure, a hospital may be made less attractive without detracting from the quality of care. For example, a period of hospitalization sometimes takes on the quality of a vacation. The patient receives personal services and has access to recreational facilities that are not available in his ordinary life. Even if the term *vacation* is something of an exaggeration, hospitalization may provide temporary relief from some unpleasant life situation. The hospital can be made less attractive in a productive way by minimizing personal services and doing for the patient only what he cannot do for himself. If the patient can make his bed and care for his personal hygiene, he should do so. Hospital facilities should be used primarily for therapeutic purposes and only secondarily for recreational purposes. Although we have stressed the importance of recreation in the total program for a patient, the point is that recreation should be a part of treatment rather than a substitute for it. Another productive way of reducing the attractiveness of the hospital is to impose limits on length of stay. In this way, the patient cannot view the hospital as a shelter provider on an indefinite basis. In relationship to this point, sometimes hospital fees can cause an aversion as they accumulate from day to day. However, many of the patients we are talking about are not at hospitals that require fees for services.

Community life can be made more attractive by aiding the patient in dealing with any problems he may have in going home. In the case of the long-term patient, the outside world can seem more pleasant if it is approached gradually through residence in a number of transitional facilities. Our experience has indicated that, if appropriate arrangements are made with family members and/or community care facilities, even long-term patients are generally pleased to leave the hospital, and may stay out on an indefinite basis.

A second way of getting the patient out of the hospital is to retrain him for life in the community. Here, we are not dealing as much with cognitive-perceptual training as we are with social skills training. Numerous techniques have been developed for such training (Bellack & Hersen, 1978), and some may be adapted for use with brain-damaged patients.

Long-term institutionalization has a double aspect. It has implications for the patient, in addition to strong implications for the family. Family members become accustomed to life without the patient. The patient's children may grow up without ever really knowing at least one of their parents. The spouse, although legally married, essentially lives the life of a single person. Parents may have to relinquish care of their children to the institution. These conditions tend to make change difficult. When the family of a long-term institutionalized patient is presented with the idea that the patient may be coming home, it may have to consider the impact of a major change in its life-style. The other aspect of the problem is that when patients are sent home, they are often not "as good as new," and the family may have to be responsible for some degree of their management. In view of these considerations, the treatment team should expect a certain degree of resistance from the family when the patient is ready to be sent home. In our experience, this resistance ranges in severity from complete refusal to minor reluctance that can eventually be overcome. One very common pattern is "excuse making," in which the family claims it would accept the patient were it not for some extenuating circumstance, such as work or the unavailability of space.

The patient without a family also has difficulties. One major problem with the long-term patient is that when he is ready to leave the institution there is often no place for him to go. It frequently becomes a matter of keeping him, or putting him out on the street without resources. More often than not we keep the patient for compassionate reasons, and he remains in the institution simply because there is no place else to go. Part of the rehabilitation program for such patients must necessarily involve seeking some form of satisfactory placement. The social worker on the team should actively pursue all possibilities for de-institutionalization. Such possibilities might include helping the patient find a job, seeking placement in various community care facilities, and working with any family members the patient may have to get them to accept the patient.

With problems related to institutionalization, there is much similarity between the brain-damaged patient and the other kinds of people who have been institutionalized for long periods of time, such as psychiatric patients, mental retardates, and prisoners. Workers at any institution that keeps people for extended time periods are probably overly familiar with these problems. One might wonder whether or not the brain-damaged patient has to be considered as a separate entity in this regard. We feel that he does, for several reasons. First, in some kinds of brain damage, chronicity may be avoidable with proper early treatment. Second, the procedures that aid in de-institutionalizing brain-damaged patients may be different from those used for de-institutionalizing other kinds of institutional residents. A case example will illustrate this point. The patient in question was a young man who sustained brain damage from cardiac arrest during surgery. He became amnesic and would have occasional episodes of assaultiveness. He was kept in a psychiatric hospital and treated primarily with phenothiazines. He remained in the hospital for many years under heavy medication of this type, although he had never been diagnosed as schizophrenic. When we started treating him, we withdrew all his medication, and shortly thereafter he brightened and became more alert. There were no episodes of assaultiveness. He was eventually discharged to his home where he has remained for over two years (at this writing), and he is doing well. In this case, the treatment of medication withdrawal was appropriate for this brain-damaged patient but probably would not have been appropriate for patients with such functional psychiatric disorders as schizophrenia. Third, the brain-damaged patient may benefit from being in a specialized setting in which he can receive cognitive-perceptual retraining. Such retraining may not be available in, or appropriate for, facilities for general psychiatric patients.

Some years ago, we became affiliated with a neuropsychiatric hospital that had a large number of chronic patients on its rolls. The hospital administration determined that, on the basis of the considerations outlined above, it might be wise to establish a separate program for brain-damaged patients only. This program was established on a 20-bed ward and was staffed with appropriate personnel. We will describe this program in some detail since it differs substantially from the Wichita program; the description should be of interest to those engaged in rehabilitation of chronic brain-damaged patients.

The Neuropsychology Center

The Neuropsychology Center, which is a 20-bed unit in a neuropsychiatric hospital, has a staffing pattern similar to that found on most psychiatric wards but with significant differences. First of all, the ward administrator is a neuropsychologist rather than a psychiatrist. Second, there are consultants from different disciplines who usually do not consult regularly on psychiatric wards: internists and neurologists. Because the ward administrator is not a physician, medical coverage is provided by physicians assigned to the ward on a part-time basis. The ward has a part-time social worker and the usual complement of nursing personnel, including 2 registered nurses and 7 nursing assistants. This staff provides 24-hour coverage, 7 days a week.

General Principles

When the Neuropsychology Center was initiated, we intended to replicate the essentials of the Wichita project in a hospital setting. In particular, we wanted to do cognitive-perceptual training utilizing behavior–therapy techniques. It soon became apparent, however, that we needed a program with a much broader scope. Numerous problems emerged: many patients were too demented or amnesic to allow for vocationally oriented retraining; many patients referred to us were chronic, and we found that we had to deal with their chronicity as well as with their organic defects; many patients also had active medical and neurological problems that required further evaluation and treatment. In our view, many referrals were being improperly, or at least not optimally, medicated and we had to become involved in the process of changing and adjusting medications. For all these reasons and more, we developed a treatment philosophy that was somewhat different from the one adopted for the Wichita project.

The concept underlying the new treatment philosophy involved behavioral and neuropsychological aspects but also included significant medical and psychosocial dimensions. One important feature of our framework is the view that every patient represents a unique behavioral-medical problem complex that requires careful evaluation. Thus, within the context of a ward milieu, we attempted to establish highly individualized programs for each patient. Although the NYU group model (Diller, 1976) and the Witchita model were appropriate, neither was sufficient. The first problem we had to deal with was diagnostic in nature; we felt that we needed a detailed and sophisticated description of the patient's condition. Even if he had been "well worked up," we wanted to do our own evaluation: we maintained an attitude of disbelief. Logically, the first component of this evaluation was medical in nature. Initially, we wanted to determine whether or not the patient's neurological difficulties could be treated with medication or some other form of somatic therapy. Sometimes neurosurgery was considered, or a trial with a new medication. We felt that, if the patient's illness could be arrested or reversed with the proper medication or other somatic treatment, our first efforts should go in that direction. We did not assume that the patient had irreversible brain damage that was incurable. An example might clarify this point. We admitted a patient who was severely impaired mentally, and who had radiological evidence of substantial cortical atrophy; the working diagnosis was Alzheimer's disease. Following examination of the patient and a review of the radiological and other laboratory material, the neurologist suggested that some of the patient's symptoms might be the result of normal-pressure hydrocephalus. To rule out this possibility, the patient was given a cisternogram to determine whether or not there was blockage in cerebrospinal fluid circulation. If such a blockage exists, it is often treatable by a neurosurgical procedure known as a *ventriculoatrial shunt.* A patient may therefore have a treatable illness rather than Alzheimer's disease. But as it turned out, this patient did not have hydrocephalus but did have Alzheimer's disease. The point is that an effort was made to determine whether or not he could be treated surgically.

Another general principle was that activity should be encouraged. In the case of institutionalized brain-damaged patients, it is generally advisable to structure this encouragement in some way through providing a milieu that stresses activity. The means for accomplishing this end were varied, but largely involved continuing efforts by the nursing staff to get the patients to scheduled activities. The ward has a daily schedule that is posted and communicated to patients at ward meetings. It includes self-care activities, housekeeping, physical exercise, and time spent off the ward. An attempt was made to structure the day on the ward much like a normal day would be structured outside a hospital. Mornings are spent on self-care and personal hygiene, the remainder of the day on a number of scheduled activities and assignments, and the evening on leisure and recreation. Sleep was permitted during the day, but it was scheduled; for example, some patients could take a nap after lunch.

The relationships between neuropsychological testing and behavior modification described elsewhere in this book were adhered to but were applied in a different way from what was done in Wichita. We continued to accept the belief that neuropsychological assessment could be productively used in the planning of behavior–modification programs. However, we found it necessary to take a broader view of neuropsychological assessment than we had in the past, particularly with regard to sole dependence on the Halstead-Reitan battery. For example, many of our patients suffered from Korsakoff's syndrome, a disorder involving moderate to severe anterograde amnesia. The Halstead-Reitan battery does not really contain tests particularly sensitive to the amnesic disorders, and so we found it necessary to depend on the interview and mental status examination. Test procedures are available for careful study of memory disorders in Korsakoff patients (Butters & Cermak, 1980), and we began using them when they became available. As most clinical psychologists know, the Wechsler Memory Scale (Wechsler, 1945) is generally accepted as the standard instrument for psychometric evaluation of memory. However, in its present form, it does not fully accomplish the task of adequately assessing patients with amnesic syndromes. First of all, it is not only a memory test but it also contains subtests that evaluate such areas as attention and mental control (e.g., reciting the alphabet). Second, one of the more diagnostic features of many amnesic disorders is an impairment of delayed recall. The recall subtests of the Wechsler Memory Scale only assess immediate recall. Finally, the Wechsler Memory Scale does not contain an adequate measure of long-term memory. For these reasons, we have developed our own series of memory tests for use on the ward, which include measures of delayed recall, list learning with an interpolated task to prevent rehearsal (Peterson & Peterson,

1959), the "Famous Faces" test for evaluation of long-term memory, and the presence of an anterograde-retrograde gradient, and a paired-associated learning task.

One significant aspect of the approach used is that of identifying maladaptive behaviors that prevent the patient from engaging in more independent, restriction-free living. Attempts are made, generally in an informal way, to alter the environmental contingencies that maintain their behaviors. Even in the case of the patient with a terminal illness, not all of the patient's behavior is directly attributable to the illness. Certain behaviors are maintained by environmental stimuli. Thus, whereas the patient's physical condition usually cannot be altered, the environmental contingencies can. The major issues we have dealt with are maladaptive attention-seeking behavior and behaviors directed toward maintaining dependence. We will give case examples for each of these areas in which informal programs were effective.

The patient for our first example was a young man with Huntington's disease who had been institutionalized for many years. His behavior in question was a tendency to form close romantic relationships with members of the nursing staff. He would develop a crush for a particular nurse and would only allow her to care for him. When this nurse was not available, he would insist, generally in the form of a temper tantrum, that she come, and would refuse to eat until she was with him. When the patient was transferred to the Neuropsychology Center, this nurse was no longer available to him, and, upon discovering this, he insisted that she continue to care for him or he would go on a "hunger strike." He did, in fact, skip a meal or two. The staff determined that the consistent attitude to be taken with this patient should be to convey to him the idea that he could eat or not as he pleased, but that we would not permit him to become malnourished. When this message was communicated to him and the attention previously paid to him when he went on a "hunger strike" was no longer forthcoming, he almost immediately returned to a normal diet. Thus, we were able to deal with a maladaptive attention-seeking behavior.

The second example of an informal program involves a patient with a multitude of illnesses, including poorly controlled diabetes and epilepsy. Prior to transfer to the Neuropsychology Center the patient had sustained an ankle injury and had to use a wheelchair. However, after he was admitted to the program, an examination by a physiatrist revealed that he was capable of walking with a brace or a walker. However, the patient claimed that he could not walk and would not do so. The program for this patient was simply, strong, consistent encouragement to get him to walk as much as possible. He was praised for walking short distances with a walker, and his many complaints about aches and pains in his back and legs did not receive much attention. The physiatrist had already indicated that, although the patient might experience some discomfort because of extended disuse of his legs, it would not be harmful but would in fact be beneficial if he experienced it. Eventually, the patient began walking, sometimes without the walker, and ultimately gave up use of his wheelchair.

We also found it necessary to extend our behavior–therapy programs beyond perceptual-cognitive retraining to remediation at more basic levels of adaptive and social behavior. It was necessary to deal with such problems as maintenance of personal hygiene, housekeeping, control of bowel and bladder function, hypersexuality, and assaultiveness. Thus, our application of behavior–therapy techniques was highly individualized, and consisted primarily of special programs tailored to the needs of patients demonstrating a wide variety of difficulties. Cognitive-perceptual training was used in some cases, particularly for patients with such specific defects as memory disorder. However, the universal application of such training in a sheltered workshoplike setting did not turn out to be appropriate or feasible. Many patients admitted to the Neuropsychology Center have terminal illnesses or conditions that are so advanced that it would be unrealistic to think in terms of their return to work or independent living. However, it is never assumed that any patient could not either leave the hospital or at least live in the hospital at a less restrictive level. Our program is time limited to a period of no more than six months. Toward the end of treatment, extensive efforts are made to place the patient outside of the hospital. If we fail to achieve this goal, an attempt is made to send the patient to as unrestricted a setting as possible within the hospital. We do not always succeed in keeping patients for no more than six months, but in general we do achieve this goal.

Typically, the patients admitted to the Neuropsychology Center have complex disorders involving physical, behavioral, and iatrogenic symptoms. Thus, it is often not helpful to characterize them by single descriptive labels such as *hemiplegic* or *Korsakoff's syndrome*. Typically, there is a combination of neurological disorders with some functional psychiatric condition. For example, one of our patients with Huntington's disease was also schizophrenic. In most cases we were convinced that we were not looking at a functional disorder mimicking an organic condition or vice versa. It usually seemed more likely that both conditions existed concurrently. Certainly, they interacted with each other even though they bore no necessary cause-and-effect relationship. In our example, whereas a certain number of patients with Huntington's disease also develop schizophrenic or schizophrenic-type conditions (Whittier, 1977), there is no evidence that Huntington's disease directly causes the symptoms of schizophrenia, such as auditory hallucinations and delusions, the symptoms that were observed in our patient. Thus, we had to deal with the dementia and movement disorder characteristic of Huntington's disease in addition to the thinking disorder associated with the schizophrenia. To illustrate further the complexities involved, we would add that the patient under discussion was also an amputee, having lost a leg in a car accident. He could not adjust to an artificial limb because of the ataxia produced by the chorea, and was therefore confined to a wheelchair.

Complexities of the type described produced numerous treatment challenges. As indicated elsewhere, brain-damaged patients do not respond in the same manner to psychotropic drugs as do nonbrain-damaged psychiatric patients. Furthermore, the use of some of these drugs is hazardous, particularly in the case of patients with cardiovascular disease. The drug issue raises the more complex problem of deciding what to treat when one has a patient with a variety of disorders. The implications of treating one of the conditions for the other conditions must be carefully considered. Thus, for example, if you treat a patient's hypertension, you might increase his depression. It is sometimes necessary to make a considered decision as to whether or not to accept some degree of risk in treatment. For example, do you treat a manic, agitated patient with lithium if he has a heart condition? If you do, it is necessary to provide a regimen of procedures to minimize this risk; for example, serial EKGs concurrent with treatment.

We did not always find that treatment of one condition had adverse effects on other conditions. Sometimes the reverse was true. We treated a number of very agitated, depressed patients, many of whom had severe sleep disturbances and other signs of chronic anxiety and discomfort. When these patients were treated for their agitation with medication or relaxation therapy, we found to our surprise that their mental functioning improved as the agitation diminished. We had one case of an aphasic patient whose speech improved significantly following treatment with lithium, a medication that was prescribed for the purpose of reducing his severe agitation. We still do not know whether the lithium had a direct effect on brain function or whether the improved speech was a secondary phenomenon related to reduce agitation (Williams & Goldstein, 1979). In either event, treatment of one symptom had clear positive implications for another apparently unrelated symptom.

These considerations can best be epitomized in two alternative ways. It can be said that the brain-damaged patient should always be viewed as a general medical, neurological, and psychiatric patient. Another way of making this point is to say that the patient should be viewed as a whole human being in a particular social situation, and his treatment should therefore not be delimited to any particular professional discipline. The first alternative reminds us that we should look at the patient's general medical, neurological, and psychiatric status, whereas the other phraseology reminds us that these evaluations should not be done in isolation but rather as parts of an interdisciplinary effort. It may be superfluous to make the point that treatment of the chronic brain-damaged patient can become exceedingly complex, particularly when we add to these problems the possible influences of long-term treatment with psychotropic medication and of other somatic therapies such as electroshock treatment (Squire, Slater, & Chace, 1975).

One major difference between the Wichita program and the Neuropsychology Center was the range of impairment. Although the Neuropsychology Center has had some patients who were comparable in impairment level to the Wichita clients, it has also taken severely demented patients. In view of the problems of these latter patients, we found it necessary to find a form of behavioral treatment that was more basic than cognitive-perceptual retraining. It occurred to us that the reality–orientation system developed by Folsom (1968) for elderly, senile patients might be appropriate for at least some of our patients. Reality orientation is essentially a system by means of which basic information is restored to the confused and demented patient through various means. It is administered largely by nursing assistants who are trained to apply its various instruments and techniques. It is both a ward milieu and a classroom form of treatment. On the ward, the patient is regularly reminded of his name, the names of those around him, the date, the location, and related basic information. Sometimes, it is necessary to assist the patient with such activities of daily living as eating and grooming. The essence of the system seems to revolve around repetition, so that eventually the patient learns certain significant pieces of basic information. In the classroom, emphasis is placed on systematically reminding the patient of his name, the time, the location, the weather, and so forth. A magnetized board is commonly used on which such items as the name of the institution, the date, the weather, the name of the therapist, the number of the ward, and related information are posted in large block letters. A typical session would last no more than a half hour, during which time each patient would introduce himself. An information review is generally held in which the patient is reminded, through the use of the board, of the posted information. Patients may be asked to write their names or produce them with anagrams. Various media, such as flash cards, large-faced clocks, picture books, calendar pads, and the like, may be used. Sometimes the therapist may give a little talk about some item of general interest, such as current events or sports.

Katz (1976) and Barnes (1974) have reported that reality orientation is of benefit to geriatric patients. Although the Barnes study utilized only six patients, Katz included 108 subjects in his study. However, he used reality orientation in combination with other techniques. Of the 108 subjects, 80 were diagnosed as having chronic brain syndrome or senile dementia. Of these 80 subjects, 35 (44%) were rated as improved following treatment.

We are conducting an ongoing study of the efficacy of reality–orientation therapy. The study is of the crossover type, in which each patient serves as his own control. One subsample receives therapy for a 3-week block of time but continues to be evaluated for an equally long time after therapy is discontinued while the other subsample is evaluated without therapy during the first time segment and given therapy for the second segment. Thus, each patient in the total sample receives therapy but is also evaluated without it. The evaluation consists of observation of activities and asking of a series of questions. Each patient is observed once a day to determine whether he is sleeping, watching television, awake but unoccupied, or engaged in some activity other than TV watching. Also, each patient is asked daily:

1. What is the day of the week; the month; the year?
2. What city; state; country are we in?
3. What is the name of the hospital?
4. What is the name of the ward?
5. What is the weather like today?

Patients in therapy were asked to answer all these questions during the therapy sessions and during daily probe periods outside the sessions. During the untreated period, patients were asked these questions only during the probe period. In order to determine whether what was taught during the sessions generalized to other aspects of orientation, they were always asked during the probe periods:

1. When is the next holiday?
2. How long have you been in this hospital? On this ward?
3. Can you name one nurse, one aide, and one patient on the ward?
4. Can you tell me about some current event?

Thus far, we have evaluated 14 patients. Preliminary analysis of their data revealed no significant group differences between the treated and the untreated segments with regard to the activities or ability to answer the questions. However, examination of the individual curves suggests that some patients did seem to benefit from the treatment. We are now trying to determine whether any particular kind of patient benefits more from reality–orientation therapy than do other kinds of patients. Thus far, we would have to conclude that reality–orientation therapy is not an efficacious treatment, as evaluated by our probe method, for demented and amnesic patients in general, but may be of benefit to certain patients, although the specific characteristics of these patients have not yet been identified.

Administrative and Policy Considerations

In the organization of programs like the Neuropsychology Center, it is necessary to develop administrative policies regarding such matters as criteria for admission, length of stay, and discharge. We have developed our policies on a tentative basis. In general, the program is open only to patients with the established diagnosis of one of the organic mental disorders (DSM-III) or the equivalent. The rationale for this rule is that it was felt to be of great benefit to consolidate patients with this group of diagnoses into one program for a number of reasons. First, a stable staff could be specifically trained to work with brain-damaged patients. Second, individuals with expertise in the area could focus their efforts onto a single program. Third, we maintained the view that the rehabilitation of the brain-damaged patient differs from the usual form of treatment of psychiatric patients in general. The differences involved require different ward milieus, different medications, and different kinds of treatment modalities. A separate ward with a separate staff appeared to be the best way of dealing with these differences. As indicated elsewhere, mainstreaming for the brain-damaged patient is a mixed affair, and putting brain-damaged patients in close proximity with chronic schizophrenic patients does not impress us as a desirable situation.

With regard to admitting patients, the final decision is made on the basis of a conference that includes a presentation of the patient by someone from the ward from which he would be coming and an interview with the patient. In the majority of cases, patients are referred to us from within the hospital. The decision to or not to admit is to some extent determined by the diagnosis, but we also take into consideration the current context of the ward. In particular, we are reluctant to admit patients demonstrating little in the way of intellectual impairment if the ward is, at the time, filled with severely demented patients. We are concerned about the impact that being on a ward with such patients would have on a substantially more intact individual. It could be perceived as a step backward in that the candidate for admission might feel that those caring for him perceived him as being like or about to be like the other patients on the ward. Thus, patients with medical histories of brain damage who do not currently exhibit prominent disturbances of intellectual function, language, or memory may not be admitted to the program. The major exception to this rule occurs when the center is the most appropriate place to conduct a comprehensive neurological and neuropsychological evaluation.

Another significant matter of policy has to do with turnover. Ideally, patients should stay no longer than three to six months in the program. There are some exceptions, but in general, when the patient has been on the ward for close to six months, we initiate a placement effort. In general, we have been successful in placing patients in community care facilities or discharging them home. A recently accomplished survey of dispositions over a one-and-a-half-year period is presented in Table 10. It should be noted that the data in Table 10 are based on a population of chronic patients, generally with long histories of institutionalization. Sometimes we fail in our efforts to de-institutionalize a patient, and it is necessary to transfer him to a long-term care facility elsewhere in the hospital. In looking back at these failures, they seem to be of two general types. In one case, the organic brain syndrome is accompanied by a severe psychiatric disorder that we cannot control. Patients who are both brain-damaged and schizophrenic fit this category. We may assist these patients in regard to activities of daily living and improving such functions as memory, but we are sometimes unable to deal with the psychosis. The second group

Table 10
Patient Disposition Data ($N = 133$)

Disposition	Percentage
Returned to an in-hospital psychiatric ward	34%
Transferred to a VA intermediate care facility	6%
Transferred to a nursing home	14%
Left hospital against medical advice	20%
Regular discharge	45%
Average length of stay 172.4 days ($SD = 148.9$)	

consists of patients with progressive terminal illnesses. Typically, these patients have either Huntington's or Alzheimer's disease. In these cases our goal is to establish contingency–management programs that allow for more satisfactory adjustments to hospital life. We try to teach these programs to those who will be caring for the patient when he is transferred to a long-term care facility.

By virtue of the policies governing the institution in which the Neuropsychology Center exists, we cannot follow up our patients after discharge. They are typically transferred to a separately staffed outpatient department that continues treatment in the community. However, we do try to follow our patients informally, and, on occasion, we are visited by patients who have been in the program. If the patient does not do well in his postdischarge situation and has to return to the hospital, we are generally willing to accept him back into the program, particularly if we feel that there is more that we can do for him. In essence, we have an open-door policy. On the basis of admittedly limited experience we would say that it is sometimes possible to rehospitalize a patient briefly, get him back on a treatment regime, and send him out again. One of the major reasons for rehospitalization is lack of compliance with the prescribed medication schedule. In these cases, it is often simply a matter of getting the patient back on his medication and readvising him and those caring for him about the importance of staying on the medication.

The Role of the Family

The family situation of the institutionalized brain-damaged patient is, in many ways, not unlike what is found for chronic institutionalized patients in general. As indicated, many long-term patients gradually lose their family attachments, and even visits or letters from family members tend to be infrequent. From the point of view of rehabilitation and placement planning, a choice must often be made with regard to whether the patient should be returned to his family or to a community care facility of some type. A common question is will the family take him back. The answer to this question is an obviously significant determinant of the patient's fate. Unfortunately, many patients remain institutionalized because the family will not take them back. In this regard the patient is generally not forced to remain in the institution on the basis of involuntary commitment, but because he has no other place to go.

In working with families of patients admitted to the Neuropsychology Center we have seen a broad spectrum of family acceptance or lack of it. It became clear that there were two significant factors usually involved: the willingness of the family to accept the patient back and the capacity of the family to care for him. The matter of capacity involves the extent of the patient's impairment in combination with the family's situation and resources. Difficulties arose when these two factors did not go well together. Either the family would refuse to accept a patient who was clearly sufficiently stable to live at home, or the family would be unrealistically willing to accept the patient back home, even though it was apparent that he was still too impaired to live at home. In the former case, we generally found it necessary to hold the patient for longer than needed, and tried to place him in a community care facility. In the latter case, if we let the patient go, he usually had to be rehospitalized after a short period of time.

It might be useful to give some examples of these two extreme situations. In one case, the patient was a young married man who entered the program because of significant intellectual impairment and an atypical dyskinesia, both of which were of unknown etiology. He also was diagnosed as schizophrenic at some time in the past, and the reason for his admission to the hospital was psychiatric in nature: an attempted sexual assault. Following treatment in the Neuropsychology Center, both the motor and intellectual symptoms remitted, and the patient began functioning in a relatively normal way. However, his wife absolutely refused to have him return home, claiming that she could not trust him around their small child. The patient, having little money and no other community resources, remained in the hospital, despite being capable of living outside, while arrangements were being made for some form of placement. In the other case, the patient was a man in his late forties who suffered severe cerebral anoxia accompanying a heart attack. He was successfully resuscitated, but came out of the episode with a severe, pervasive dementia. The dementia was accompanied by an agitated affective state, sometimes resulting in verbal abusiveness or assaultiveness. Although some progress was made in treating him, he remained impaired after several months of hospitalization. Since the verbal abusiveness and assaultiveness diminished greatly, the patient's very devoted wife expressed the desire to take him home. Although we had some doubts about what would happen, we went along with her wishes and discharged the patient. Several weeks later, however, he was readmitted to the hospital. After what seemed to be a sincere effort on the part of the wife to manage the patient at home, it became apparent to her that she was unable to do so. Subsequently, the patient was placed in a nursing home.

One obvious suggestion in regard to the latter case in particular would be to train the family to care for the patient. The NYU group has developed a program to train wives of aphasic patients to help their husbands with their language disabilities. The problem here is that often we do not have the substantive basis for this kind of training. However, when patients have been placed on effective programs of medication and contingency management, we have been able to instruct family members in how to continue these programs. A good example involves the patients effectively treated with lithium. Since lithium ingestion does not generally lead to any perceptible changes in subjective experience, we instructed family members that it was critical that the patient take the lithium on a regular basis. They were told that lithium does not work for immediate symptomatic relief, and that for it to be effective, a steady blood level must be maintained for what can be a long period of time.

In our view, both of these cases have their unfortunate aspects. In the first case, the patient has to remain institutionalized for no clinically relevant reason, and since he does have anxieties about going out of the hospital, there is some problem with maintaining his motivation to leave. In the second case, premature discharge of the patient may have hindered his returning home again at some later time. While his prognosis is poor, he just may recover sufficient function in years to come to be able to make it at home. However, his wife may have been so discouraged by the first attempt that she is probably much less likely to accept the patient back again in the future.

In our experience, the role of the social worker in regard to establishing contact with the family cannot be overestimated. However, we have also found it useful to have the family visit the hospital to talk to various members of the treatment team. During such meetings, it is possible to discuss the patient's illness, the treatment program, and the likely outcome. In some cases, it is necessary to communicate to the family the tragic nature of the patient's disorder, particularly when a terminal illness is diagnosed. Sometimes, however, the family visit can be a source of encouragement when the patient is found to be responding to treatment and improving. We also found that it was sometimes useful to have the patient attend a family meeting for purposes of negotiating discharge arrangements. Concrete matters, such as the family's finances and the conditions under which the patient can return home, are often high on the agenda for such meetings.

The rehabilitation process has its practical aspects, many of which, at least in part, relate to the family situation. One of these important practicalities involves the economic status of the patient. The two basic issues are whether or not the patient can afford continued treatment, and whether or not his finances are sufficient to support independent living in the community. These matters get us into such intricacies as pensions, disability compensation, and guardianship arrangements for management of funds. The issue of competency to handle funds often becomes a major consideration. In our experience, these matters often become a source of substantial conflict among the patient, the family, and the treatment team. It is unpleasant to comment on the darker side of human nature, but sometimes the family may appear to become unduly mercenary with regard to management of the patient's funds. Sometimes there are fiscal difficulties because the family is simply unable to provide for the support of an unemployable patient. The major practical suggestion we have is to have an aggressive social worker on the team who can serve as an advocate for the patient. The patient's rights to his property and money should be protected, and efforts should be made to seek funding when it is needed. For example, the patient may be entitled to veterans or social security benefits and not be aware of it. Alternatively, family resources may exist to which the patient is entitled, but which are not made accessible to him for one reason or another.

Rehabilitation of brain-damaged patients does not generally connote a return to normalcy. Rather, the term *rehabilitation* is generally taken to mean restoration of function sufficient to allow for a more independent level of living than was previously possible. A major problem with regard to the family relates to whether or not the patient has recovered sufficiently, in its judgment, to be able to return home. If there is any question, we handled this problem through the use of trial visits. Initially, the patient may go home for a weekend. If he does well he may go out for increasingly lengthy stays until he is fully accepted back. If things do not work out well, then alternative plans are made, such as placing the patient in a nursing home or community-care facility. In some cases, we experienced reluctance on the part of the family when in our judgment the patient was capable of living at home and the family had the resources to care for him. In these cases, we tended to adopt a firm attitude, and exerted some pressure on the family to accept the patient. To do so, the principle arguments directed to the family were that a hospital is not a place for housing people simply because they are not wanted, and the responsibility for the care of an impaired family member could not simply be completely transferred to an institution. Family members frequently made remarks to the effect that the patient is doing so well in the hospital that it would be best if he just stayed there. Such comments as, "He looks so good" and "He seems happy in the hospital," were frequently made by family members. Our responses to these remarks were to the effect that long-term institutionalization is ultimately detrimental to patients and should be avoided whenever possible, and it is unfair to other people who seriously require hospitalization to have a bed occupied on an indefinite basis by someone who no longer belongs in a hospital. In essence, our attitude was that the family should not be permitted to arbitrarily abrogate its social responsibility or its responsibility to the patient.

When the patient does go home, the question arises as to what he should do there. The Wichita clients tended to return to school or to work, but this happened far less frequently in the case of the Neuropsychology Center patients. Members of the latter group tended to be older, more impaired, and institutionalized for lengthier periods of time. Therefore, the treatment goal was generally de-institutionalization rather than return to gainful employment. Some of our patients did go, or attempted to go, back to work, but the number that did so was small. In most cases, if the patient was referred to the Neuropsychology Center, he was not viewed as potentially employable. The center was primarily established for chronic, severely demented patients. Thus the problem becomes primarily one of maintaining the patient at an optimal level without the necessity of rehospitalization. Wherever possible, we attempted to continue in the community the program that was established in the hospital. The success of this endeavor was informally evaluated through the outpatient facility located on the grounds of the hospital. Most of our discharged patients were followed as outpatients. Of course, if the patient had to be rehospitalized shortly after discharge, we were not successful. In many cases, we discharged those patients who were stabilized with medications that appeared to do an optimal job of symptom relief. We were insistent with the patient and the family that the medication schedule should be complied with. It was important to emphasize that even if the patient feels better for an extended period of time, he should continue to take his medication unless a doctor tells him to discontinue it. In some cases, we were able to counsel the family members concerning the nature of the patient's difficulties and how to handle certain situations. For example, we worked on a program in which the mother of a patient was instructed in how to continue a behavior–therapy program started in the hospital. We unfortunately do not have the resources available for home visiting, but obviously such a capability would be highly desirable. The concept of hospital-based home care seems to be an excellent one, but it was not implemented in our case. Indeed, the whole issue of continuity of care is one that we have not yet adequately addressed and represents a weakness of the program.

A Preliminary Assessment of the Neuropsychology Center Program

The Neuropsychology Center is an ongoing program and it would be premature to attempt to accomplish an extensive program evaluation. However, if such an evaluation were to be accomplished, it might address itself to these questions:

1. Is it advantageous to place chronic brain-damaged patients on their own ward within the organizational structure of a neuropsychiatric hospital?
2. Does neuropsychological assessment provide data to be used in formulating rehabilitation programs that are more useful than the usually obtained clinical data?
3. Are techniques derived from behavior therapy effective in the treatment and management of adults with severe, chronic brain disorders? If so, what are the limitations of such treatments?
4. Are there any pharmacological treatments that are particularly effective in treating brain-damaged patients?
5. What can be said at this point about treatment of the various patient groups described at the beginning of this chapter?

It would seem to us that these five questions represent key areas in regard to determining the effectiveness of our program and programs similar to ours. We will briefly address ourselves to each of them.

1. Obviously, the question of whether it is better to treat brain-damaged patients on a special ward can be determined only by doing something like the Ellsworth, Collins, Casey, Schoonover, Hickey, Hyer, Twemlow, and Nesselroade (1979) study in which patients on special and mixed wards are contrasted in terms of outcome variables. Practically speaking, however, the decision to create special programs is generally made on an administrative basis without reference to research findings of that type. In such decision making, rational considerations involving establishing or not establishing special programs may be invoked. In the absence of an Ellsworth-type study we have

offered such considerations throughout this book. In a special program it is possible to establish a treatment team with the expertise and training needed for sophisticated rehabilitative efforts. As has been pointed out, basic knowledge of neurology and neuropsychology is essential. We have indicated our preference for including individuals on the team with expertise in behavior therapy. We also pointed out that the treatment programs for brain-damaged patients may be different from those commonly used for psychiatric patients. The stress on perceptual and cognitive skills that is so important in treating the brain-damaged individual generally is not nearly as crucial in the treatment of patients with emotional disorders. Thus, whereas it would be our opinion that the establishment of a special program would be in the best interest of the brain-damaged patient, the proof that such programs are better than general psychiatric programs remains to be established.

2. Our answer regarding the question concerning the neuropsychological tests would have to be a qualified one. One major difficulty is that the severely impaired chronic patient is often unable to take the complex type of tests contained in the Halstead-Reitan battery. A substantial percentage of patients in the Neuropsychology Center were not testable with this procedure. Some of them could take the Luria-Nebraska battery (Golden, 1981), which is generally less demanding than the Halstead-Reitan. There appears to be a need for the development of simpler assessment procedures for this kind of patient. The other qualification would be that the neuropsychological test results, although often useful, should not be viewed as the sole consideration in the planning of a rehabilitation program. On the basis of our experience, we cannot concur with Luria's presentation (1948/1963) of cases in which rapid recovery of function often came apparently solely as a result of neuropsychologically oriented retraining. Somehow, our patients seemed to be more complicated than that and reflected complex mixtures of neurological, general medical, psychiatric, and socioeconomic difficulties.

With these qualifications, the considerations offered in Chapter 5 concerning the use of neuropsychological tests in rehabilitation planning were found to be helpful. For each testable patient a full neuropsychological assessment report was written that provided a series of recommendations for retraining, where applicable, and for management. The reports were read by the ward physicians, the nursing staff, the rehabilitation specialists, and the consultants to the program. Attempts were made to integrate the recommendations contained in the report with the other assessments done, and often the recommendations would be presented to our consultant in behavior therapy, who would make suggestions concerning program design and implementation.

The specific ways in which the tests were used in the design of programs are illustrated elsewhere in this book. Although the use of the concept of hemisphere asymmetries was somewhat helpful, it was not as pertinent as it was in the case of the Wichita program. The majority of the Neuropsychology Center patients did not have unilateral or focal lesions, and so specific language or visual-spatial training was not recommended with great frequency. The most frequently observed deficit area was memory, and we continue to make extensive use of memory training. Whereas the existence of memory disorder is often clinically obvious (as was pointed out earlier), there are many types of memory disorder, and neuropsychological assessment is useful in regard to identifying the type of disorder as a basis for rational treatment.

Since most of the patients admitted to the Neuropsychology Center had a chronic disorder, it is clear that the major neuropsychological deficit found among our patients was dementia. That is, most of the patients had generalized impairment of intellectual function, and the existence of specific deficits was generally accompanied by dementia. The major exceptions were the amnesic patients, who often had average intellectual levels despite their severe impairments of recent memory. Most of these patients had alcoholic Korsakoff's syndrome. In view of the nature of the population, then, the major consideration became general level of performance, since there was great variability in regard to degree of dementia. Level of performance, as estimated by the IQ or average impairment rating values, tended to be used as a prognostic indicator of what functional level the patient might achieve and level of performance was also of assistance in determining how much structure the patient required for his management on the ward. The level of performance indices often helped the staff to not underestimate the patient's capacities. Initially, they were surprised to learn that patients with specific disorders in such areas as language or memory maintained average IQs, and could do things that did not involve those specific deficits. Thus, the tests were useful in demonstrating that neuropsychological deficits could be specific, and that normal level intellectual demands could reasonably be placed on patients in areas that did not involve those deficits.

3. The application of behavior–therapy principles and methods to adult patients with acquired brain damage is still in its infancy. With regard to the chronic, institutionalized patient without a specific neu-

ropsychological deficit, we would have to say that its major contributions have been in two areas: memory training and treatment of maladaptive behavior. We have reviewed some of the literature on memory training elsewhere in this book, and have also presented some of our own cases. We have also tried to show instances in which formal or informal programs could significantly alter such maladaptive behaviors as incontinence and inappropriate verbalizations. For research purposes, and ideally for clinical purposes as well, it is useful to implement these programs as formal single-subject design studies, but in an actual clinical situation, particularly when staff is not plentiful, this procedure is often not practical. Our experience has been that when staff members are trained in general behavior therapy principles, they can apply those principles in an informal but often effective way in treating patients. Formal programs tend to require large staffs and large amounts of staff time, resources that we generally did not have available. However, informal interventions based on such principles as differential attention, selective reinforcement, avoidance of unscheduled reinforcement, and overcorrection can often be effective.

The limitations of behavior therapy in this area remain undefined. Clearly it does not influence the underlying neurological disorders of these patients, but no one has suggested that it does or should. It would appear to us that two reasonable questions can be asked: Can behavior therapy assist in the remediation of such neuropsychological deficits as language, memory, and visual-spatial disorders? Can it assist in modifying the maladaptive behaviors of brain-damaged patients that are apparently maintained by environmental contingencies? There are no definitive answers as yet. There have been reports of successful memory training, notably the oft-cited study of Lewinsohn, Danaher, and Kikel (1977), and we have provided a small number of reasonably successful cases in which maladaptive behaviors were effectively altered.

The application of behavior therapy has frequently been associated with ethical issues (Bellack & Hersen, 1977; Kazdin, 1977), and we might comment on this matter here. Ethical considerations related to institutional treatment of the long-term chronic patient are matters of great concern, typically when continued institutionalization is not on a voluntary basis. In the case of the Neuropsychology Center, very few of the patients were in the hospital on an involuntary basis (e.g., through legal commitment), and several of those who were, remained in the hospital after their involuntary status was removed. However, some might argue that from an ethical standpoint, not only should patients have to stay in a hospital on an involuntary basis for as brief a period as possible, but active efforts should be made to return patients to the community even if they are in the hospital on a voluntary basis. As we have pointed out, these patients sometimes remain in the hospital simply because they have no place to go. Thus, although some of our programs involved mildly aversive treatment, such as control of a patient's cigarettes, the goal of that treatment was that of assisting the patient in being able to adapt to a less restrictive environment. It may be pointed out, however, that we never used severely aversive procedures, such as shock, restriction, or food deprivation, and cannot recommend the use of such procedures with brain-damaged patients.

4. With regard to pharmacological treatment of adult brain-damaged patients, the results thus far are somewhat discouraging. We have reported on our successes with lithium (Williams & Goldstein, 1979) and we are continuing to make effective use of this medication on a clinical basis. However, the research with vasodilators and such other agents as lecithin and piracetam showed the rather typical pattern of early modest success but discouraging results later on (Corkin, Growdon, Sullivan, & Shedlack, 1981; Gershon, 1979). However, clinical pharmacologists (Perel, 1981) have pointed out that research with such compounds as lecithin, choline and piracetam has suffered from the lack of adequate preliminary pharmacokinetic investigation. These investigations determine the optimal dose ranges and the schedule for optimal administration of the agent. Thus, we do not know yet what optimal dose level is for such substances as lecithin, and we do not know how often to administer it. Thus, there is some possibility that following these detailed investigations, the subsequent clinical trials with these various agents may yield more positive results.

5. The reader who works with patients of the various types described here may wish to know what our impressions and findings are for the individual types. We will briefly review them here.

The Amnesic Alcoholic. The basic positive finding with these patients and other amnesic patients is that they are able to learn new material if it is specific. Our general approach with these patients involved the development of procedures to teach them specific items of information on a need-to-know basis. This teaching was accomplished through utilization of the Premack recency principle or through simple repetition. Thus, we were able to have patients learn and retain such items as the number of the ward, the names of staff, various times at which activities were scheduled, such as distribution of cigarettes, and related material. We have no evidence that it is necessary to associate a preserved long-term memory with the material to be learned in order to produce retention, but when this method was used we were able to demonstrate that retention was achieved. However, in the case of a severely amnesic patient, we were able to achieve significant improvement in retention on the basis of simple repetition of the same material over a long series of sessions. A detailed description of this case was presented in Chapter 4.

Several of the successful lithium cases were amnesic alcoholics. Those who responded well tended to be agitated and tense, in addition to exhibiting their memory difficulties. Based on our findings, the best candidate for lithium therapy among our patients was the agitated Korsakoff patient. The mechanism underlying our findings with these patients is unknown, but if there are no complicating issues, such as abnormal cardiac or thyroid function, the use of lithium might be considered on the basis of our success with it.

The amnesic alcoholics we treated who were not also severely demented tended to be good candidates for community placement. They did not remain in the program for long periods of time, and were generally acceptable to their families or community care facilities. In many cases, the issue of continued drinking remained a problem and it was recommended that these patients might require further treatment for their alcoholism. However, for some of them, excessive alcohol use was no longer a major problem.

The Nonamnesic Alcoholic. We did not develop a specific training regime for the patient with alcoholic dementia. These patients were reviewed very carefully from a medical standpoint, since they frequently had alcohol-related physical difficulties, notably liver disease. Since these patients typically were not severely impaired and did not have major sensory or motor disabilities, they tended to do well at work assignments off the ward. We noted much variability among them in regard to motivation. One group of our alcoholic dementia patients was extremely apathetic in relation to our other patients and saw no need to maintain a normal activity level or to change in any way. Typically, they denied or minimized their alcoholic histories. Generally, we were unable to remotivate these patients and they were not among our treatment successes. However, other alcoholic patients were quite motivated to do something about their lives. They willingly attended activities, went to their assignments, and generally cooperated for their assessment and treatment. In the case of these patients, the best course seemed to involve active milieu therapy, ideally involving some kind of work assignment within their capacity. We were generally able to send these patients out of the hospital in good physical health and with a high level of motivation to abstain from drinking and, in some cases, to go back to some form of employment.

The Atherosclerotic Patient. We treated very few patients with diffuse cerebral vascular disease, and those we did were so variable that nothing definitive can be said. Actually, it is now thought that diffuse cerebral arterisclerosis is a relatively rare condition that does not generally occur in the absence of significant cardiovascular disease. This association was clearly the case for one of our patients whose hypertension and other medical problems greatly complicated our treatment efforts. However, we did treat a number of poststroke patients, and can comment definitively about them. In many ways, the stable poststroke patient can be treated much like the clients in the Wichita program. The patients who had left-hemisphere strokes and still needed speech therapy were seen by a speech pathologist and assigned to a therapist. Although most students of this area point out that therapy is most effective shortly after the stroke, most of our patients had their strokes in the remote past but continued to require hospitalization. Our philosophy was to attempt treatment regardless of the chronicity of the condition. We will not go into details about the speech therapy done, since it varied with the type of speech or language disorder that each patient had. We saw very few patients with right-hemisphere strokes and did not develop an organized program for those we did see. Of course the NYU group has developed elaborate training systems for right-hemisphere stroke patients, and there is nothing we can contribute at this point to supplement their work in that area. However, our recommendation would clearly be to use these methods for the appropriate patients.

It should be pointed out that rehabilitation of the chronic stroke patient is a somewhat different matter from what is described as rehabilitation for stroke in the standard texts. For example, Hirschberg, Lewis, and Thomas (1964), in their discussion of the hemiplegic patient, stress helping the patient get out of bed, stand up, ambulate, and engage in self-care activities. Most of the chronic stroke patients we have seen in the Neuropsychology Center were well beyond that stage, to the extent that it was sometimes difficult to determine the side of the hemiparesis without doing neuropsychological tests or a neurological examination. Typically, the factors contributing to these patients' need for continuing hospitalization were mainly intellectual impairment leading to difficulties with independent living in the community, residual language or memory difficulties, and a variety of problems of a psychiatric or psychosocial nature. The residual motor or sensory weakness of one side of the body was generally the least of their problems. For example, one of our left-hemisphere stroke patients had a significant degree of residual aphasia, but almost no sensory or motor dysfunction of his right side. Most of the hemiparesis had disappeared long before we saw him. In any event, it is important to distinguish between the treatment needs of the acute stroke patient and those of the patient who had a stroke during the relatively remote past, but who continues to need treatment.

The Terminally Ill Patient. In recent years, there have been extensive efforts made to determine the etiology of the various progressive neurological diseases and to develop treatments that, although perhaps not influencing the ultimate outcome, could at least provide some symptomatic relief during the course of the illness. Extensive progress has been made in our understanding of the basis of such disorders as multiple sclerosis, Huntington's disease, and Alzheimer's disease. It is hoped that these findings in neurochemistry and virology will ultimately lead to methods of prevention and treatment. What appears to have happened, however, is that advances in medicine and the quality of nursing care have significantly increased the life span of individuals with progressive neurological disorders. In the older texts, for example, it was indicated that the expected life span for Alzheimer's disease patients was about five years from the time of appearance of the first symptoms. Alzheimer's disease patients now often live much longer than that. Indeed, even if the symptoms appear during the presenile period, the patient may live a relatively normal life span with good care. Recently, Butler (1978) gives the 5-year survival figure but qualifies it by indicating that this is the case in the absence of artificial survival methods. However, with intensive nursing care, control of infection, and related life preserving procedures, survival may go on for substantially longer than five years.

As indicated, the use of such medications as lecithin and piracetam to improve cognitive abilities in dementing patients has not yet led to impressive results. Our successful efforts to treat these patients behaviorally have been limited to modifying behaviors that appeared to be mediated by environmental contingencies. The "hunger strike" episode of a Huntington's disease patient that we described would be a good example of this type of treatment. We have not noted cognitive improvement in these patients with lecithin, nor was lithium effective with them. The use of very small doses of neuroleptics, notably Haldol, seems to be effective in relieving nocturnal agitation (sundowning).

In regard to management of these patients, in addition to close medical attention, we have noted that the Alzheimer's disease patients in particular do well by having a companion who can be a volunteer, staff member, or more intact patient. The companions for our patients typically spent a great deal of time with them, taking them to various activities, some of which were recreational in nature. In our program, we did not attempt to do anything structured in the form of behavior–therapy programs, but simply provided the patients with sufficient structure for them to engage in stimulating activities which they would have great difficulty in doing independently. With regard to Alzheimer's disease patients living in the community, Eisdorfer (1980) has described a program in which patients are transported to relatives of other Alzheimer's disease patients so that family members can continue to work and/or engage in recreational activities. Thus, for example, if a family wishes to go out for an evening, the patient can be brought to another Alzheimer's disease family who cares for him while his family is out. This simple but innovative system appears to be very effective in regard to forestalling institutionalization for as long as possible.

With regard to the Huntington's disease patient, there is the double problem of the dementia and the movement disorder. We use the standard medications for control of movement disorder with our Huntington's patients, but the reader who has not had experience with these patients should be aware that the degree of control that can be achieved with medication is generally only partial, and many patients will continue to have a great deal of chorea even with very high dosages of Haldol or related agents.

In general, although the outlook continues to be grim for patients with progressive neurological disorders, medical treatment and management of these patients has progressed to the point that they can now be more free, content, and productive during their remaining years than before. Our own program has stressed contingency management for maladaptive behaviors, maintenance of activity and stimulation, where possible with the aid of a patient companion, and careful medical attention to avoid the kinds of complications these patients often develop, and which are frequently the immediate cause of death. Medication, although of limited value, appears to have some beneficial effects in regard to reducing agitation and controlling to some extent the choreiform movements of the Huntington's disease patient.

The Head-Injured Patient. Long-term institutionalization of head-injured patients is somewhat unusual, and generally connotes one of two things. In some cases the head injury is so severe and disabling that it is difficult for the patient to function outside of an institutional setting. Alternatively, the patient may have acquired significant psychiatric difficulties in association with the head–injury experience, or previous psychiatric difficulties were so exacerbated by the head injury that a very severe, functional, psychiatric disorder emerges. As often occurs, although these conditions can and do improve while the patient is in the hospital, the process of institutionalization itself may promote continued hospitalization beyond the point at which it is really required.

In the Neuropsychology Center, and we suspect in most facilities for chronic patients, it is unlikely that patients with recent head injuries who are still in the natural recovery period will be seen. Certainly the patients we saw were beyond the acute phase of their injuries, and it was clear that the treatment called for would have to be of a somewhat different nature from what is typically done for the patient who is still recovering from recent trauma. Even the young head-injured patients we saw generally had their injuries several years before they came to our program. However, certain of these patients continued to have specific neuropsychological deficits in such areas as language and memory, which we attempted to treat directly. For example, one of our young patients had a persistent, chronic aphasia marked by slow, labored speech and handwriting. We made specific attempts to help him accelerate his speech and writing, with some success. One of his desires was to be able to sign his name more rapidly, and with regular practice he was able to do so. Another patient had a specific language difficulty in the area of word finding, although his speech was otherwise fluent and his comprehension was normal. The speech pathologist had him make up lists of names of things he had difficulty with, and he was asked to refer to the list when he got into trouble. Again, the method helped him with word findings, even though he had the condition for some time before he entered our program.

Although we were able to work effectively with the neuropsychological difficulties of these patients, as was also the case in the Wichita program, we often had major problems in regard to getting these patients out of the hospital after we did as much as we could for them. The problems in this area were found to be particularly complex. For example, one young patient was highly motivated to leave, but because he continued to have difficulties with memory and judgment and he had no family resources, the staff was rightfully reluctant to simply discharge him. Efforts had to be made not only to find an appropriate community–care facility, but to convince the patient of the need for him to be in such a facility. A great deal of time had to be spent counseling with the patient and finding a setting that he found satisfactory. It appeared to us that we received a particularly high level of resistance to reaccept patients from families of young head-injured individuals. This pattern led us to examine the reasons why these patients had head injuries. In several instances, these patients had preinjury difficulties with alcohol or drug abuse and were injured while intoxicated. In these cases, the families feared that the patient would begin drinking or abusing drugs again, which, in combination with the impaired judgment associated with the brain trauma, could result in dire consequences. In one case, the patient was a war hero who was injured while trying to save some comrades during combat in Vietnam. He sustained a severe disability affecting his ability to walk, his language, and his intellectual processes in general. Whereas this patient's mother was initially reluctant to accept him back home, she did so after it was apparent that he had achieved substantial recovery. However, in the cases of the patients who were injured while intoxicated, this reacceptance appeared to be far less forthcoming. In essence, whereas our program was somewhat successful in regard to restoration of function in head-injured patients, getting these patients out of the hospital was a complex matter involving numerous issues other than head injury itself. In this regard, we found it very important to examine the circumstances of the head injury, particularly in regard to its relationship to problems of impulse control and abuse of alcohol or drugs.

Conclusions

Rehabilitation of the hospitalized patient with chronic brain damage is a major clinical issue, but there is little research or specialized clinical practice in the area. Centers of the kind described here are few in number, and those that do exist must operate on a more or less ad hoc basis, because of the dearth of research involving treatment of patients of this type. It seems clear that there is a commonly shared implicit assumption that these patients are untreatable. We have tried to point out that, while the underlying neuropathologies for many of the chronic brain diseases are incurable, a great deal can be done in the management of patients with these disorders. As we have indicated, although it may not be possible to change the condition of the patient, it is possible to alter the environment in which he or she lives.

It is clear that clinical rehabilitative issues regarding the chronic patient go far beyond the neurological disability. Problems related to institutionalization on a long-term basis, family matters, the iatrogenic effects of treatment, secondary psychiatric disorders, and related matters all have to be dealt with in some fashion. All these issues are closely related to where the patient can go after leaving the institution and the level of care that has to be provided on a continuing basis. Modern psychiatric treatment now uses the concept of the least restrictive alternative to plan treatment. Often, this concept is not applied to the chronically brain-damaged patient, largely because of the inertia generated by many years of institutionalization. If the patient has been on a closed psychiatric ward for many years, there seems to be a tendency to have him stay there. It is therefore crucial to regularly review the status of the chronic patient so that he may gradually be moved to less restrictive facilities and more independent living, as warranted by changes in his condition. Much of the problem of chronicity revolves around the self-fulfilling prophecy that since there is no cure for the patient's condition, he will never change.

Rehabilitation planning for chronic patients requires carefully considered goal setting. Unrealistically ambitious goals lead to disappointment, while overly modest goals lead to little if any progress. Patients admitted to the Neuropsychology Center had goals formulated for them very early in the rehabilitation process. Sometimes, the goal was simply placement in a nursing home; sometimes we aimed at restoring the patient to independent life in the community. The nature of the patient's pathology, the related symptoms, the neuropsychological test results, the family situation, the patient's finances, and many additional factors may all relate to goal formation. The matter of whether or not the patient has a progressive illness is an obviously crucial factor.

One does not generally think in terms of active treatment for such disorders as multiple sclerosis, Huntington's disease, Korsakoff's syndrome, and other chronic brain diseases. There is no medicine to cure these illnesses, nor is there any surgery that will remove the pathological agent. Medical treatment of illnesses of this type generally has the aims of relieving symptoms and retarding progression. Thus, for example, the Korsakoff's patient is given thiamine in the hope that it will prevent further hemorrhaging. The Huntington's disease patient is given phenothiazine or other medications to reduce the severity of the movement disorder. There is, of course, hope that the cause and cure for these illnesses will be discovered. However, there is also hope that behavioral therapy can be productively used with people with these illnesses.

Our experience indicates that the inpatient Neuropsychology Center is a productive method for assessing, reevaluating, and treating chronically brain-damaged patients. Such a center can provide the focus for multidisciplinary efforts at de-institutionalization and restoration of function. The concept should ultimately be extended into the community, but it must of necessity begin in the hospital, where the patients are. From a behavior–therapy standpoint, having a center allows for the acquisition of stimulus control, and provides a stable setting for single-subject design studies. It also provides a controlled environment for doing laboratory assessments of various types and for conducting medication trials. In our clinical experience, we have been able to restore many of our patients to community life with a minimum of recidivism. We have also been convinced that whereas one might maintain a largely behavioral orientation, the use of chemical treatments should certainly not be rejected out of hand. Our experience with lithium was pleasantly surprising in that regard. Thus, a genuine rather than a nominal interdisciplinary approach can be of great benefit.

10

CONCLUDING COMMENTS

Time has passed between the early years of the Wichita program and the completion of this book. Still, little progress has been made in regard to its topic. A work originally written by Luria (1948/1963) has gained in popularity, and a book on the role of neuropsychology in relation to rehabilitation has been published (Golden, 1978). Diller and his colleagues at NYU are continuing their impressive work with stroke and head-injured patients and publishing their findings (Diller & Gordon, 1981). To the best of our knowledge, no other major contributions have been made to our understanding of how to rehabilitate brain-damaged adults, nor have there been major advances with regard to techniques or technologies. However, in our perception, there has been an increased public and professional demand for such rehabilitation efforts. Perhaps the aura of pessimism has been lifted to some extent, and people are beginning to seek active treatment for various brain disorders. Perhaps this movement is largely attributable to changes in health care philosophy, particularly in regard to the developing belief that patients should no longer be indefinitely, and perhaps permanently, institutionalized because there is no medical cure for their illness. It is now mandatory in accredited treatment facilities to have a treatment plan for each patient and to comment on each patient's rehabilitation potential. Indeed, an assessment of rehabilitation potential must be made even for patients with terminal illnesses. Additionally, the relatively positive experience with de-institutionalization of psychiatric patients may ultimately have had an influence on the ways in which we deal with brain-damaged patients.

It is hoped that this book may add to the impetus of rehabilitation. Much of our discussion concerned outcomes of rehabilitation programs. In general, these outcomes, when not dramatic, were at least favorable. Other investigators working in this area also report favorable outcomes. What do we mean by favorable? In this regard, we can be fairly specific in terms of what "favorable" does and does not mean. The data derived from the Wichita program clearly indicated that there was improvement on neuropsychological tests in individuals who went through the rehabilitation program, with little improvement found in the control group that did not go through the program. Perhaps more significantly, it was also shown that many of the subjects who went through the program were either institutionalized or unemployed and living at home before the program, and working or in school after it. The clear implication of this finding is that natural recovery of function does not appear to be enough to allow for an optimal adjustment after brain damage. It would appear that rehabilitation is not simply keeping the patient busy while he gets better by himself. Whereas the Wichita program seemed to be associated with better adjustment in a general way, the behavior–modification material presented in Chapter 4 also shows specific learning phenomena in our brain-damaged subjects. These findings strongly suggest that the laws of learning are operative in brain-damaged patients (of the type participating in the study) and contradict the view that the brain-damaged patient does not learn from experience. It is interesting to note that although the role of training in recovery of function has been recognized and documented at least since the time of Lashley (1933), there has been little recognition of this principle in the clinical realm.

The preliminary findings coming out of the Neuropsychology Center suggest that much can also be done for the chronic, institutionalized patient. For these patients, such modalities as reality orientation, single-subject design behavior–therapy programs, informal programs, and more formal cognitive-perceptual retraining can be effective. However, in the case of these patients, it is particularly important to establish a proper milieu and to carry on programs of a psychosocial nature involving the family and reintegration into the community. We also found that whereas certain medications used with these patients can be counterproductive, other drugs, particularly lithium in our case, can be effective.

We learned from the Wichita and the Neuropsychology Center programs that brain-damaged patients frequently have significant psychiatric difficulties associated with their neurological disorders. Some aspects of the nature of these difficulties were explored in Chapter 8. Although this finding was certainly not surprising, it represents an area that is not always dealt with effectively in the case of the brain-damaged patient. Depending on the setting in which he finds himself, the patient may be treated only for the traditionally psychiatric aspects of his disorder, or these aspects may be essentially neglected. The approach we would favor involves giving the patient emotional support while focusing on the crucial adaptive difficulties induced by his brain damage. Thus, for example, it is not sufficient to treat the agitated amnesic patient just for the agitation. When this approach is taken, the nature of the treatment may further impair the patient's memory. In our view, both the memory disorder and the agitation should be treated. Perhaps the agitation can be treated with medication, and memory can be improved with some form of retraining. It may also be mentioned that the effectiveness of such retraining can be greatly compromised when one tries to do it with an untreated, very agitated patient.

The term *favorable outcome* should not be taken to mean that the present authors or anyone else has discovered some treatment that "cures" brain damage in the sense of restoration of structural loss. Perhaps through chemical or behavioral methods remaining neurons begin to work harder, but it does not seem prudent at this point to speculate about the mechanisms underlying the behavioral changes noted. Thus, although we appear to have demonstrated some degree of recovery of function through behavioral and chemical means, we would certainly not claim that we have, in any sense, reversed the structural damage experienced by our clients. On the other hand, we feel that our rehabilitative efforts were not merely palliative or epiphenomenal in nature. The attempt was not simply to induce the client to accept or feel more comfortable with his impairments. Nor was it an effort directed toward reducing demands on the client. Rather, particularly in the case of the Wichita program, we attempted to deal with the adaptive deficits directly, and, in doing so, had some success in restoring some degree of function. In this regard, we would take issue with the simplistic notion that the brain-damaged patient should not be subject to excessive demands. Such a treatment philosophy could encourage the development of levels of expectation that are insufficiently high. Instead, we would encourage adoption of the idea of a gradient of adaptive challenges in which the patient is not put into situations in which he will develop a catastrophic reaction, but is put into situations that require full exercise of the adaptive capacities he still has available. In those cases in which function could not be restored, as may occur in the case of the patient with a progressive disease of the brain, attempts were made to alter environmental contingencies to allow for more optimal functioning with what adaptive resources remain available. Even in these cases, however, a lessening of demand is not always the wisest course to follow, and these patients often respond to reasonable challenges. In essence, rehabilitation efforts neither reverse brain pathology, nor do they merely induce the patient to comfortably accept his situation.

It would appear that the approach being suggested here is a holistic or comprehensive one that emphasizes neuropsychology and behavioral methods, but that also calls for backups in the psychiatric, pharmacological, and psychosocial realms. It requires expertise in neuropsychological assessment, behavioral therapy, and cognitive-perceptual retraining methods, but it also leans heavily on sensitivity to emotional and social problems. It is also a practical approach, in that it calls for a pragmatic assessment of what methods are available and using those that are successful. Perhaps the best example of this feature of the approach is the lithium experience. Starting with a somewhat antipharmacological bias, and being aware that since the Wichita program was nonmedical (it made essentially no active use of medication), we ultimately became convinced that lithium was effective, and we continue to use it. Similarly, we had to shift from cognitive-perceptual retraining to other methods when we became faced with the problems of long-term chronic patients. It is our view that the future of the area of rehabilitation of brain-damaged adults lies in advances in a number of areas, including neuropsychological assessment, cognitive-perceptual retraining, behavior therapy, psychopharmacology, and psychiatric and psychological intervention.

REFERENCES

Adams, G. F., & McComb, S. G. Assessment and prognosis in hemiplegia. *Lancet*, 1953, *2*, 266–269.

Adams, H. E., & Unikel, I. P. *Issues and trends in behavior therapy*. Springfield, Ill.: Charles C Thomas, 1973.

Adams, R. D., & Victor, M. *Principles of neurology*. New York: McGraw-Hill, 1977.

Agras, W. S., Barlow, D. H., Chapin, H. N., Abel, G. G., & Leitenberg, H. Behavior modification of anorexia nervosa. *Archives of General Psychiatry*, 1974, *30*, 279–286.

Agulnik, P. L., DiMascio, A., & Moore, P. Acute brain syndrome associated with lithium therapy. *American Journal of Psychiatry*, 1972, *129*, 621–623.

Alajouanine, T. H., & L'hermitte, F. Des agnosies électives. *Encephale*, 1957, *46*, 505.

Albert, M. L., Goodglass, H., Helm, N. A., Rubens, A. B., & Alexander, M. P. *Clinical aspects of dysphasia*. New York: Springer-Verlag, 1981.

Alford, G. S., Blanchard, E. B., & Buckley, T. M. Treatment of hysterical vomiting by modification of social contingencies: A case study. *Journal of Behavior Therapy and Experimental Psychiatry*, 1972, *3*, 209–212.

Allport, G. W. *Personality, a psychological interpretation*. New York: Holt, 1937.

Avedon, E. M. *Therapeutic recreation service: An applied behavioral science approach*. Englewood Cliffs, N.J.: Prentice-Hall, 1974.

Ayllon, T., & Azrin, N. H. *The token economy: A motivational system for therapy and rehabilitation*. New York: Appleton-Century-Crofts, 1968.

Ayllon, T., & Michael, J. The psychiatric nurse as behavioral engineer. *Journal of Experimental Analysis of Behavior*, 1959, *2*, 323–334.

Azrin, N. H., & Foxx, R. M. A rapid method of toilet training the institutionalized retarded. *Journal of Applied Behavioral Analysis*, 1971, *4*, 89–99.

Azrin, N. H., & Nunn, R. G. Habit reversal: A method of eliminating nervous habits and tics. *Behavior Research & Therapy*, 1973, *11*, 619–628.

Azrin, N. H., Nunn, R. G., & Frantz-Renshaw, S. Habit reversal treatment of thumb sucking. *Behavior Research & Therapy*, 1980, *18*, 395–399.

Babcock, H. An experiment in the measurement of mental deterioration. *Archives of Psychology*, 1930, *18*, 117.

Baer, D. M., & Wolf, M. M. Recent examples of behavior modification in preschool settings. In C. Neuringer & J. L. Michael (Eds.), *Behavior modification in clinical psychology*. New York: Appleton-Century-Crofts, 1970.

Baer, D. M., Wolf, M. M., & Risley, T. R. Some current dimensions of applied behavior analysis. *Journal of Applied Behavior Analysis*, 1968, *1*, 91–97.

Ball, J. A. C., & Taylor, A. R. Effect of cyclandelate on mental function and cerebral blood flow in elderly patients. *British Medical Journal*, 1967, *3*, 525–578.

Ban, T. A. Vasodilators, stimulants and anabolic agents in the treatment of geropsychiatric patients. In M. A. Lipton, A. DiMascio, & K. F. Killam (Eds.), *Psychopharmacology: A generation of progress*. New York: Raven Press, 1978.

Ban, T. A., & Lehmann, H. E. Nicotinic acid in the treatment of schizophrenia. *Canadian Mental Health Association, Collaborative Study—Progress Report, I*. Toronto: Canadian Mental Health Association, 1970.

Bandura, A. *Principles of behavior modification*. New York: Holt, Rinehart & Winston, 1969.

Bandura, A. Psychotherapy based on modeling principles. In A. E. Bergin & S. L. Garfield (Eds.), *Handbook of psychotherapy and behavior change*. New York: Wiley, 1971.

Barker, R. G. The social psychology of physical disability. *Journal of Social Issues*, 1948, *4*, 28–38.

Barker, R. G., & Wright, B. A. The social psychology of adjustment to physical disability. In J. F. Garret (Ed.), *Psychological aspects of physical disability*. Department of Health, Education, and Welfare, Office of Vocational Rehabilitation, Rehabilitation Service Series, 1952, *210*, 18–32.

Barker, R. G., Wright, B. A., & Gonick, H. R. Adjustment to physical handicap and illness: A survey of the social psychology of physique and disability. New York: *Social Science Research Council Bulletin*, 1946, *55*.

Barker, R. G., & Wright, H. F. *Midwest and its children*. New York: Harper & Row, 1954.

Barnes, J. A. Effects of reality orientation classroom on memory loss, confusion, and disorientation in geriatric patients. *The Geronologist*, 1974, *14*, 138–142.

Barton, E. S., Guess, D., Garcia, E., & Baer, D. M. Improvement of retardates' mealtime behaviors by timeout procedures using multiple-baseline techniques. *Journal of Applied Behavior Analysis*, 1970, *3*, 77–84.

Barton, R., & Hurst, L. Unnecessary use of tranquilizers in elderly patients. *British Journal of Psychiatry*, 1966, *112*, 989–990.

Basso, A., Capitani, E., & Vignolo, L. A. Influence of rehabilitation on language skills in aphasic patients. *Archives of Neurology*, 1979, *36*, 190–196.

Basso, A., Faglioni, P., & Vignolo, L. A. Etude controlée de la réducation du langage dans l'aphasie: comparison entre aphasiques traités et non–traités. *Revue Neurologie*, 1975, *131*, 607–614.

Beaumont, J. G., & Dimond, S. J. The clinical assessment of interhemispheric psychological functioning. *Journal of Neurology, Neurosurgery and Psychiatry*, 1973, *36*, 445–447.

Bellack, A. S., & Hersen, M. *Behavior modification: An introductory textbook*. Baltimore: Williams & Wilkins, 1977.

Bellack, A. S., & Hersen, M. Assessment and single-case research. In M. Hersen & A. S. Bellack (Eds.), *Behavior therapy in the psychiatric setting*. Baltimore: Williams & Wilkins, 1978.

Bellack, A. S., & Hersen, M. (Eds.). *Research and practice in social skills training*. New York: Plenum Press, 1979.

Bellack, A. S., Hersen, M., & Turner, S. M. Effects of social disruption, stimulus interference and aversive conditioning on auditory hallucinations. *Behavior Modification*, 1977, *1*, 249–258.

Bender, M. B. *Disorders in perception*. Springfield, Ill.: Charles C Thomas, 1952.

Bentin, S., & Gordon, H. W. Assessment of cognitive asymmetries in brain-damaged and normal subjects: Validation of a test battery. *Journal of Neurology, Neurosurgery and Psychiatry*, 1979, *42*, 715–723.

Benton, A. Visuoperceptive, visuospatial and visuoconstructive disorders. In K. M. Heilman & E. Valenstein (Eds.), *Clinical neuropsychology*. New York: Oxford University Press, 1979.

Benton, A. L. *Right-left discrimination and finger localization: Development and pathology*. New York: Hoeber-Harper, 1959.

Benton, A. L. (Ed.). *Behavioral change in cerebrovascular disease*. New York: Harper & Row, 1970.

Ben-Yishay, Y., Diller, L., & Mandelberg, I. The ability to profit from cues as a function of initial competence in normal and brain-injured adults: A replication of previous findings. *Journal of Abnormal Psychology*, 1970, *76*, 378–379.

Ben-Yishay, Y., Diller, L., Gerstman, L., & Gordon, W. Relationship between initial competence and ability to profit from cues in brain-damaged individuals. *Journal of Abnormal Psychology*, 1970, *78*, 248–259.

Ben-Yishay, Y., Gerstman, L., Diller, L., & Haas, A. Prediction of rehabilitation outcomes from psychometric parameters in left hemiplegics. *Journal of Consulting and Clinical Psychology*, 1970, *34*, 436–441.

Ben-Yishay, Y., Rattok, J., & Diller, L. (Eds.). A clinical strategy for the systematic amelioration of attentional disturbances in severe head trauma patients. In *Working approaches to remediation of cognitive deficits in brain damaged*. Unpublished document, Institute of Rehabilitation Medicine, New York, 1979.

Bergin, A. E., & Strupp, H. H. *Changing frontiers in the science of psychotherapy*. New York: Aldine, 1972.

Bigger, T. J., Jr., Kantor, S. J., Glassman, A. H., & Perel, J. M. Cardiovascular effects of tricyclic antidepressant drugs. In M. A. Lipton, A. DiMascio, & K. F. Killam (Eds.), *Psychopharmacology: A generation of progress*. New York: Raven Press, 1978.

Binder, L. M., & Schreiber, V. Visual imagery and verbal mediation as memory aids in recovering alcoholics. *Journal of Clinical Neuropsychology*, 1980, *2*, 71–73.

Bleuler, M. *Die schizophrenen Geistesstörungen im lichte Langjähriger kranken und Familiengeschichten*. Stuttgart: Thieme, 1972.

Bogen, J. E., & Vogel, J. P. Cerebral commissurotomy in man. *Bulletin of the Los Angeles Neurological Society*, 1962, *27*, 169–172.

Boll, T. J. Diagnosing brain impairment. In B. B. Wolman (Ed.), *Clinical diagnosis of mental disorders*. New York: Plenum Press, 1978.

Boll, T. J. The Halstead-Reitan neuropsychology battery. In S. B. Filskov & T. J. Boll (Eds.), *Handbook of clinical neuropsychology*. New York: Wiley-Interscience, 1981.

Boll, T. J., Heaton, R., & Reitan, R. M. Neuropsychological and emotional correlates of Huntington's Chorea. *The Journal of Nervous and Mental Disease*, 1974, *158*, 61–69.

Boring, E. G. *A history of experimental psychology* (2nd ed.). New York: Appleton-Century-Crofts, 1950.

Browne, W. A. F. *What asylums were, are, and ought to be*. Edinburgh: Adam and Charles Black, 1837.

Bruell, J. H., & Simon, J. I. Development of objective predictors of recovery in hemiplegic patients. *Archives of Physical Medicine and Rehabilitation*, 1960, *41*, 564–569.

Butler, R. Alzheimer's disease, senile dementia and related disorders: The role of NIA. In R. Katzman, R. D. Terry, & K. L. Bick (Eds.), *Alzheimer's disease: Senile dementia and related disorders*. New York: Raven Press, 1978.

Butters, N. Amnesic disorders. In K. M. Heilman & E. Valenstein (Eds.), *Clinical neuropsychology*. New York: Oxford University Press, 1979.

Butters, N., & Cermak, L. S. Neuropsychological studies of alcoholic Korsakoff patients.

In G. Goldstein & C. Neuringer (Eds.), *Empirical studies of alcoholism.* Cambridge, Mass.: Ballinger, 1976.

Butters, N., & Cermak, L. S. *Alcoholic Korsakoff's syndrome.* New York: Academic Press, 1980.

Butters, N., Rosen, J., Soeldner, C., & Stein, D. A comparison of one-stage and serial ablations of the middle third of sulcus principalis on delayed-alternation performance. *Journal of Comparative and Physiological Psychology,* 1975, *89,* 1077–1082.

Caffey, E. M., Galbrecht, C. R., & Klett, C. J. Brief hospitalization and aftercare in the treatment of schizophrenia. *Archives of General Psychiatry,* 1971, *24,* 81–86.

Cantwell, J. D. Running. *Journal of the American Medical Association,* 1978, *240,* 1409–1410.

Carey, M. E., Young, H. F., Rish, B. L., & Mathis, J. L. Follow-up study of 103 American soldiers who sustained a brain wound in Vietnam. *Journal of Neurosurgery,* 1974, *41,* 542–549.

Carmon, A., Gordon, H. W., Bental, E., & Harness, B. Z. Retraining in literal alexia: Substitution of impaired left-hemispheric processing by a right-hemisphere perceptual strategy. *Bulletin of the Los Angeles Neurological Societies,* 1977, *42,* 41–50.

Cermak, L. S. Imagery as an aid to retrieval for Korsakoff patients. *Cortex,* 1975, *11,* 163–169.

Chapman, L. F., & Wolff, H. G. Studies of human cerebral hemisphere function: Adaptive capacity after loss of hemisphere tissue. *Transactions of the American Neurological Academy,* 1956, 175–178.

Chapman, L. F., & Wolff, H. G. The cerebral hemispheres and the highest integrative functions of man. *Archives of Neurology,* 1959, *1,* 357–424.

Christensen, A. L. *Luria's neuropsychological investigation.* New York: Spectrum, 1975.

Ciminero, A. R., Calhoun, K. S., & Adams, H. E. *Handbook of behavioral assessment.* New York: Wiley, 1977.

Clausen, J. P. Effect of physical training on cardiovascular adjustments to exercise in man. *Physiological Review,* 1977, *57,* 779–815.

Collins, F. L., Jr. Behavioral medicine. In L. Michelson, M. Hersen, & S. M. Turner (Eds.), *Future perspectives in behavior therapy.* New York: Plenum Press, 1981.

Cone, J. D., & Hawkins, R. P. (Eds.). *Behavioral assessment: New directions in clinical psychology.* New York: Brunner/Mazel, 1977.

Conners, C. K. (Ed.). *Clinical use of stimulant drugs in children.* The Hague: Excerpta Medica, 1974.

Corkin, S., Growdon, J. H., Sullivan, E. V., & Shedlack, K. Lecithin and cognitive function in aging and dementia. In A. D. Kidman, J. K. Tomkins, & R. A. Westerman (Eds.), *New approaches to nerve and muscle disorders.* The Netherlands: Excerpta Medica, 1981.

Corte, H. E., Wolf, M. M., & Locke, B. J. A comparison of procedures for eliminating self-injurious behavior of retarded adolescents. *Journal of Applied Behavior Analysis,* 1971, *4,* 201–213.

Critchley, M. *The parietal lobes.* London: Arnold, 1953.

Damasio, A. The frontal lobes. In K. M. Heilman & E. Valenstein (Eds.), *Clinical neuropsychology.* New York: Oxford University Press, 1979.

Davis, J. M. Antipsychotic drugs. In H. I. Kaplan, A. M. Freedman, & B. J. Sadock (Eds.), *Comprehensive textbook of psychiatry, III.* Baltimore: Williams & Wilkins, 1980.

DeJong, R. N. *The neurologic examination.* New York: Harper & Row, 1967.

DeLemos, G. P., Clement, W. R., & Nickels, E. Effects of diazepam suspension in geriatric patients hospitalized for psychiatric illnesses. *Journal of the American Geriatric Society,* 1965, *13,* 355–359.

Denckla, M. B., & Heilman, K. M. Childhood learning disabilities. In K. M. Heilman & E. Valenstein (Eds.), *Clinical Neuropsychology.* New York: Oxford University Press, 1979.

Dewhurst, K., Oliver, J. E., & McKnight, A. L. Socio-psychiatric consequences of Huntington's Disease. *British Journal of Psychiatry,* 1970, *116,* 255–258.

Diagnostic and Statistical Manual of Mental Disorders (DSM-III) (3rd ed.). Washington, D.C.: American Psychiatric Association, 1980.

Diller, L. Presentation II. In A. L. Benton (Ed.), *Behavioral change in cerebrovascular disease.* New York: Harper & Row, 1970.

Diller, L. A model for cognitive retraining in rehabilitation. *The Clinical Psychologist,* 1976, *29,* 13–15.

Diller, L., & Gordon, W. A. Rehabilitation and clinical neuropsychology. In S. B. Filskov & T. J. Boll (Eds.), *Handbook of clinical neuropsychology.* New York: Wiley, 1981.

Diller, L., & Weinberg, J. Accidents in hemiplegia. *Archives of Physical Medicine and Rehabilitation,* 1970, *51,* 358–363.

Diller, L., & Weinberg, J. Studies in scanning behavior in hemiplegia. In L. Diller & J. Weinberg (Eds.), *Studies in cognition & rehabilitation in hemiplegia.* New York: New York University Medical College, Institute of Medicine, 1971.

Diller, L., Ben-Yishay, Y., Gerstman, L. J., Goodkin, R., Gordon, W., & Weinberg, J. *Studies in cognition and rehabilitation. Rehabilitation Monograph,* 1974, *50.*

Dimond, S. J., & Beaumont, J. G. *Hemisphere function in the human brain.* London: Elek Science, 1974.

Doehring, D. G., & Reitan, R. M. MMPI performance of aphasic and nonaphasic brain-damaged patients. *Journal of Clinical Psychology,* 1960, *16,* 307–309.

Doehring, D. G., & Reitan, R. M. Concept attainment of human adults with lateralized cerebral lesions. *Perceptual and Motor Skills,* 1962, *14,* 27–33.

Dohrenwend, B. S., & Dohrenwend, B. P. Some issues in research on stressful life events. *Journal of Nervous and Mental Disease,* 1978, *166,* 7–15.

Dolan, M. P., & Norton, J. C. A programmed training technique that uses reinforcement to facilitate acquisition and retention in brain-damaged patients. *Journal of Clinical Psychology,* 1977, *33,* 496–501.

Drabman, R. S., Jarvie, G. J., & Hammer, D. Residential child treatment. In M. Hersen & A. S. Bellack (Eds.), *Behavior therapy in the psychiatric setting.* Baltimore: Williams & Wilkins, 1978.

Drachman, D. A. Central cholinergic system and memory. In M. A. Lipton, A. DiMascio, & K. F. Killam (Eds.), *Psychopharmacology: A generation of progress.* New York: Raven Press, 1978.

Edelwich, J., & Brodsky, A. *Burnout: Stages of disillusionment in the helping professions.* New York: Human Science Press, 1980.

Eisdorfer, C. *Clinical approach to dementia.* Lecture presented at Western Psychiatric Institute and Clinic, Pittsburgh, Pa., February 1980.

Eisenson, J. *Examining for aphasia: A manual for the examination of aphasia and related disturbances.* New York: Psychological Corporation, 1954.

Ellsworth, R. B., Collins, J. F., Casey, N. A., Schoonover, R. A., Hickey, R. H., Hyer, L., Twemlow, S. W., & Nesselroade, J. R. Some characteristics of effective psychiatric treatment programs. *Journal of Consulting and Clinical Psychology,* 1979, *47,* 799–817.

Epstein, L. J. Anxiolytics, antidepressants, and neuroleptics in the treatment of geriatric patients. In M. A. Lipton, A. DiMascio, & K. F. Killam (Eds.), *Psychopharmacology: A generation of progress.* New York: Raven Press, 1978.

Epstein, L. H., & Hersen, M. Behavioral control of hysterical gagging. *Journal of Clinical Psychology*, 1974, *30*, 102–104.

Epstein, L. H., Katz, R. C., & Zlutnick, S. Behavioral medicine. In M. Hersen, R. M. Eisler, & P. M. Miller (Eds.), *Progress in behavior modification* (Vol. 7). New York: Academic Press, 1979.

Fairweather, G. W. (Ed.). *Social psychology in treating mental illness: An experimental approach*. New York: Wiley, 1964.

Filskov, S. B., & Boll, T. J. (Eds.). *Handbook of clinical neuropsychology*. New York: Wiley-Interscience, 1981.

Finger, S. *Recovery from brain damage: Research and theory*. New York: Plenum Press, 1978.

Finlayson, M. A. J., Johnson, K. A., & Reitan, R. M. Relationship of level of education to neuropsychological measures in brain-damaged and nonbrain-damaged adults. *Journal of Consulting and Clinical Psychology*, 1977, *45*, 402–411.

Flor-Henry, P., & Yeudall, L. T. Neuropsychological investigation of schizophrenia and manic-depressive psychoses. In J. Gruzelier & P. Flor-Henry (Eds.), *Hemisphere asymmetries of function in psychopathology*. Amsterdam: Elsevier/North Holland, 1979.

Folsom, J. C. Reality orientation for the elderly mental patient. *Journal of Geriatric Psychiatry*, 1968, *1*, 291–307.

Folstein, M. F., Folstein, S. E., & McHugh, P. R. "Mini-mental status," a practical method for grading the cognitive state of patients for the clinician. *Journal of Psychiatric Research*, 1975, *12*, 189–198.

Gainotti, G. Emotional behavior and hemispheric side of the lesion. *Cortex*, 1972, *8*, 41–55.

Gardner, H. *The shattered mind*. New York: Knopf, 1974.

Gardner, H., Zurif, E., Berry, T., & Baker, E. Visual communication in aphasia. *Neuropsychologia*, 1976, *14*, 275–292.

Gardner, J. M. Teaching behavior modification to nonprofessionals. *Journal of Applied Behavior Analysis*, 1972, *5*, 517–521.

Garfield, S. L. Psychological treatments of abnormal behavior. In A. E. Kazdin, A. S. Bellack, & M. Hersen (Eds.), *New perspectives in abnormal psychology*. New York: Oxford University Press, 1980.

Garron, D. C. Huntington's chorea and schizophrenia. In A. Barbeau, T. N. Chase, & G. W. Paulson (Eds.), *Advances in neurology, Volume 1: Huntington's chorea 1872–1972*. New York: Raven Press, 1973.

Gasparrini, B., & Satz, P. A treatment for memory problems in left-hemisphere CVA patients. *Journal of Clinical Neuropsychology*, 1979, *1*, 137–150.

Gershon, S. *Pharmacotherapy of dementia*. Paper presented at the 69th Annual Meeting of the American Psychopathological Association, New York, March 1979.

Geschwind, N. Disconnexion syndromes in animals and man. *Brain*, 1965, *88*, 237–294; 585–644.

Gianutsos, R., & Gianutsos, J. Rehabilitating the verbal recall of brain-injured patients by mnemonic training: An experimental demonstration using single-case methodology. *Journal of Clinical Neuropsychology*, 1979, *1*, 117–135.

Gibbs, C. J., Jr., & Gajdusek, D. C. Subacute spongiform virus encephalopathies: The transmissible virus dementias. In R. Katzman, R. D. Terry, & K. L. Bick (Eds.), *Alzheimer's disease: Senile dementia and related disorders*. New York: Raven Press, 1978.

Giles, D. K., & Wolf, M. M. Toilet-training institutionalized, severe retardates: An application of operant behavior modification techniques. *American Journal of Mental Deficiency*, 1966, *70*, 766–780.

Goffman, E. *Asylums: Essay on social situation of mental patients and other inmates.* New York: Aldine, 1961.

Goldberg, S. C., Schooler, N. R., Hogarty, G. E., & Roper, M. Prediction of relapse in schizophrenic outpatients treated by drug and sociotherapy. *Archives of General Psychiatry,* 1977, *34,* 171–184.

Golden, C. J. *Diagnosis and rehabilitation in clinical neuropsychology.* Springfield, Ill: Charles C Thomas, 1978.

Golden, C. J. A standardized version of Luria's neuropsychological tests. In S. B. Filskov & T. J. Boll (Eds.), *Handbook of clinical neuropsychology.* New York: Wiley-Interscience, 1981.

Golden, C. J., Hammeke, T. A., & Purisch, A. D. *The Luria-Nebraska battery manual.* Los Angeles: Western Psychological Services, 1980.

Goldman, P. S. Functional development of the prefrontal cortex in early life and the problem of neuronal plasticity. *Experimental Neurology,* 1971, *32,* 366–387.

Goldman, P. S. Developmental determinants of cortical plasticity. *Acta Neurobiologiae Experimentalis,* 1972, *32,* 495–511.

Goldman, P. S. An alternative to developmental plasticity: Heterology of CNS structures in infants and adults. In D. G. Stein, J. J. Rosen, & N. Butters (Eds.), *Plasticity and recovery of function in the central nervous system.* New York: Academic Press, 1974.

Goldstein, G. The use of clinical neuropsychological methods in the lateralization of brain lesions. In S. J. Dimond & J. G. Beaumont (Eds.), *Hemisphere function in the human brain.* London: Elek Science, 1974.

Goldstein, G. Cognitive and perceptual differences between schizophrenics and organics. *Schizophrenia Bulletin,* 1978, *4,* 160–185.

Goldstein, G. Methodological and theoretical issues in neuropsychological assessment. *Journal of Behavioral Assessment,* 1979, *1,* 23–41.

Goldstein, G. In defense of the computer: Remarks on objective neuropsychological interpretation. In *Quantitative and qualitative approaches to clinical neuropsychology.* Symposium presented at the 88th Annual Convention of the American Psychological Association, Montreal, Canada, September 1980.

Goldstein, G., & Shelly, C. H. Univariate vs. multivariate analysis in neuropsychological test assessment of lateralized brain damage. *Cortex,* 1973, *9,* 204–216.

Goldstein, G., & Shelly, C. H. Statistical and normative studies of the Halstead neuropsychological test battery relevant to a neuropsychiatric hospital setting. *Perceptual and Motor Skills,* 1972, *34*(2), 603–620.

Goldstein, G. & Shelly, C. H. Field dependence and cognitive, perceptual and motor skills in alcoholics: A factor-analytic study. *Quarterly Journal of Studies on Alcohol,* 1971, *32*(1), 29–40.

Goldstein, G., Neuringer, C., & Olson, J. Impairment of abstract reasoning in the brain damaged: Qualitative or quantitative? *Cortex,* 1968, *4,* 372–388.

Goldstein, K. *The organism.* New York: American Book, 1939.

Goldstein, K. The two ways of adjustment of the organism to cerebral defects. *Journal of Mt. Sinai Hospital,* 1942, *9,* 4.

Goldstein, K. *Language and language disturbances.* New York: Grune, 1948.

Goldstein, K. *Human nature in the light of psychopathology.* Cambridge: Harvard University Press, 1951.

Goldstein, K. The organismic approach. In S. Arieti (Ed.), *American handbook of psychiatry* (Vol. 2).New York: Basic Books, 1959.(a)

Goldstein, K. Functional disturbances in brain damage. In S. Arieti (Ed.), *American handbook of psychiatry.* New York: Basic Books, 1959.(b)

Goldstein, K., & Scheerer, M. Abstract and concrete behavior: An experimental study with special tests. *Psychological Monographs,* 1941, *53,* (2, Whole No. 239).

Goodglass, H., & Kaplan, E. *The assessment of aphasia and related disorders.* Philadelphia: Lea & Febiger, 1972.

Goodwin, F. K., Cowdry, R. W., & Webster, M. H. Predictors of drug response in the affective disorders: Towards an integrated approach. In M. A. Lipton, A. DiMascio, & K. F. Killam (Eds.), *Psychopharmacology: A generation of progress.* New York: Raven Press, 1978.

Groves, I. D., & Carroccio, D. F. A self-feeding program for severely and profoundly retarded. *Mental Retardation,* 1971, *9,* 10–12.

Growdon, J. H., & Corkin, S. Neurochemical approaches to the treatment of senile dementia. In J. O. Cole & J. E. Barrett (Eds.), *Psychopathology in the aged.* New York: Raven Press, 1980.

Guess, D., Sailor, W., Rutherford, G., & Baer, D. M. An experimental analysis of linguistic development: The productive use of the plural morpheme. *Journal of Applied Behavior Analysis,* 1968, *1,* 297–306.

Gur, R. E. Left hemisphere dysfunction and left hemisphere overactivation in schizophrenia. *Journal of Abnormal Psychology,* 1978, *87,* 226–238.

Gur, R. E. Cognitive concomitants of hemispheric dysfunction in schizophrenia. *Archives of General Psychiatry,* 1979, *36,* 269–274.

Gurland, B., Dean, L., Cross, P., & Golden, R. The epidemiology of depression and dementia in the elderly: The use of multiple indicators of these conditions. In J. O. Cole & J. E. Barrett (Eds.), *Psychopathology in the aged.* New York: Raven Press, 1980.

Guthrie, M. A personal view of genetic counseling. In Y. E. Hsia, K. Hirschhorn, R. L. Silverberg, & L. Godmilow (Eds.), *Counseling in genetics.* New York: Alan R. Liss, 1979. (Reprinted by Committee to Combat Huntington's Disease Inc.)

Hagen, C. Communication abilities in hemiplegia: Effect of speech therapy. *Archives of Physical Medicine and Rehabilitation,* 1973, *54,* 454–463.

Halstead, W. C. Preliminary analysis of grouping behavior in patients with cerebral injury by the method of equivalent and non-equivalent stimuli. *American Journal of Psychiatry,* 1940, *96,* 1263–1294.

Halstead, W. C. *Brain and intelligence.* Chicago: University of Chicago Press, 1947.

Halstead, W. C., & Rennick, P. Toward a behavioral scale for biological age. *Social and psychological aspects of aging.* New York: Columbia University Press, 1962.

Halstead, W. C., & Wepman, J. M. The Halstead-Wepman aphasia screening test. *Journal of Speech and Hearing Disorders,* 1949, *14,* 9–15.

Hammer, M., Makiesky-Barrow, S., & Gutwirth, L. Social networks and schizophrenia. *Schizophrenia Bulletin,* 1978, *4,* 522–545.

Hécaen, H. Aphasic, apraxic and agnostic syndromes in right and left hemisphere lesions. In P. J. Vinken & G. W. Bruyn (Eds.), *Handbook of clinical neurology* (Vol. 4). Amsterdam: North Holland, 1969.

Hécaen, H., & Albert, M. L. *Human neuropsychology.* New York: Wiley-Interscience, 1978.

Hécaen, H., & Angelergues, R. Agnosia for faces. *Archives of Neurology,* 1962, *7,* 92–100.

Heilman, K. M. Neglect and related disorders. In K. M. Heilman & E. Valenstein (Eds.), *Clinical neuropsychology.* New York: Oxford University Press, 1979.

Heilman, K. M., & Valenstein, E. (Eds.). *Clinical neuropsychology.* New York: Oxford University Press, 1979.

Hersen, M., & Barlow, D. H. *Single-case experimental designs: Strategies for studying behavior change.* New York: Pergamon Press, 1976.

Hersen, M., & Bellack, A. S. *Behavior therapy in the psychiatric setting.* Baltimore: Williams & Wilkins, 1978.

Hersen, M., & Bellack, A. S. *Behavioral assessment: A practical handbook* (2nd ed.). New York: Pergamon Press, 1981.

Herz, M. I., Endicott, J., & Spitzer, R. L. Brief hospitalization of patients with families: Initial results. *American Journal of Psychiatry,* 1975, *132,* 413–418.

Herz, M. I., Endicott, J., & Spitzer, R. L. Brief hospitalization: A two year follow-up. *American Journal of Psychiatry,* 1977, *134,* 502–507.

Hilgard, E. R. *Theories of learning* (2nd ed.). New York: Appleton-Century-Crofts, 1956.

Himmelhoch, J. M., Mulla, D., Neil, J. F., Detre, T. P., & Kupfer, D. J. Incidence and significance of mixed affective states in a bipolar population. *Archives of General Psychiatry,* 1976, *33,* 1062–1066.

Himmelhoch, J. M., Neil, J. F., May, S. J., Fuchs, C. Z., & Licata, S. M. Age, dementia, dyskinesias, and lithium response. *American Journal of Psychiatry,* 1980, *137,* 941–945.

Hirschberg, G. G., Lewis, L., & Thomas, D. *Rehabilitation: A manual for the care of the disabled and elderly.* Philadelphia: J. B. Lippincott, 1964.

Hirschberg, G. G., Lewis, L., & Vaughan, P. *Rehabilitation: A manual for the care of the disabled and elderly* (2nd ed.). New York: J. B. Lippincott, 1976.

Hogarty, G. E., Goldberg, S. C., Schooler, N. R., & The Collaborative Study Group. Drug and sociotherapy in the aftercare of schizophrenic patients: Part III. Adjustment of nonrelapsed patients. *Archives of General Psychiatry,* 1974, *31,* 609–618.

Hogarty, G. E., Ulrich, R. F., Goldberg, S. C., & Schooler, N. R. Sociotherapy and the prevention of relapse among schizophrenic patients: An artifact of drug? In R. L. Spitzer & D. F. Klein (Eds.), *Evaluation of psychological therapies.* Baltimore: Johns Hopkins University Press, 1976.

Holland, A. L. Some current trends in aphasia rehabilitation. *Asha,* 1969, *11,* 3–7.

Holland, A. L. Case studies in aphasia rehabilitation using programmed instruction. *Journal of Speech and Hearing Disorders,* 1970, *35,* 377–390.

Holland, A. L. Some practical considerations in aphasia rehabilitation. In M. Sullivan & M. S. Kommers (Eds.), *Rationale for adult therapy.* Omaha: University of Nebraska Medical Center, 1977.

Holland, A. L. *Current developments in the treatment of aphasia.* Paper presented at Conference on Advances in Clinical Neuropsychology, Pittsburgh, Pa., September 1979.

Holland, A. L. *CADL Communicative Abilities in Daily Living. A test of functional communication for aphasic adults.* Baltimore: University Park Press, 1980.

Holland, A. L. Observing functional communication of aphasic adults. *Journal of Speech and Hearing Disorders,* 1982, *47,* 50–56.

Holland, A. L., & Harris, A. Aphasia rehabilitation using programmed instruction: An intensive care history. In H. Sloane & B. D. Macaulay (Eds.), *Operant procedures in remedial speech and language training.* Boston: Houghton Mifflin, 1968.

Holland, A. L.,& Levy, C. The development and evaluation of programmed instruction techniques for aphasia rehabilitation. *Final Report Social Rehabilitation Service of the Department of Health, Education, and Welfare,* 1969.

Holland, A. L., & Levy, C. Syntactic generalization in aphasics as a function of retraining an active sentence. *Acta Symbolica,* 1971, *2,* 34–41.

Holland, A. L., & Sonderman, J. Effects of a program based on the Token Test. *Journal of Speech and Hearing Research,* 1974, *17,* 589–598.

Holmes, T. H., & Rahe, R. H. The social readjustment rating scale. *Journal of Psychosomatic Research,* 1967, *2,* 213–218.

Homme, L. E. Perspectives in psychology: XXIV. Control of coverants, the operants of the mind. *Psychological Record,* 1965, *15,* 501–511.

Horenstein, S. Presentation 17: Effects of cerebrovascular disease on personality and emotionality. In A. L. Benton (Ed.), *Behavioral change in cerebrovascular disease.* New York: Harper & Row, 1970.

Huntington, G. On chorea. *The Medical and Surgical Reporter,* 1872, *26,* 317–321.

Jacobs, E. A., Winter, P. M., Alvis, H. J., & Small, S. M. Hyperoxygenation effect on cognitive functioning in the aged. *New England Journal of Medicine,* 1969, *281,* 753–757.

Jervis, G. A. The mental deficiencies. In S. Arieti (Ed.), *American handbook of psychiatry* (Vol. 2). New York: Basic Books, 1959.

Kallman, W. M., Hersen, M., & O'Toole, D. H. The use of social reinforcement in a case of conversion reaction. *Behavior Therapy,* 1975, *6,* 411–413.

Kaplan, E. *Presidential Address.* Second European Conference of the International Neuropsychological Society, Noordwijkerhout, Holland, June 1979.

Katz, M. M. Behavioral change in the chronicity pattern of dementia in the institutional geriatric resident. *Journal of the American Geriatrics Society,* 1976, *24,* 522–528.

Katzman, R., Terry, R. D., & Bick, K. L. (Eds.), *Alzheimer's disease: Senile dementia and related disorders.* New York: Raven Press, 1978.

Kazdin, A. E. The effect of response cost and aversive stimulation in suppressing punished and nonpunished speech disfluencies. *Behavior Therapy,* 1973, *4,* 73–82.

Kazdin, A. E. *The token economy: A review and evaluation.* New York: Plenum Press, 1977.

Keith, R. L., & Darley, F. L. The use of a specific electric board in rehabilitation of the aphasic patient. *Journal of Speech and Hearing Disorders,* 1967, *32,* 148–153.

Kertesz, A. Recovery and Treatment. In K. M. Heilman & E. Valenstein (Eds.), *Clinical neuropsychology.* New York: Oxford University Press, 1979.

Kimura, D., & Durnford, M. Normal studies on the function of the right hemisphere in vision. In S. J. Dimond & J. G. Beaumont (Eds.), *Hemisphere function in the human brain.* London: Elek Science, 1974.

Kinsbourne, M., & Smith, W. L. *Hemispheric disconnection and cerebral function.* Springfield, Ill.: Charles C Thomas, 1974.

Klonoff, H., Fibiger, C. H., & Hutton, G. H. Neuropsychological patterns in chronic schizophrenia. *Journal of Nervous and Mental Disease,* 1970, *150,* 291–300.

Kolb, L. C. Disturbances of the body-image. In S. Arieti (Ed.), *American handbook of psychiatry* (Vol. 1). New York: Basic Books, 1959.

Kubler-Ross, E. K. *On death and dying.* New York: Collier MacMillan, 1969.

Lahey, B. B. *The modification of language behavior.* Springfield, Ill.: Charles C Thomas, 1973.

Langer, E. J., Rodin, J., Beck, P., Weinman, C., & Spitzer, L. Environmental determinants of memory improvement in late adulthood. *Journal of Personality and Social Psychology,* 1979, *37,* 2003–2013.

Lashley, K. S. Integrative functions of the cerebral cortex. *Physiological Review,* 1933, *13,* 1–42.

Lawton, M. P. Psychosocial and environmental approaches to the care of senile dementia patients. In J. O. Cole & J. E. Barrett (Eds.), Psychopathology in the aged. New York: Raven Press, 1980.

Lehmann, H. E., & Ban, T. A. Central nervous system stimulants and anabolic substances in geropsychiatric therapy. In S. Gershon & A. Raskin (Eds.), *Aging, Volume 2, Genesis and treatment of psychologic disorders in the elderly.* New York: Raven Press, 1975.

Leitenberg, H. (Ed.). *Handbook of behavior modification and behavior therapy.* Englewood Cliffs, N.J.: Prentice-Hall, 1976.

Leitenberg, H., & Callahan, E. J. Reinforced practice and reduction of different kinds of fears in adults and children. *Behavior Research and Theory,* 1973, *11,* 10–30.

Levin, H. S. The acalculias. In K. M. Heilman & E. Valenstein (Eds.), *Clinical neuropsychology.* New York: Oxford University Press, 1979.

Lewinsohn, P. M. *Depression: A social learning theory perspective.* Lecture: Western Psychiatric Institute and Clinic, April 1979.

Lewinsohn, P. M., Danaher, B. G., & Kikel, S. Visual imagery as a mnemonic aid for brain-injured persons. *Journal of Consulting and Clinical Psychology,* 1977, *45,* 717–723.

Lezak, M. D. *Neuropsychological Assessment.* New York: Oxford University Press, 1976.

Lezak, M. D. *"Context", the key to understanding the patient.* Symposium-Case Analysis in Neuropsychology. Second European Conference of the International Neuropsychological Society, Noordwijkerhout, Holland, June, 1979.

Lieb, J., Lipsitch, I. T., & Slaby, A. E. *The crisis team: A handbook for the mental health professional.* New York: Harper & Row, 1973.

Lieberman, M. Institutionalization of the aged: Effects on behavior. *Journal of Gerontology,* 1969, *24,* 330–340.

Lieberman, M., & Lakin, M. On becoming an institutionalized person. In R. H. Williams, C. Tibbits, & W. Donahue (Eds.), *Processes of aging, Volume 1: Social and Psychological Perspectives.* New York: Atherton Press, 1963.

Lieberman, M., Prock, V. N., & Tobin, S. S. Psychological effects of institutionalization. *Journal of Gerontology,* 1968, *23,* 342–353.

Lindsley, O. R. Operant conditioning methods applied to research in chronic schizophrenia. *Psychiatric Research Reports,* 1956, *5,* 118–139.

Lindsley, O. R., & Skinner, B. F. A method for the experimental analysis of the behavior of psychotic patients. *American Psychologist,* 1954, *9,* 419–420.

Lipton, M. A., DiMascio, A., & Killam, K. F. (Eds.). *Psychopharmacology: A generation of progress.* New York: Raven Press, 1978.

Loeber, R. Engineering the behavioral engineer. *Journal of Applied Behavioral Analysis,* 1971, *4,* 321–326.

Logue, P. E. *Understanding and living with brain damage.* Springfield, Ill.: Charles C Thomas, 1975.

Lorayne, H., & Lucas, J. *The memory book.* Briarcliff Manor, N. Y.: Stein & Day, 1974.

Lovaas, O. I., & Newsom, C. D. Behavior modification with psychotic children. In H. Leitenberg (Ed.), *Handbook of behavior modification and behavior therapy.* Englewood Cliffs, N.J.: Prentice-Hall, 1976.

Lovaas, O. I., & Simmons, J. Manipulation of self-destruction in three retarded children. *Journal of Applied Behavioral Analysis,* 1969, *2,* 143–157.

Luria, A. R. *Restoration of function after brain injury.* New York: Macmillan 1963. (Originally published in Russian, 1948.)

Luria, A. R. *Higher cortical functions in man.* New York: Basic Books, 1966/Consultants Bureau Enterprises, 1980. (Originally published in Russian, 1962.)

Luria, A. R. *Traumatic aphasia (its syndromes, psychology & treatment).* The Hague: Mouton, 1970.

Luria, A. R. *The man with a shattered world.* New York: Basic Books, 1972.

Luria, A. R. *The working brain.* New York: Basic Books, 1973.

Luria, A. R., & Majovski, L. V. Basic approaches used in American and Soviet clinical neuropsychology. *American Psychologist,* 1977, *32,* 959–968.

Lutzker, J. R. *Behavior therapy and rehabilitation.* Symposium presented at the 14th Annual Convention of the Association for Advancement of Behavior Therapy, New York, November 1980.

Lutzker, J. R., & Sherman, J. Producing generative sentence usage by imitation and reinforcement procedures. *Journal of Applied Behavior Analysis,* 1974, *7,* 447–460.

McKeown, D., Adams, H. E., & Forehand, R. Generalization to the classroom of principles of behavior modification taught to teachers. *Behavioral Research and Therapy,* 1975, *13,* 85–92.

Mahoney, F. I., & Bartel, S. J. Rehabilitation of the hemiplegic patient: A clinical evaluation. *Archives of Physical Medicine and Rehabilitation,* 1954, *35,* 359–362.

Mandell, A. J. Asymmetry and mood emergent properties of serotonin regulation: A mechanism of action of lithium. Lecture to the Department of Psychiatry, University of Pittsburgh, October 1978.

Maslow, A. H. *Toward a psychology of being.* New York: Van Nostrand Reinhold, 1968.

Matarazzo, J. D. *Wechsler's measurement and appraisal of adult intelligence* (5th ed.). Baltimore: Williams & Wilkins, 1972.

Matarazzo, J. D., Weins, A. N., Matarazzo, R. G., & Goldstein, S. G. Psychometric and clinical test-retest reliability of the Halstead impairment index in a sample of healthy, young, normal men. *Journal of Nervous and Mental Disease,* 1974, *158,* 37–49.

Matarazzo, J. D., Matarazzo, R. G., Weins, A. N., Gallo, A. E., & Klonoff, H. Retest reliability of the Halstead impairment index in a normal, a schizophrenic and two samples of organic patients. *Journal of Clinical Psychology,* 1976, *32,* 338–349.

Matarazzo, J. D., Carmody, T. P., & Jacobs, L. D. Test-retest reliability and stability of the WAIS: A literature review with implications for clinical practice. *Journal of Clinical Neuropsychology,* 1980, *2,* 89–105.

Meehl, P. *Clinical versus statistical prediction.* Minneapolis: University of Minnesota Press, 1954.

Meier, M. J. Objective behavioral assessment in diagnosis and prediction. In A. L. Benton (Ed.), *Behavior changes in cerebrovascular disease.* New York: Harper & Row, 1970.

Meier, M. J. Some challenges for clinical neuropsychology. In R. M. Reitan & L. A. Davison (Eds.), *Clinical neuropsychology: Current status and applications.* New York: Winston Wiley, 1974.

Meier, M. J. Education for competency assurance in human neuropsychology: Antecedents, models and directions. In S. B. Filskov & T. J. Boll (Eds.), *Handbook of clinical neuropsychology.* New York: Wiley, 1981.

Meier, M. J., & Resch, J. A. Behavioral prediction of short-term neurologic change following acute onset of cerebrovascular symptoms. *Mayo Clinic Proceedings,* 1967, *42,* 641–647.

Merikangas, J. R. Common neurologic syndromes in medical practice. *Medical Clinics of North America,* 1977, *61,* 723–735.

Michaux, W. W., *et al. The first year out: Mental patients after hospitalization.* Baltimore: Johns Hopkins Press, 1969.

Miller, E. Psychological intervention in the management and rehabilitation of neuropsychological impairments. *Behavioral Research and Therapy,* 1980, *18,* 527–535.

Miller, P. M. Behavioral treatment of alcoholism. In M. Hersen & A. S. Bellack (Eds.), *Behavior therapy in the psychiatric setting.* Baltimore: Williams & Wilkins, 1978.

Milner, B. Some effects of frontal lobectomy in man. In J. M. Warren & K. Akert (Eds.), *The Frontal Granular Cortex and Behavior.* New York: McGraw-Hill, 1964.

Mitchell, R. A. Reality orientation for brain damaged patients. *Staff,* 1966, *3,* 3–64.

Monakow, C. *Die Lokalisation im Grosshirn und der Abbau der Funktionen durch cortical Herde.* Wiesbaden: Bergmann, 1914.

Morrow, L., Vrtunski, P., Kim, Y., & Boller, F. *Arousal responses to emotional stimuli in unilateral hemisphere damaged patients.* Paper presented at Second European Conference of the International Neuropsychological Society, Noordwijkerhout, Holland, June 1979.

Moss, C. S. *Recovery with aphasia: Aftermath of my stroke.* Chicago: University of Illinois Press, 1973.

Myslobodsky, M. S., & Weiner, M. Clinical psychology in the chemical environment. *Psychological Reports,* 1978, *43,* 247–276.

Nathan, P. E. Management of the chronic alcoholic: A behavioral viewpoint. In J. P. Brady & H. K. Brodie (Eds.), *Controversy in psychiatry.* New York: Saunders, 1977.

Newcombe, F. *Missile wounds of the brain: A study of psychological deficits.* Oxford: Clarendon Press, 1969.

Newcombe, F. Selective deficits after focal cerebral injury. In S. J. Dimond & J. G. Beaumont (Eds.), *Hemisphere function in the human brain.* London: Elek Science, 1974.

Ojemann, G., & Fedio, P. Effect of stimulation of the human thalamus and parietal and temporal white matter on short-term memory. *Journal of Neurosurgery,* 1968, *29,* 51–59.

Ojemann, G., & Mateer, C. Human language cortex: Localization of memory, syntax, and sequential motorphoneme identification systems. *Science,* 1979, *205,* 1401–1403.

Olsen, C. W., & Ruby, C. Anosognosia and autotopagnosia. *Archives of Neurology,* 1941, *46,* 340–345.

Oltman, J. E., & Friedman, S. Comments on Huntington's chorea. *Diseases of the Nervous System,* 1961, *22,* 313–319.

Parsons, O. A., & Farr, S. P. The neuropsychology of alcohol and drug use. In S. B. Filskov & T. J. Boll (Eds.), *Handbook of clinical neuropsychology.* New York: Wiley, 1981.

Parsons, O. A., & Prigatano, G. P. Methodological considerations in clinical neuropsychological research. *Journal of Consulting and Clinical Psychology,* 1978, *46,* 608–619.

Patterson, G. R. The aggressive child: Victim and architect of a coercive system. In E. J. Marsh, E. J. Hamerlynck, & L. C. Handy (Eds.), *Behavior modification and families.* New York: Brunner/Mazel, 1976.

Paul, G. L. Strategy of outcome research in psychotherapy. *Journal of Consulting Psychology,* 1967, *31,* 104–118.

Paul, G. L., & Lentz, R. J. *Psychosocial treatment of chronic mental patients. Milieu versus social-learning programs.* Cambridge, Harvard University Press, 1977.

Penfield, W., & Roberts, L. *Speech and brain mechanisms.* Princeton, N.J.: Princeton University Press, 1959.

Perel, J. Personal communication, June, 1981.

Peszczynski, M. Prognosis for rehabilitation of older adults and the aged hemiplegic patient. *American Journal of Cardiology,* 1961, *7,* 365–369.

Peterson, L. R., & Peterson, M. J. Short-term retention of individual verbal items. *Journal of Experimental Psychology,* 1959, *58,* 193–198.

Pichot, P., Lempérière, T., & Perse, J. *Le test de Rorschach et la personnalité épileptique.* Paris: Presses Universitaires de France, 1955.

Pincus, J. H., & Tucker, G. J. *Behavioral neurology* (2nd ed.). New York: Oxford University Press, 1978.

Plotnikoff, N. Learning and memory enhancement by pemoline and magnesium hydroxide (PMH). *Recent Advances in Biological Psychiatry,* 1968, *10,* 102–120.

Plotnikoff, N. Pemoline: Review of performance. *Texas Reports in Biology and Medicine,* 1971, *29,* 467–479.

Poon, L. W., Fozard, J. L., Cermak, L. S., Arenberg, D., & Thompson, L. W. *New directions in memory and aging*. Hillsdale, N.J.: Lawrence Erlbaum Associates, 1980.

Porch, B. E. Disorders of Communication. In G. G. Hirschberg, L. Lewis, & P. Vaughan (Eds.), *Rehabilitation: A manual for the care of the disabled and elderly*. Philadelphia: J. B. Lippincott, 1964.

Premack, D. Reinforcement theory. In D. Levin (Ed.), *Nebraska Symposium on Motivation*. Lincoln, Nebraska: University of Nebraska Press, 1965.

Prigatano, G. P., & Parsons, O. A. Relationship of age and education to Halstead test performance in different patient populations. *Journal of Consulting and Clinical Psychology*, 1976, *44*, 527–533.

Reed, H. B. C., & Reitan, R. M. A comparison of the effects of the normal aging process with the effects of organic brain-damage on adaptive abilities. *Journal of Gerontology*, 1963, *18*, 177–179.

Reed, T. E., & Chandler, J. H. Huntington's chorea in Michigan I. Demography and Genetics. *American Journal of Human Genetics*, 1958, *10*, 201–225.

Reisberg, B., Ferris, S. H., & Gershon, S. Pharmacotherapy of senile dementia. In J. O. Cole & J. E. Barett (Eds.), *Psychopathology in the aged*. New York: Raven Press, 1980.

Reischel, K. L. Psychological benefits of exercise. *Sports Medicine*, 1977, 16–17.

Reitan, R. M. The relation of the trail making test to organic brain damage. *Journal of Consulting Psychology*, 1955, *19*, 393–394.

Reitan, R. M. Validity of the trail making test as an indicator of organic brain damage. *Perceptual and Motor Skills*, 1958, *8*, 271–276. (a)

Reitan, R. M. Intellectual functions in aphasic and non-aphasic brain injured subjects. *Neurology*, 1958, *3*(3), 202–212. (b)

Reitan, R. M. Qualitative versus quantitative mental changes following brain damage. *The Journal of Psychology*, 1958, *46*, 339–346. (c)

Reitan, R. M. Impairment of abstraction ability in brain damage: Quantitative versus qualitative changes. *The Journal of Psychology*, 1959, *48*, 97–102. (a)

Reitan, R. M. The comparative effects of brain damage on the Halstead impairment index and the Wechsler-Bellevue scale. *The Journal of Clinical Psychology*, 1959, *25*, 281–285. (b)

Reitan, R. M. Effects of brain-damage on a psychomotor problem-solving task. *Perceptual Motor Skills*, 1959, *9*, 211–215. (c)

Reitan, R. M. A research program on the psychological effects of brain lesions in human beings. In N. R. Ellis (Ed.), *International Review of Research in Mental Retardation* (Vol. 1). New York: Academic Press, 1966. (a)

Reitan, R. M. Problems and prospects in studying the psychological correlates of brain lesions. *Cortex*, 1966, *2*, 127–154. (b)

Reitan, R. M., & Davison, L. A. *Clinical neuropsychology: Current status and applications*. Washington, D.C.: V. H. Winston, 1974.

Reitan, R. M., & Tarshes, E. L. Differential effects of lateralizing brain lesions on the trail making test. *Journal of Nervous and Mental Disease*, 1959, *129*, 257–262.

Rogers, C. *Client-centered therapy*. Boston: Houghton Mifflin, 1951.

Rusk, H. A. *Rehabilitation medicine*. St. Louis: C. V. Mosby, 1977.

Russell, E. W., Neuringer, C., & Goldstein, G. *Assessment of brain damage: A neuropsychological key approach*. New York: Wiley, 1970.

Russell, W. R. Cerebral involvement in head injury. *Brain*, 1932, *55*, 549–603.

Russell, W. R., & Smith, A. Post-traumatic amnesia in closed head injury. *Archives of Neurology*, 1961, *5*, 16–29.

Rylander, G. *Personality changes after operations on the frontal lobes*. London: Oxford University Press, 1939.

Saan, R. J., & Schoonbeek, H. R. Problemen bij de behandelig en de evaluatie van een patient met rekenstoornissen. In A. P. Cassee, P. E. Boeke, & J. J. Barendregt (Eds.), *Klinische psychologie in Nederland, Deel 2.* Deventer: Van Loghum Slaterus, 1973.

Safar, P., Bleyaert, A., Nemoto, E. M., Moossy, J., & Snyder, J. V. Resuscitation after global brain ischemia-anoxia. *Critical Care Medicine,* 1978, *6,* 215–226.

Sands, E., Sarno, M. T., & Shankweiler, D. Long-term assessment of language function in aphasia due to stroke. *Archives of Physical Medicine and Rehabilitation,* 1969, *50,* 202–222.

Sarno, M. T., Silverman, M., & Sands, E. Speech therapy and language recovery in severe aphasia. *Journal of Speech and Hearing Research,* 1970, *13,* 607–623.

Satterfield, J. H., Atoian, G., Brashears, G. C., Burleigh, A. C., & Dawson, M. E. Electrodermal studies of minimal brain dysfunction children. In C. K. Connors (Ed.), *Clinical use of stimulant drugs in children.* The Hague: Excerpta Medica, 1974.

Satz, P. Specific and nonspecific effects of brain lesions in man. *Journal of Abnormal Psychology,* 1966, *71,* 65–70.

Scheerer, M. Problems of performance analysis in the study of personality. *Annals of the New York Academy of Science,* 1946, *46,* 653–678.

Schwartz, R. M., & Gottman, J. M. Toward a task analysis of assertive behavior. *Journal of Consulting and Clinical Psychology,* 1976, *44,* 910–920.

Scoville, W. B., & Milner, B. Loss of recent memory after bilateral hippocampal lesions. *Journal of Neurology & Neurosurgery Psychiatry,* 1957, *20,* 11–21.

Searleman, A. A review of right hemisphere linguistic capabilities. *Psychological Bulletin,* 1977, *84,* 503–528.

Seligman, M. E. P. *Helplessness: On depression development and death.* San Francisco: W. H. Freeman, 1975.

Semmes, J., Weinstein, S., Ghent, L., & Teuber, H. L. *Somatosensory changes after penetrating brain wounds in man.* Cambridge: Harvard University Press, 1960.

Shagass, C. Twisted thoughts, twisted brain waves? In C. Shagass, S. Gershon, & A. J. Friedhoff (Eds.), *Psychopathology and brain dysfunction.* New York: Raven Press, 1977.

Shallice, T. Case study approach in neuropsychological research. *Journal of Clinical Neuropsychology,* 1979, *1,* 183–211.

Shure, G. H., & Halstead, W. C. Cerebral localization of intellectual processes. *Psychological Monographs,* 1958, *72,* (12, Whole No. 465).

Siller, J. *Changing attitudes towards the disabled.* Symposium: Presented at the 87th Annual American Psychological Association Convention, New York, September 1979.

Simmel, M. L., & Counts, S. Some stable response determinants of perception, thinking and learning: A study based on the analysis of a single test. *Genetic Psychology Monographs,* 1957, *56,* 3–157.

Skinner, B. F. *Science and human behavior.* New York: Macmillan, 1953.

Sloane, H. N., & Macaulay, B. D. (Eds.). *Operant procedures in remedial speech and language training.* Boston: Houghton Mifflin, 1968.

Small, L. *Neuropsychodiagnosis in psychotherapy.* New York: Brunner/Mazel, 1973.

Smith, A. *Changes in psychological test performances of brain-operated schizophrenics after an eight-year interval.* Unpublished doctoral dissertation, New York, Yeshiva University, 1958.

Smith, A. The duration of impaired consciousness as an index of severity in closed head injury: A review. *Diseases of the Nervous System,* 1961, *2,* 1–6.

Smith, A. Principles underlying human brain functions in neuropsychological sequelae of different neuropathological processes. In S. B. Filskov & T. J. Boll (Eds.), *Handbook of clinical neuropsychology.* New York: Wiley-Interscience, 1981.

Smith, A., & Kinder, E. F. Changes in psychological test performance of brain-operated schizophrenics after eight years. *Science*, 1959, *129*, 149–150.

Sovner, R., & DiMascio, A. Extrapyramidal syndromes and other neurological side effects of psychotropic drugs. In M. A. Lipton, A. DiMascio, & K. F. Killam (Eds.), *Psychopharmacology: A generation of progress*. New York: Raven Press, 1978.

Sparks, R. W., Helm, N., & Albert, M. Aphasia rehabilitation resulting from melodic-intonation therapy. *Cortex*, 1974, *10*, 303–316.

Sparks, R. W., & Holland, A. Method: Melodic intonation therapy for aphasia. *Journal of Speech and Hearing Disorders*, 1976, *41*, 287–297.

Spitzer, R. L., & Endicott, J. *Schedule for affective disorders and schizophrenia (SADS)*. New York: Biometrics Research, New York State Department of Mental Hygiene, 1973.

Squire, L. R., Slater, P. C., & Chace, P. M. Retrograde amnesia: Temporal gradient in very long term memory following electroconvulsive therapy. *Science*, 1975, *187*, 77–79.

Stanton, A. H., & Schwartz, M. S. *The mental hospital: A study of institutional participation in psychiatric illness and treatment*. New York: Basic Books, 1954.

Stedman's Medical Dictionary (22nd ed.). Baltimore: Williams & Wilkins, 1972.

Stein, D., Rosen, J., & Butters, N. (Eds.), *Plasticity and recovery of function in the central nervous system*. New York: Academic Press, 1974.

Stephens, L. P. Reality orientation. *American Psychiatric Association Hospital and Community Psychiatry Service*, 1969, 1–11.

Stern, R. S. Behavior therapy and psychotropic medication. In M. Hersen & A. S. Bellack (Eds.), *Behavior therapy in the psychiatric setting*. Baltimore: Williams & Wilkins, 1978.

Stricker, E. *Animal models of recovery of brain function*. Paper presented at Conference on Advances in Clinical Neuropsychology. Pittsburgh, Pa.: September, 1979.

Tarter, R. Neuropsychological investigations of alcoholism. In G. Goldstein and C. Neuringer (Eds.), *Empirical studies of alcoholism*. Cambridge, Mass.: Ballinger, 1976.

Taulbee, L. R., & Folsom, J. C. Reality orientation for geriatric patients. *Hospital and Community Psychiatry*, 1966, *17*, 133–135.

Taylor, A. R. Speculation about site of action of cyclospasmol on cerebral metabolism. In G. Stocker, R. A. Kuhn, P. Hall, G. Becker, & E. van der Veen (Eds.), *Assessment in cerebrovascular insufficiency*. Stuttgart: Thieme, 1971.

Taylor, M. L. Language therapy. In H. G. Burr (Ed.), *The aphasic adult: Evaluation and rehabilitation*. Charlottesville, Va.: Wayside Press, 1964.

Teuber, H. L. Some alterations in behavior after cerebral lesions in man. In A. D. Bass (Ed.), *Evolution of nervous control from primitive organisms to man*. Washington, D.C.: American Association for the Advancement of Science, 1959.

Teuber, H. L. The riddle of frontal lobe function in man. In J. M. Warren & K. Akert (Eds.), *The frontal granular cortex and behavior*. New York: McGraw-Hill, 1964.

Teuber, H. L., & Liebert, R. S. Specific and general effects of brain injury in man; evidence of both from a single task. *Archives of Neurology and Psychiatry*, 1958, *80*, 403–404.

Teuber, H. L., & Mishkin, M. Judgement of visual and postural vertical after brain injury. *Journal of Psychology*, 1954, *38*, 161–175.

Teuber, H. L. & Weinstein, S. Performance on a formboard task after penetrating brain injury. *Journal of Psychology*, 1954, *38*, 177–190.

Trower, P., Bryant, B., & Argyle, M. *Social skills and mental health*. Pittsburgh: University of Pittsburgh Press, 1978.

Ullmann, L. P., & Krasner, L. (Eds.). *Case studies in behavior modification.* New York: Holt, Rinehart & Winston, 1965.

Valenstein, E. S. *Brain control: A critical examination of brain stimulation and psychosurgery.* New York: Wiley, 1973.

Valenstein, E., & Heilman, K. M. Emotional disorders resulting from lesions of the central nervous system. In K. M. Heilman & E. Valenstein (Eds.), *Clinical neuropsychology.* New York: Oxford University Press, 1979.

Vaughan, C. E., & Leff, J. P. The influence of family and social factors on the course of psychiatric illness: A comparison of schizophrenic and depressed neurotic patients. *British Journal of Psychiatry,* 1976, *129,* 125–137.

Walker, A. E. Prognosis in post-traumatic epilepsy—a ten year follow-up of craniocerebral injuries of World War II. *Journal of the American Medical Association,* 1957, *164,* 1636–1641.

Walker, A. E., Caveness, W. F., & Critchley, M. (Eds.). *Late effects of head injury.* Springfield, Ill.: Charles C Thomas, 1969.

Walker, A. E., & Erculei, F. *Head-injured man fifteen years later.* Springfield, Ill.: Charles C Thomas, 1969.

Walker, A. E., & Jablon, S. A follow-up of head injured men of World War II. *Journal of Neurosurgery,* 1959, *16,* 600–610.

Walsh, K. W. *Neuropsychology: A clinical approach.* Edinburgh: Churchill Livingstone, 1978.

Wechsler, D. *The measurement of adult intelligence* (3rd ed.). Baltimore: Williams & Wilkins, 1944.

Wechsler, D. A standardized memory scale for clinical use. *Journal of Psychology,* 1945, *19,* 87–95.

Wechsler, D. *Wechsler adult intelligence scale.* New York: Psychological Corporation, 1955. (a)

Wechsler, D. *Manual for the Wechsler adult intelligence scale.* New York: Psychological Corporation, 1955. (b)

Wechsler, D. *The measurement and appraisal of adult intelligence* (4th ed.). Baltimore: Williams & Wilkins, 1958.

Weed, L. L. Medical records, medical education & patient care. Chicago: Year Book Medical Publications, 1970.

Weinberg, J., Diller, L., Gordon, W. A., Gerstman, L. J., Lieberman, A., Lakin, P., Hodges, G., & Ezrachi, O. Visual-scanning training effect on reading related tasks in acquired right brain damage. *Archives of Physical Medicine and Rehabilitation,* 1977, *58,* 479–486.

Weinstein, E. A., Cole, M., Mitchell, M. S., & Lyerly, O. G. Agnosia and aphasia. *Archives of Neurology,* 1964, *10,* 376–386.

Weinstein, E. A., & Kahn, R. L. *Denial of illness: Symbolic and physiological aspects.* Springfield, Ill.: Charles C Thomas, 1955.

Weinstein, E. A., & Kahn, R. L. Symbolic reorganization in brain injuries. In S. Arieti (Ed.), *American handbook of psychiatry* (Vol. 1). New York: Basic Books, 1959.

Wells, C. E. Pseudodementia. *American Journal of Psychiatry,* 1979, *136,* 895–900.

Wells, C. E. The differential diagnosis of psychiatric disorders in the elderly. In J. O. Cole & J. E. Barrett (Eds.), *Psychopathology in the aged.* New York: Raven Press, 1980.

Wepman, J. P. *Recovery from aphasia.* New York: Ronald Press, 1951.

Whittier, J. R. Hereditary chorea (Huntington's chorea): A paradigm of brain dysfunction with psychopathology. In C. Shagass, S. Gershon, & A. J. Friedhoff (Eds.), *Psychopathology and brain dysfunction.* New York: Raven Press, 1977.

Wilcox, M. J., & Davis, G. A. A method for promoting aphasic communicative effectiveness. *Journal of Speech and Hearing Disorders,* in press.

Williams, K. H., & Goldstein, G. Cognitive and affective responses to lithium in patients with organic brain syndrome. *American Journal of Psychiatry,* 1979, *136,* 800–803.

Wilson, R. S., & Garron, D. C. Cognitive and affective aspects of Huntington's disease. *Advances in Neurology,* 1979, *23,* 193–201.

Wilson, R. S., Rosenbaum, G., & Brown, G. The problem of premorbid intelligence in neuropsychological assessment. *Journal of Clinical Neuropsychology,* 1979, *1,* 49–53.

Wincze, J. P., Leitenberg, H., & Agras, W. S. The effects of token reinforcement and feedback on the delusional verbal behavior of chronic paranoid schizophrenics. *Journal of Applied Behavioral Analysis,* 1972, *5,* 247–262.

Wing, J. K., & Brown, G. W. *Institutionalism and schizophrenia.* London: Cambridge University Press, 1970.

Wolfe, G., Stricker, E. M., & Zigmond, M. J. Brain lesions: Induction, analysis and the problem of recovery of function. In S. Finger (Ed.), *Recovery from brain damage: Research and theory.* New York: Plenum Press, 1978.

Wolpe, J. *Psychotherapy by reciprocal inhibition.* Stanford, Calif.: Stanford University Press, 1958.

Wolpe, J. *The practice of behavior therapy* (2nd ed.). New York: Pergamon Press, 1973.

Wright, B. A. *Physical disability: A psychological approach.* New York: Harper & Row, 1960.

Yacorzynski, G. K. An evaluation of the postulates underlying the Babcock deterioration test. *Psychological Review,* 1941, *48,* 261–267.

Yates, A. *Behavior therapy.* New York: Wiley, 1970.

Yates, A. *Theory and practice in behavior therapy.* New York: Wiley, 1975.

Young, J., & Wincze, J. The effects of the reinforcement of compatible and incompatible alternative behaviors on the self-injurious and related behaviors of a profoundly retarded family adult. *Behavior Therapy,* 1974, *5,* 614–623.

Young, L. D., Taylor, I., & Holmstrom, V. Lithium treatment of patients with affective illness associated with organic brain symptoms. *American Journal of Psychiatry,* 1977, *234,* 1405–1407.

Zaidel, E. Unilateral auditory language comprehension on the token test following cerebral commissurotomy hemispherectomy. *Neuropsychologia,* 1977, *15,* 1–13.

Zubin, J., & Spring, B. Vulnerability: A new view of schizophrenia. *Journal of Abnormal Psychology,* 1977, *86,* 103–126.

INDEX